VACUUM TECHNOLOGY

Vacuum Technology

Andrew Guthrie, Professor of Physics

ALAMEDA STATE COLLEGE, HAYWARD, CALIFORNIA

John Wiley and Sons, Inc.

New York · London ·· Sydney

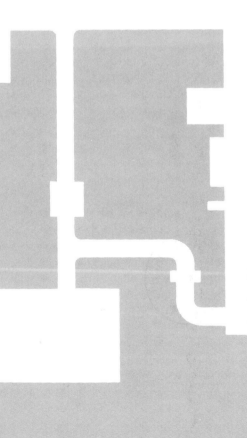

Preface

Developments in the vacuum field have been rapid during the last few years. These developments have included much larger systems, new techniques and equipment, and operation at extremely low pressures. More people are continually becoming involved in vacuum work. There are a number of excellent books on the subject of vacuum; however, in general, they assume a fair degree of scientific knowledge on the part of the reader and treat the broader principles and techniques of vacuum. It is the purpose of this book to consolidate the vast amount of material on this subject, both published and unpublished, into sets of working rules primarily for technicians engaged in setting up, operating, and maintaining vacuum systems.

The treatment is nonmathematical, with the arithmetic used extending only to the application of decimals, powers of ten, and logarithms. Where simple arithmetic is involved, numerous examples are given. However, tables, graphs, and simple rules are employed extensively. Using the material in this book, the technician should be able to choose the appropriate commercial vacuum components, fabricate various parts for vacuum use, set up a complete system, test components and complete systems, perform routine maintenance, and recognize and correct malfunctions. Chapter 14 is devoted exclusively to very high and ultra-high vacuum systems, but some information pertaining to such systems is contained in other chapters.

Although this book is intended primarily for the technician, it is hoped that it will be useful to a certain degree to the engineer and the scientist because of its tables and graphs, and the physical data and design formulas included in the appendices.

The terminology used generally follows that recommended by the American Vacuum Society in its *Glossary of Terms Used in Vacuum Technology* (Pergamon Press, London, 1958). Extensive use has been made of data from various manufacturers, and every effort has been made to acknowledge the source of such information in the text.

I wish to express my appreciation to Carole Abbott for typing the manuscript, and for subsequent help on details which had to be worked out while the book was in various stages of publication.

ANDREW GUTHRIE

Berkeley, California
August 1963

Contents

Chapter Nine   Cleaning Techniques   262

Chapter Ten   Some Fabrication Techniques   279

Chapter Eleven   Baffles, Traps, and Valves   322

Chapter Fifteen   Finding and Repairing Leaks   456

Appendices   503

Index   523

ONE

The Nature of Vacuum

1.1 What Is Vacuum?

The word vacuum is derived from the Greek word meaning empty. In practice some type of vessel (vacuum enclosure, chamber, or container) which is open to the surrounding air is used. As air is removed by some pumping means, a vacuum is obtained. Clearly various degrees of vacuum can be obtained, depending on how much air is removed from the enclosure. Practically, a vacuum vessel which is empty, i.e., free of all matter, is never obtained. If this were possible the vacuum would be called *perfect* or *absolute*. Many vacuum terms are defined in the text as they are used. The subject of vacuum is treated in general in the references listed in Appendix E.

1.2 General Nature of a Gas

The word gas is often used loosely in vacuum practice to denote both *noncondensable* gases and *vapors*. The noncondensable gases, sometimes called *permanent* gases, cannot be compressed to liquid or solid form at ordinary room temperature. Dry air is an example of a noncondensable gas, and water vapor is an example of a vapor.

1

In order to produce liquid air it is necessary not only to use a high degree of compression but also to do so at low temperatures.

A gas is made up of many, small, invisible particles called molecules which are moving about rapidly in all directions. It is possible to visualize these molecules by imagining tennis balls moving about rapidly in all directions in a room. They collide with each other and rebound from the walls of the room. Most of them move with about the same speed, although occasionally it will be observed that one moves considerably faster or slower than the others. Figure 1.1 is a schematic representation of a vessel containing a gas. The arrows show the directions in which the molecules are moving, and it is found that, on the average, as many molecules are moving in one direction as in any other direction.

If the enclosure shown in Fig. 1.1 were heated, it would be found that the molecules had a greater mean speed than before the enclosure was heated. Again an occasional molecule would be found to be moving considerably faster or slower than the bulk of the molecules. The speeds of molecules are quite high, even at room temperature (20° centigrade or 68° Fahrenheit). At this temperature a hydrogen molecule has an average speed of about 5700 feet per second whereas an air molecule has an average speed of about 1500 feet per second. Actually air consists of several gases, the most prominent being nitrogen and oxygen (see Table 1.1). The average speed for an air molecule given above is an average for the molecules of the various gaseous constituents. The distance between molecules is, on the average, much greater than the diameter of a molecule. This is true for the air

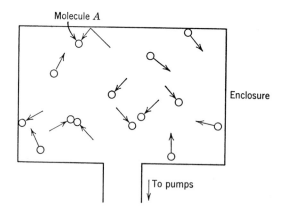

Fig. 1.1   Molecules moving about in an enclosure.

Table 1.1 Composition of Dry Air at Sea Level

Gas	Percent by Volume
Nitrogen	78.08
Oxygen	20.95
Argon	0.93
Carbon dioxide	0.03
Neon	0.0018
Helium	0.0005
Methane	0.0002
Krypton	0.0001
Hydrogen	0.00005
Xenon, etc.	Traces

Handbook of Chemistry and Physics, Chemical Rubber Publishing Co., Cleveland, Ohio, 43rd ed., 1961.

around us, and if some of the air is removed from such an enclosure as shown in Fig. 1.1, the average distance increases since there are fewer molecules distributed through the volume.

Referring again to Fig. 1.1, it is seen that molecule *A* has struck the inside wall of the enclosure and rebounded. Actually many molecules strike the wall in a short time, and they strike from all directions up to grazing incidence. The striking and rebounding of the molecule results in a "push" or force on the wall. The average force on a unit area of the wall, say one square inch, is called the *pressure*. This is a fundamental quantity in vacuum work. The pressure of a gas depends on how many molecules are striking the unit area in any given time (a second, a minute, etc.) and on how fast they are moving. Consequently, the pressure will drop as the gas is pumped out (fewer molecules) and will rise as the gas is heated (molecules move faster).

It has been stated that a gas exerts a pressure on the inside walls of the enclosure. Actually there is a gas pressure anywhere in the gas. Suppose an object, such as a sheet of metal, is placed in the enclosure. Then each side of the sheet will be subjected to a pressure equal to that exerted on the inside walls of the enclosure. This is illustrated in Fig. 1.2. Although many units of pressure are used in

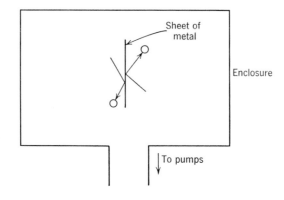

Fig. 1.2   Molecules exert a pressure on an object in the gas.

practice, they are all derived from one basic unit, which depends on the characteristics of the air around us (the atmosphere). Consequently, this unit is discussed next.

1.3   The Atmosphere

Since in most vacuum work some type of vessel which is open to the air (the atmosphere) is used, gases making up the atmosphere are of considerable importance. Table 1.1 shows the proportions of the various gases making up dry air at sea level. The proportions change (very slowly) with altitude, but this change becomes significant only at extremely high altitudes.

The atmosphere exerts a pressure on any object immersed in it. Normally reference is made to the pressure at sea level and at a temperature of 0° centigrade (0°C) or 32° Fahrenheit (32°F), called the *standard* or *normal atmosphere* (atm). The relationships between the various temperature scales are discussed in Section 1.8. To measure the actual pressure of the atmosphere, a glass U-tube partially filled with mercury is used, with one end connected to some type of vacuum pump. This is illustrated in Fig. 1.3*b*. If both ends of the U-tube are exposed to air, the mercury levels in both sides will be the same (Fig. 1.3*a*). This is simply an example of a liquid seeking its own level. If one side of the U-tube is connected to the vacuum pump, the mercury will rise in that side and fall in the other side, as is shown in Fig. 1.3*b*. This difference in level is caused by the atmosphere exerting a pressure on the mercury in side *B*. There is no balancing pressure on side *A*

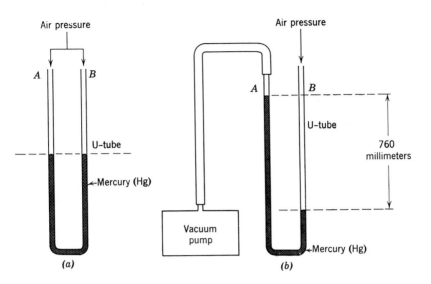

Fig. 1.3 (a) U-tube with both sides exposed to the atmosphere. (b) U-tube with one side pumped.

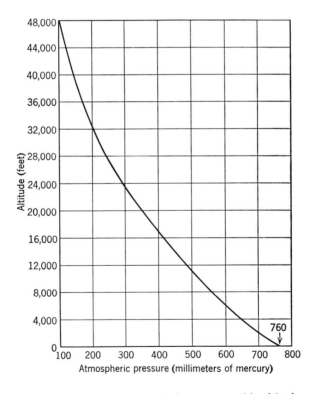

Fig. 1.4 Change in atmospheric pressure with altitude.

since the amount of air left is negligible. Therefore, the atmospheric pressure is balanced by the mercury column (side A). The mercury in this column can be expressed in terms of its weight or its height, either of which is the atmospheric pressure. Such a U-tube is a form of mercury manometer.

It must be emphasized that the standard or normal atmosphere refers to the atmospheric pressure at sea level and at 0°C. As the altitude increases, the height of the mercury column decreases. Most vacuum systems are operated at relatively low altitudes, and the range of atmospheric pressure involved has little effect, as a rule, on design considerations. A knowledge of how the pressure changes with altitude becomes particularly important in connection with various space studies. Figures 1.4 and 1.5 show the relationship between pressure and altitude. The basic unit of pressure used in vacuum work comes from the height of the mercury column (at sea level and at 0°C), which is 760 millimeters. It is often stated that this pressure is 760 millimeters of mercury, and the pressure in a vacuum system is often specified in terms of millimeters of mercury (mm Hg). For example, if the pressure in a vacuum system is 1 mm Hg, this pressure corresponds to $\frac{1}{760}$ th of an atmosphere.

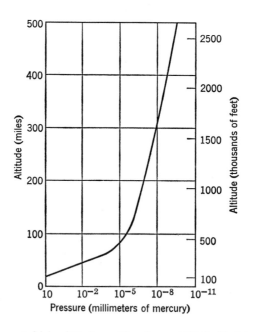

Fig. 1.5 Pressure at high altitudes. (Based on *ARDC Model Atmosphere*, a report of the Air Research and Development Command, 1959.)

1.4   Units of Pressure

Table 1.2 lists some common units of length, area, volume, and weight for reference purposes. Some of these relationships are needed in converting from one pressure unit to another. Table 1.3 shows the relationships between various pressure units used in vacuum practice. In examining this table it will be noted that the units Torr and millitorr are shown as being very nearly the same as the millimeter of mercury

Table 1.2   Some Common Units

Length

millimeter = mm
centimeter = cm
meter = m
inch = in.
foot = ft
yard = yd
1 mm = 0.1 cm = 0.04 in.
1 cm = 10 mm = 0.4 in.
1 m = 100 cm = 39.4 in.
1 in. = 2.54 cm
1 ft = 12 in. = 30.48 cm
1 yd = 3 ft = 36 in. = 81.4 cm

Area

square millimeter = sq mm = mm^2
square centimeter = sq cm = cm^2
square meter = sq m = m^2
square inch = sq in. = $in.^2$
square foot = sq ft = ft^2
square yard = sq yd = yd^2
$1 mm^2 = 0.01 cm^2 = 0.0015 in.^2$
$1 cm^2 = 100 mm^2 = 0.15 in.^2$
$1 m^2 = 10^4 cm^2 = 1535 in.^2$
$1 in.^2 = 6.45 cm^2$
$1 ft^2 = 144 in.^2 = 929 cm^2$
$1 yd^2 = 9 ft^2 = 8360 cm^2$

Volume

cubic millimeter = cu mm = mm^3
cubic centimeter = cu cm = cm^3 = cc
cubic meter = cu m = m^3
cubic inch = cu in. = $in.^3$
cubic foot = cu ft = ft^3
cubic yard = cu yd = yd^3
liter = l
$1 mm^3 = 10^{-3} cc = 0.00006 in.^3$
$1 cc = 10^3 mm^3 = 10^{-3} l = 0.06 in.^3$
$1 m^3 = 10^6 cc = 1000 l = 34.7 ft^3$
$1 l = 1000 cc = 61 in.^3$
$1 in.^3 = 16.4 cc$
$1 ft^3 = 1728 in.^3 = 28.4 l$
$1 yd^3 = 27 ft^3 = 7668 l$

Weight

gram = g
kilogram = kg
ounce = oz
pound = lb
ton = ton
1 oz = 28.6 g
1 lb = 16 oz = 454 g
1 ton = 2000 lb = 909 kg
1 g = 0.0350 oz
1 kg = 1000 g = 2.205 lb

Table 1.3   Some Useful Pressure Relationships

standard or normal atmosphere = atmospheric pressure (sea level,
0°C) = 1 atmosphere = 1 A.P. = 1 atm
millimeter of mercury = mm of Hg = mm Hg
micron of mercury = μ of Hg = μ Hg (or μ)
1 atm = 760 mm Hg = 760,000 μ Hg (or μ) = 29.92 in. Hg =
14.7 lb/in.² = 14.7 psi = 1,013,250 dynes/cm²
1 mm Hg = 0.0013 atm (roughly one one-thousandth of an atmosphere) = 1000 μ Hg (or μ)
1 μ Hg (or 1 μ) = 0.0000013 atm (roughly one millionth of an atmosphere) = 0.001 mm Hg
1 Torr = 1 mm Hg (very nearly)
1 millitorr = 0.001 Torr = 1 μ Hg (or μ) (very nearly)
1 bar = one million (10⁶) dynes/cm²
1 millibar (1 mb) = one thousand (10³) dynes/cm²
1 microbar (1 μb) = 1 dyne/cm² = 0.745 μ Hg (or μ)

and the micron of mercury (μ Hg or μ) respectively. For all practical
purposes these units can be used interchangeably. Actually the Torr
and millitorr are coming into more general use and have been recommended for adoption as standards by the American Vacuum Society.
However, commercial equipment in the United States is calibrated in
millimeters of mercury or microns of mercury. Consequently these
latter units are used in the early chapters of this book. In many cases
both millimeters of mercury and Torr are listed. In the later chapters
Torr and microns are used exclusively. The following examples show
how one pressure unit is converted to another.

EXAMPLE 1

A vacuum gauge, e.g., a mercury manometer, shows the pressure in
a vacuum system to be 28 inches of Hg (in. Hg). What is the pressure
in the system in millimeters of Hg? Gauges which read in in. Hg are
normally calibrated so that zero corresponds to an atmosphere. Since
an atmosphere corresponds to 29.92 in. Hg, when the gauge reads this
value, it will correspond to a very good vacuum. Consequently, a
reading of 28 in. Hg will correspond to a pressure of 29.92 − 28 = 1.92
in. Hg. Now 1 in. = 2.54 cm or 25.4 mm. Therefore, 1.92 in. =
1.92 × 25.4 = 48.8 mm Hg. In using this type of gauge, a value of
30 in. instead of 29.92 in. is often used as an approximation for a very
good vacuum.

EXAMPLE 2

The pressure in a vacuum system is found to be 10 mm Hg. What is the pressure in psi?

In Table 1.3 it is seen that 1 atm = 760 mm Hg or 14.7 psi. Therefore,

$$10 \text{ mm Hg} = \frac{10}{760} \times 14.7 = 0.19 \text{ psi}$$

or

$$1 \text{ mm Hg} = 0.0013 \text{ atm} = 0.0013 \times 14.7 \text{ psi}$$

Therefore,

$$10 \text{ mm Hg} = 10 \times 0.0013 \times 14.7 = 0.19 \text{ psi}$$

It should be noted that 1 atm = 14.7 psi corresponds to about 1 ton/ft². This is an important consideration in the design of vacuum vessels.

In practice the most common type of computation encountered involves the use of decimals and powers of ten. Because of its importance, this type of manipulation will be treated separately.

1.5   Decimals and Powers of Ten

Table 1.4 lists a number of powers of ten for reference purposes. Simple rules can be stated for using the table and also for extending

Table 1.4   Powers of Ten

$10^6 = 1,000,000$	$10^{-6} = 0.000001 = \dfrac{1}{1,000,000}$
$10^5 = 100,000$	$10^{-5} = 0.00001 = \dfrac{1}{100,000}$
$10^4 = 10,000$	$10^{-4} = 0.0001 = \dfrac{1}{10,000}$
$10^3 = 1000$	$10^{-3} = 0.001 = \dfrac{1}{1000}$
$10^2 = 100$	$10^{-2} = 0.01 = \dfrac{1}{100}$
$10^1 = 10$	$10^{-1} = 0.1 = \dfrac{1}{10}$
$10^0 = 1$	

it to other powers. The number 10^3 can be expressed as 1000, where there are three zeros after the 1, i.e., the number of zeros is equal to the power. Then 10^9 would have nine zeros after the 1 (1,000,000,000). A number like 4000 is expressed as 4×10^3, and, in reverse, a number like 6×10^3 is 6000. For a negative number such as 10^{-3} the decimal point is moved three places to the left, i.e., $10^{-3} = 0.001$. This means that the number of zeros after the decimal is one less than the power, two in this case. The quantity 10^{-6} would be 0.000001, with five zeros after the decimal. A number such as 5×10^{-4} would be 0.0005, and, in reverse, a number like 0.006 would be 6×10^{-3}.

Some other simple rules can be stated as follows.

Dividing by a power of ten where the power is positive is the same as multiplying by the same power of ten but with the power negative. Dividing by 10^3 is the same as multiplying by 10^{-3}, i.e., $1/10^3 = 10^{-3}$ or 0.001. Similarly, dividing by 5×10^2 is the same as multiplying by $1/(5 \times 10^2) = \frac{1}{5} \times 10^{-2}$ or 0.2×10^{-2} or 2×10^{-3}. Dividing by a power of ten with the power negative is the same as multiplying by the same power of ten with the power positive. Dividing by 10^{-3} is the same as multiplying by $1/10^{-3} = 10^3$ or 1000. Also, dividing by 2×10^{-4} is the same as multiplying by $1/(2 \times 10^{-4}) = \frac{1}{2} \times 10^4$ or 5×10^3.

In multiplying powers of ten together simply add the powers. Of course, if one of the powers is negative and the other positive, the sum is the difference between the two. Suppose 10^{-4} is multiplied by 10^3. The answer is 10^{-1} ($-4 + 3 = -1$). Multiplying 10^{-1} by 10^3 gives 10^2 ($3 - 1 = 2$) and multiplying 10^{-4} by 10^{-3} gives 10^{-7} ($-3 - 4 = -7$).

EXAMPLE 1

The pressure in a vacuum system is read as 10^{-2} μ. What is the pressure in mm Hg?

$$1 \text{ mm Hg} = 1000 \ \mu \quad \text{or} \quad 1 \ \mu = \frac{1}{1000} = 10^{-3} \text{ mm Hg}$$

Therefore,

$$10^{-2} \ \mu = 10^{-2} \times 10^{-3} = 10^{-5} \text{ mm Hg}$$

EXAMPLE 2

The pressure in a vacuum system is read as 2×10^{-4} mm Hg. What is the pressure in microns?

$$1 \text{ mm Hg} = 1000 \ \mu$$

Therefore,

$$2 \times 10^{-4} \text{ mm Hg} = 2 \times 10^{-4} \times 1000 = 2 \times 10^{-4} \times 10^{3}$$
$$= 2 \times 10^{-1} = 0.2 \ \mu$$

The converting of millimeters of mercury to microns and vice versa is a common procedure in high vacuum practice. The thing to remember is that a millimeter of mercury is one thousand times as large as a micron. Therefore, any given pressure is expressed as fewer millimeters of mercury than microns.

1.6 Degrees of Vacuum

Vacuum systems are often classified according to their operating pressures. One method of classifying which is being widely accepted is the following:

Low vacuum	760 to 25 mm Hg
Medium vacuum	25 to 10^{-3} mm Hg

or

Rough vacuum	760 to 1 mm Hg
Fine vacuum	1 to 10^{-3} mm Hg

and

High vacuum	10^{-3} to 10^{-6} mm Hg
Very high vacuum	10^{-6} to 10^{-9} mm Hg
Ultra-high vacuum	10^{-9} mm Hg and less

The pressure ranges in Torr would be very nearly the same numerically.

1.7 Mean Free Path of a Gas

As the molecules move about they collide with each other. The distance any molecule travels before colliding with another molecule will be different from collision to collision. However, we can speak of an average distance that a molecule travels before colliding with another molecule. This average distance between collisions is called the *mean free path*, a very important quantity since it determines the design of various types of pumps and enters into many aspects of vacuum practice. As the pressure of a gas is decreased, there will be fewer molecules available to make collisions with each other. Consequently, the average distance between collisions (the mean free path) will increase as the pressure decreases. For any vacuum vessel, it is possible to reduce the pressure to the point where molecules will go from one side of the container to the other without hitting other molecules. This is often called the *free molecule* pressure region. There is also a pressure region called the *viscous* region, where molecules make many collisions with other molecules in passing across a container. Generally speaking, vacuum gauges are designed to operate in one or the other of these pressure regions. Also, the nature of the flow of a gas is different in these two pressure regions. Rules for handling the flow of gas through pipes, liquid nitrogen traps, etc., are given in later chapters.

The mean free path of one gas will not be the same as the mean free path of another gas (same pressure). This is because the mean free path depends on the size of a molecule, and different gases have molecules of different sizes. The smaller the molecules, the larger will be the mean free path. In spite of this fact, it is possible to give a relationship between mean free path and pressure, which is intended for air but will still give useful approximate values for other gases. This relationship is

$$\text{Mean free path, } L \text{ (in centimeters)} = \frac{5}{\text{pressure (in microns)}}$$

Mean free path is often written MFP or mfp. Consider three cases to get some idea of the magnitudes of mean free paths.

EXAMPLE 1

Consider air at a pressure of 1 atm (atmospheric pressure at sea level and 0°C). This pressure = 760 mm Hg = 760 × 1000 = 760,000 μ. Therefore

$$\text{Mean free path} = \frac{5}{760,000} = \frac{5}{0.76} \times 10^{-6} = 6.6 \times 10^{-6} \text{ cm}$$

or   approx. 2.6×10^{-6} in.

Consequently, at 1 atm a molecule will make about half a million collisions in traveling a distance of 1 in. Of course the 1 in. distance is not in a straight line since the molecule travels a very erratic path because of the collisions.

EXAMPLE 2

Take air at a pressure of 1 μ. In this case the mean free path = $\frac{5}{1}$ = 5 cm. Here the molecule travels an average distance of 5 cm or about 2 in. between collisions.

EXAMPLE 3

Take air at a pressure of 10^{-6} mm Hg. Now 10^{-6} mm Hg = $10^{-6} \times 10^3 = 10^{-3}$ μ. Therefore,

$$\text{Mean free path} = \frac{5}{10^{-3}} = 5 \times 10^3 = 5000 \text{ cm}$$

or

$$\frac{5000}{2.54} = 1970 \text{ in.}$$

or

$$\frac{1970}{12} = 164 \text{ ft}$$

At this pressure the molecules will cross most vacuum vessels several times per second.

1.8   Temperature Scales

The "hotness" or "coldness" of a body is usually expressed in terms of a temperature. A common method of measuring temperature is with a thermometer consisting of an evacuated glass tube partially filled with mercury. The mercury expands as it gets hotter and contracts as it cools so that the length of the mercury column gives a measure of the temperature of the body with which it is in contact. In most cases the freezing point and boiling point of water are used as "fixed points" to determine a temperature scale.

The common temperature scale used in the home, in weather forecasting, and in most engineering work is the Fahrenheit (F) scale. Here the boiling point of water is taken as 212°F and the freezing point of water is taken as 32°F so that there are 180° (212° − 32°) between these temperatures. In most vacuum work other temperature scales are more commonly used, particularly the centigrade (C) and absolute (A) or Kelvin (K) scales. On the centigrade scale the boiling and freezing points of water are 100°C and 0°C, respectively, whereas on the absolute or Kelvin scale these temperature points are 373°K and 273°K. Consequently there are 100° between these fixed points on

Table 1.5   Temperature Scales

Centigrade Scale, °C	Fahrenheit Scale, °F	Absolute (Kelvin) Scale, °A or °K	Rankine Scale, °R	
100	212	373	671.4	Boiling point of water
90	194	363	653.4	water
80	176	353	635.4	
70	158	343	617.4	
60	140	333	599.4	
50	122	323	581.4	
40	104	313	563.4	
30	86	303	545.4	
20	68	293	527.4	
10	50	283	509.4	Freezing
0°	32°	273° 491.4°		point of water
−10	14	263	473.4	water
−20	−4	253	455.4	
−30	−22	243	437.4	
−40	−40	233	419.4	
−50	−58	223	401.4	
−60	−76	213	383.4	
−70	−94	203	365.4	
−80	−112	193	347.4	Temperature
−90	−130	183	329.4	of dry ice
−100	−148	173	311.4	
−110	−166	163	293.4	
−120	−184	153	275.4	
−130	−202	143	257.4	
−140	−220	133	239.4	
−150	−238	123	221.4	
−160	−256	113	203.4	
−170	−274	103	185.4	
−180	−292	93	167.4	
−190	−310	83	149.4	Temperature
−200	−328	73	131.4	of liquid nitrogen
−210	−346	63	113.4	
−220	−364	53	95.4	
−230	−382	43	77.4	
−240	−400	33	59.4	
−250	−418	23	41.4	
−260	−436.	13	23.4	
−270	−454	3	5.4	Absolute
−273°	−459.4°	0°	0°	zero

both these scales. A final temperature scale sometimes used in vacuum work is the Rankine scale (R). Here the boiling and freezing points of water are 671.4°R and 491.4°R, respectively, with 180° between these points. Therefore, the magnitude of 1°R is the same as that of 1°F.

Table 1.5 provides convenient information for converting from one of these temperature scales to any other. Also shown in this table are the temperatures of dry ice and liquid nitrogen, which are commonly used in cold traps. If the table is not used, conversion from one temperature to another can readily be done by using these simple conversion formulas:

$$\text{Number of } °C = \tfrac{5}{9} \times \text{number } °F - 17.8 = \text{number } °K - 273$$

$$= \tfrac{5}{9} \times \text{number } °R - 273$$

$$\text{Number of } °F = \tfrac{9}{5} \times \text{number } °C + 32 = \tfrac{9}{5} \times \text{number } °K - 459.4$$

$$= \text{number } °R - 459.4$$

$$\text{Number of } °K = \text{number } °C + 273$$

$$= \tfrac{5}{9} \times \text{number } °F + 255.2 = \tfrac{5}{9}°R$$

$$\text{Number of } °R = \tfrac{9}{5} \times \text{number } °C + 491.4$$

$$= \text{number } °F + 459.4 = \tfrac{9}{5}°K$$

1.9   Effect of Pressure on a Gas

It is well known that when a gas is compressed it occupies less space. Conversely if the pressure on a gas is decreased, it will expand. It is of interest to note how the volume occupied by a gas depends on the pressure which is exerted on the gas. The law governing this behavior is known as Boyle's law, which can be expressed so: The volume of a gas varies inversely as the pressure of the gas as long as the temperature remains constant. It is important to note that the temperature must remain unchanged. As will be seen later, heating or cooling a gas will cause it to expand or contract.

An example of what happens to the volume of a gas when it is compressed is given in Fig. 1.6. Here the gas initially occupies 10 liters (10 l) in a cylinder at a pressure of 50 mm Hg (Fig. 1.6a). When the pressure is increased to 100 mm Hg the gas (Fig. 1.6b) then occupies a volume of 5 l, i.e., when the pressure is doubled the volume is reduced

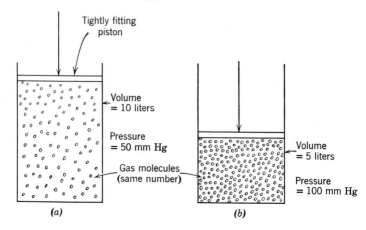

Fig. 1.6   Effect of pressure on a gas.

to one-half its original value. For the relationship to hold, the temperature of the gas before compression must equal the temperature after compression. One way of achieving this would be to thermostatically control the temperature of the air in a vessel containing the cylinder. One other point to note is that there must be no loss or gain of gas during this compression process.

If several gases are contained in a vessel, each gas will contribute its own pressure, and the total pressure will be the sum of the pressures of all the gases. The pressure of each gas is its pressure when it occupies the vessel by itself. This is called the *partial pressure* of the gas. The pressure of each individual gas in a mixture obeys Boyle's law just as does the pressure of all the gases. In each case the volume is taken as the total volume available in the container. The term partial pressure has the same meaning in the case of vapors. However, Boyle's law can be applied only when vapors are not near their liquefaction points. Often Boyle's law is expressed as: The volume of a gas times its pressure is constant when the temperature remains unchanged. This can also be stated: The initial pressure times the initial volume equals the final pressure times the final volume (temperature constant).

EXAMPLE 1

A gas at a pressure of 10^{-4} mm Hg occupies a volume of 2000 cc. If the volume is increased to 8000 cc, what is then the pressure of the gas? Assume the temperature of the gas does not change.

By Boyle's law, initial pressure $\times$ initial volume = final pressure $\times$ final volume. Therefore,

$$10^{-4} \times 2000 = \text{final pressure} \times 8000$$

or

$$\text{Final pressure} = \frac{10^{-4} \times 2000}{8000} = \frac{1}{4} \times 10^{-4}$$

$$= 0.25 \times 10^{-4} \text{ mm Hg} = 2.5 \times 10^{-5} \text{ mm Hg}$$

EXAMPLE 2

A mixture of nitrogen and helium occupies a volume of 100 l, and the total pressure of the mixture is 20 mm Hg. Suppose the partial pressures of the nitrogen and helium are 15 mm Hg and 5 mm Hg respectively. If the volume is reduced to 25 l, with the temperature constant, what then is the partial pressure of the helium?

The initial volume of helium is 100 l, and its final volume is 25 l. By Boyle's law, $5 \times 100 =$ final partial pressure of helium $\times$ 25. Therefore,

$$\text{Final partial pressure of helium} = \frac{5 \times 100}{25} = 20 \text{ mm Hg}$$

Similarly, the final partial pressure of the nitrogen can be found, thus:

$$15 \times 100 = \text{final partial pressure of nitrogen} \times 25$$

or

$$\text{Final partial pressure of nitrogen} = \frac{15 \times 100}{25} = 60 \text{ mm Hg}$$

The final pressure of the mixture will be the sum of the final partial pressures of the nitrogen and the helium or $20 + 60 = 80$ mm Hg. This could also be found directly from Boyle's law, thus:

$$20 \times 100 = \text{final pressure of mixture} \times 25$$

or

$$\text{Final pressure of mixture} = \frac{20 \times 100}{25} = 80 \text{ mm Hg}$$

1.10   Effect of Temperature on a Gas

As has been mentioned, heating a gas causes it to expand, whereas cooling it causes it to contract. The law governing this behavior is known as Charles' law and can be stated: The volume of a gas varies directly as the absolute temperature when the pressure is kept constant. Note that the temperature scale used is the absolute or Kelvin scale and that the pressure must not change. In addition, the quantity of gas must not change. This law is often expressed as: The volume of a gas divided by its absolute temperature is constant as long as the pressure does not change. Another way of stating the law is: The initial volume of a gas divided by its initial absolute temperature equals the final volume divided by its final absolute temperature (pressure constant).

Figure 1.7 illustrates Charles' law. Here a gas is contained in a cylinder with a movable piston which seals off the gas. Assume that the piston is weightless so that it exerts no pressure on the gas. Initially the gas, occupying a volume of 10 l, is at a temperature of 0°C (273°K) and is subject to 1 atm pressure. Suppose now that the gas is heated to a temperature of 273° (546°K). It will be found that the piston will rise until the volume occupied by the gas becomes 20 l. Of course the pressure is still the same, viz., 1 atm. Therefore, when the temperature is doubled on the absolute scale (273°K to

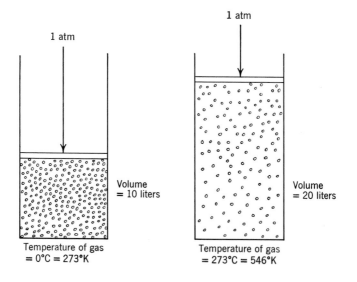

Fig. 1.7   Effect of heating a gas.

546K°) the volume also doubles. Again, no gas is lost or gained in the process.

EXAMPLE 1

A gas has a volume of 5 ft³ at a temperature of 20°C. What will the volume of the gas be at 100°C (pressure constant)?

By Charles' law, initial volume divided by initial absolute temperature equals final volume divided by final absolute temperature.

$$\text{Initial absolute temperature} = 20° + 273° = 293°\text{K}$$

$$\text{Final absolute temperature} = 100° + 273° = 373°\text{K}$$

Therefore,

$$\frac{5}{293} = \frac{\text{final volume}}{393}$$

or

$$\text{Final volume} = \frac{5 \times 393}{293} = 6.7 \text{ ft}^3$$

EXAMPLE 2

Suppose a gas occupies a volume of 1000 cc at 200°C. What temperature will be required to reduce the volume to 800 cc (pressure constant)?

The initial absolute temperature is 200° + 273° = 473°K. Therefore,

$$\frac{1000}{473} = \frac{800}{\text{final absolute temperature}}$$

or

$$\text{Final absolute temperature} = \frac{800 \times 473}{1000} = 378.4°\text{K}$$

$$= 378.4° - 273° = 105.4°\text{C}$$

1.11   The General Gas Law

In Sections 1.9 and 1.10 the effect of pressure on volume (temperature constant) and the effect of temperature on volume (pressure constant) have been discussed. In practice changes might occur simultaneously in all three quantities—temperature, pressure, and volume. However, there is a relationship governing the changes in these quantities which is known as the General Gas Law. This law states that the product of the pressure and the volume when divided by the absolute temperature gives a quantity which is always constant. The amount of gas involved must not change. Another way of expressing the law is: The initial pressure of a gas times its initial volume, divided by its initial absolute temperature, equals the final pressure times the final volume, divided by the final absolute temperature.

Referring again to Fig. 1.7, if the pressure on the gas were increased at the same time that the temperature was increased, the gas would not have expanded as much as is shown. The pressure could have been increased by adding weights on the piston. Consider the special case where the pressure is doubled and the absolute temperature is also doubled. Then it would be found that the volume of the gas would remain unchanged, i.e., it would still be 10 l. However, in most cases changes will occur in all three quantities—temperature, pressure, and volume.

EXAMPLE 1

A gas occupies a volume of 200 l at a pressure of 10^{-4} mm Hg and a temperature of 20°C. If the pressure is increased to 10^{-3} mm Hg and the temperature raised to 50°C, what is the volume of the gas?

The initial absolute temperature is $20° + 273° = 293°$K and the final absolute temperature is $50° + 273° = 323°$K. Therefore, by the General Gas Law,

$$\frac{10^{-4} \times 200}{293} = \frac{10^{-3} \times \text{final volume}}{323}$$

or

$$\text{Final volume} = \frac{10^{-4} \times 200 \times 323}{293 \times 10^{-3}} = \frac{200 \times 323 \times 10^{-4} \times 10^{3}}{293}$$

$$= \frac{200 \times 323 \times 10^{-1}}{293} = \frac{20 \times 323}{293} = 22.0 \text{ l}$$

EXAMPLE 2

A gas has a volume of 1200 cc, a temperature of 0°C, and a pressure of 100 μ. If the volume changes to 2000 cc and the pressure becomes 50 μ, what is the temperature of the gas?

The initial absolute temperature of the gas is $0° + 273° = 273°$K. By the General Gas Law

$$\frac{100 \times 1200}{273} = \frac{50 \times 2000}{\text{final absolute temperature}}$$

or

$$\text{Final absolute temperature} = \frac{50 \times 2000 \times 273}{100 \times 1200} = \frac{5 \times 273}{6} = 227.5°\text{K}$$

$$= 227.5° - 273° = -45.5°\text{C}$$

In vacuum practice a fixed amount of gas is not involved, as a rule. The interest is in removing gas (and vapor) from some vessel which has a fixed volume. How rapidly molecules are being removed will be indicated by how the pressure drops, and the final pressure reading will be an indication of the number of molecules remaining. Methods for treating the pumping of vacuum vessels and related phenomena will be considered in subsequent chapters. Although it is not spelled out there, these methods are derived from the General Gas Law.

1.12   Some Other Characteristics of Gases

In vacuum work there are some other characteristics of gases which are quite important in certain circumstances. Characteristics which will be discussed in more detail later, particularly in connection with leak detectors and vacuum gauges, are *diffusion, viscosity,* and *thermal conductivity.*

Everyone is familiar with the fact that if a bottle of ammonia is opened at one corner of a room, the odor of ammonia is soon detected throughout the room. It is said that the ammonia vapor has diffused through the air in the room because of the relatively high speed of the ammonia molecules. In vacuum work, where the pressures are analogously low, gases and vapors will diffuse quite rapidly because there are fewer collisions.

If a stream of gas moves through a stationary gas, the moving gas tends to "drag" along some of the stationary gas. At the same time the moving gas is slowed down. The amount of "drag" involved is expressed in terms of a quantity called the viscosity, which is a characteristic of each gas or vapor. This property of a gas or vapor is important in leak detection since it determines the ease with which the gas or vapor will pass through a small leak.

It is well known that copper is a better conductor of heat than is wood. Similarly gases and vapors differ in their ability to conduct heat. It is possible to speak of the thermal conductivity of a particular gas or vapor. In the micron range of pressure, the conductivity of a gas or vapor depends on the pressure, so this property becomes the basis on which certain classes of vacuum gauges work, viz., the Pirani and thermocouple gauges.

The material in the sections above has been concerned largely with gases. However, much of it can be applied to vapors as long as the temperature is well above the liquefaction temperature. For reference purposes, the important formulas governing the behavior of gases are summarized in Appendix C. Also, the properties of some gases and vapors important in vacuum work are listed in Appendix A.

TWO

Vacuum Systems

2.1 Elements of a Vacuum System

In spite of rather large differences among various vacuum systems, all of them possess some elements in common. Figure 2.1 shows a vacuum system intended for operation in the 10^{-6} mm Hg range. The actual lower limit of pressure which will be reached with such a system will depend on the choice and design of components as well as on the care taken in fabrication and cleaning. The major components involved in this system are the vacuum vessel, the pumps, and the piping connecting the vessel and the pumps. Other components include vacuum gauges for assessing the vacuum conditions, valves, a baffle, a cold trap, and miscellaneous hardware including seals, protective devices, etc.

The vacuum vessel and the various pumping lines are constructed of glass or metal. The use of glass is confined to relatively small systems, because of its fragility and the cost and difficulty of fabricating large glass parts. Either glass tubing or piping can be used for pumping lines, the latter being more rugged. Metals commonly used are copper, brass, steel, aluminum, and stainless steel. Tubing is often used for pumping lines in small systems, but piping is usually used in the larger units. The vapor pump can be an oil or a mercury diffusion

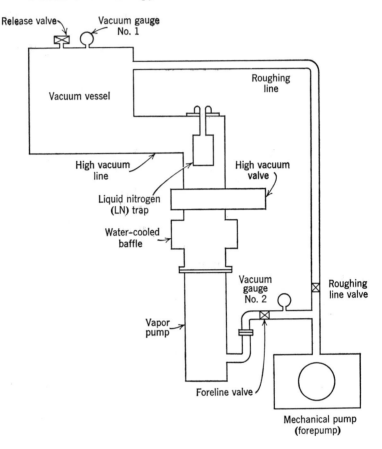

Fig. 2.1 Typical vacuum system involving mechanical and vapor pumps.

pump, although sometimes ejector stages are used either separately or in a combination diffusion-ejector pump. If a mercury pump is used, a cold trap (usually a water-cooled baffle) should be included between this pump and the mechanical pump so that mercury doesn't get into the mechanical pump or into the air. Mercury is quite toxic and precautions should be taken to keep its vapor out of the air. The type of mechanical pump used for a system operating in this pressure range is usually a rotary oil-sealed pump. Sometimes a Roots blower-type pump is used in conjunction with a rotary oil-sealed pump.

It will be noted that two vacuum gauges are included in the system— vacuum gauge No. 1 on the vacuum vessel for measuring the pressure in this vessel and vacuum gauge No. 2 in the foreline for measuring the forepressure. Gauge No. 1 will be more sensitive than gauge No. 2, that is, it will measure lower pressures. It is usually desirable to include a roughing line, as shown, so that the vessel can be pumped down without the vapor pump having to be cooled down. In this particular system four valves are included. The high vacuum valve makes it possible to isolate the pumps from the rest of the system. The foreline valve is closed when pumping with the forepump through the roughing line. Often a separate roughing pump is used. After the vacuum vessel has been pumped down, the roughing line valve is closed and the foreline and high vacuum valves are opened. The release valve shown on the vacuum vessel is used to let the whole system or the part of the system up to the high vacuum valve down to air (atmospheric pressure). In many cases a second release valve is included in the foreline, so arranged that the foreline valve closes and the release valve opens in case of power failure.

A water-cooled baffle is normally used with oil diffusion pumps to keep oil out of the system. In the case of mercury pumps lower temperatures are required so that a dry ice trap or Freon refrigeration is used. The liquid nitrogen trap pumps vapors from the system. Often a manifold is included between the vapor pump and the vessel and various items of auxiliary equipment, including cold traps and valves, can be included in such a manifold. Of course, automatic and protective devices can be incorporated in any vacuum system.

2.2   Some Types of Vacuum Systems

Vacuum systems are used, of course, in many different applications, including vacuum impregnation of metals, vacuum casting of metals, freeze drying of foods and pharmaceuticals, evaporation of metals, nuclear particle accelerators, research apparatus, etc. The exact nature of the system will depend on its application. A principal distinction is the pressure range in which the system is to be operated. Suggested classifications of pressure range have been included in Section 1.6. Operating in the millimeter of mercury or micron range generally requires the use of mechanical pumps only, or in the cases of very large installations, steam ejector pumps. As better vacuum conditions are needed, i.e., lower pressures, other types of pumps have to be included and more attention has to be given to the choice of components and construction materials. In addition it may be necessary to include cold traps. Obviously, suitable vacuum gauges for measuring the vacuum conditions will have to be provided. As the operating pressure is reduced, not only must greater attention be given to the choice of components and construction materials but, also, fabrication and cleaning techniques become more critical.

Vacuum systems are often divided into "static" and "dynamic" systems. A static system is one that has been pumped down and then sealed off. Radio and television tubes are examples of static vacuum systems. In spite of the most careful efforts to remove gases and vapors from the construction materials, there will still be a gradual evolution of gases and vapors, causing an increase in the pressure. To avoid this, gettering materials are usually included in static systems. These are chemically active materials which are usually deposited in the form of a thin film on the inside of the system and which combine with the gases and vapors. Dynamic systems are continually pumped systems. Essentially all industrial vacuum processes involve the use of dynamic systems and these processes often involve a cycling operation.

2.3   Quantity of Gas and Throughput

The *quantity of a gas* is a measure of the number of molecules in the gas; this measure is the product of the pressure times the volume, as determined at the temperature of the gas. Common units are: micron liters (μl), millimeters of mercury liters (mm Hg l), atmospheric cubic centimeters (cc of volume at atmospheric pressure or std cc), micron cubic feet (μft^3), etc. Actually there is more concern with the rate of flow of gas, commonly called *throughput*, in vacuum practice. Throughput is the quantity of gas at a specified temperature (usually the gas temperature) passing an open cross section of the vacuum system (usually a pipe) in unit time (a second, a minute, an hour, etc.). This is illustrated by Fig. 2.2, which shows a segment of pipe that might

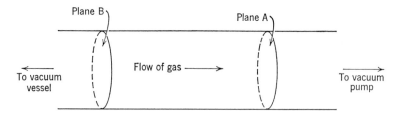

Fig. 2.2   Rate of flow of a gas.

connect a pump and a vacuum vessel. In order for gas to flow from left to right, the pressure on the left side must be higher than the pressure on the right side. If it were possible to directly measure the quantity of gas passing a plane A (cross section of pipe) in a specified time, then this throughput could be found by dividing the quantity by the time. In practice it is possible to measure the pressure in plane A directly with a vacuum gauge and obtain the volume of gas passing through plane A in unit time by other means, to be discussed later. Suppose the gas in the pipe between planes A and B passes plane A in unit time, say 1 second (sec). Suppose, further, that this volume is 100 l and the pressure in plane A is 10^{-5} mm Hg. Then the throughput of gas is 10^{-5} mm Hg times 100 l divided by the time (1 sec) or

$$\frac{10^{-5} \times 100}{1} = 10^{-3} \text{ mm Hg l/sec}$$

The symbol Q is normally used to designate the throughput. Various units are used for Q, the relationship among several of them being listed in Table 2.1. Often standard leaks used with helium leak detectors are calibrated in standard cc per second (std cc/sec) or micron cubic feet per hour (μcfh). A standard cubic centimeter is a volume of 1 cc at normal atmospheric pressure (760 mm Hg) and at normal temperature (0°C). It is easy to convert from one unit to another by

Table 2.1   Some Units of Throughput

1 standard cubic centimeter per second = 1 std cc per sec or 1 atm cc per sec = 1 cubic centimeter of gas at atmospheric pressure (often written as 1 cc/sec)

1 micron cubic foot per hour = 1 micron cu ft per hr = 1 μ ft^3/hr = 1 μcfh = 1 cubic foot of gas per hour at 1 micron Hg pressure

1 micron liter per second = 1 μl per sec = 1 liter of gas per second at 1 micron Hg pressure

1 std cc/sec = 96,700 μcfh (approximately 100,000) = 760 μl/sec

1 μcfh = 0.0079 μl/sec = 0.0000104 std cc/sec (roughly one hundred thousandth or 10^{-5} of a std cc/sec)

1 μl/sec = 127 μcfh = 0.0013 std cc/sec

Note: / stands for per

1 ft^3 = 28.3 l

1 μ Hg = 1 millitorr (very nearly)

referring to Table 2.1. It should be noted that the value of the through-put will increase as the time involved increases. For example, a throughput of 10^{-5} std cc/sec would be

$$60 \times 10^{-5} = 6 \times 10^{-4} \text{ std cc/min}$$

or

$$60 \times 60 \times 10^{-5} = 3600 \times 10^{-5} = 3.6 \times 10^{-2} \text{ std cc/hr}$$

In going from units per hour to units per minute or units per second the procedure is, of course, reversed. Since a cubic foot is larger than a liter, one would need fewer cubic feet than liters at a given pressure and in a given time to equal the same throughput. It should also be noted that a leak into a vacuum system represents a certain through-put. However, in this case one usually refers to a leak or a leak rate.

EXAMPLE

Air from the atmosphere is observed to flow through a leak in a system at the rate of 3×10^{-10} std l/sec. What is the rate of leak (throughput) in μcfh?

From Table 2.1 it is seen that

$$1 \text{ std cc/sec} = 96,700 \ \mu\text{cfh}$$

Now

$$1 \text{ std cc/sec} = \frac{1}{1000} \text{ std l/sec}$$

so

$$1 \text{ std l/sec} = 1000 \times 96,700 \ \mu\text{cfh}$$

Therefore,

$$3 \times 10^{-10} \text{ std l/sec} = 3 \times 10^{-10} \times 1000 \times 96,700 \ \mu\text{cfh}$$

$$= 2.9 \times 10^{-2} \ \mu\text{cfh}$$

2.4   Comparison with an Electric Circuit

Sometimes it is convenient to compare a vacuum system with a simple electric circuit to help remember some vacuum terms. But it must be kept in mind that this comparison cannot be carried too far. It is possible to think of the vacuum system as constituting a circuit where air enters the system through leaks and is pumped through the system by vacuum pumps and ejected back into the atmosphere. The quantity of air passing any plane of the system per unit time (throughput) is always the same throughout the system. This flow of gas can be compared to an electrical current in a wire. Of course, if vapors should enter the system with the air, these vapors could be removed by a cold trap and therefore would not contribute to the throughput farther down the system. In addition, various parts of the vacuum system could contribute gases and vapors.

Comparing the vacuum circuit with a simple electric circuit, electric batteries would correspond to the vacuum pumps and the flow of electricity (electric current) would correspond to the flow of gas. When an electric current flows through a wire, the wire is heated and there is a drop in electrical voltage (potential) across the wire. One can compare the pressure drop in a vacuum system with the voltage drop in the wire. Furthermore, in the case of the electric circuit the wire is said to offer resistance to the flow of electricity. Similarly, in the case of the vacuum circuit the various parts of the vacuum system (piping, liquid nitrogen traps, etc.) offer *resistance* to the flow of gas.

In the case of the electric circuit there is a relationship between the electric potential drop (voltage drop), the electric current and the resistance of the circuit. This relationship is: potential drop equals electric current times resistance. The corresponding relationship for a vacuum circuit is: pressure drop equals throughput times gas flow resistance. Here, the pressure drop corresponds to potential drop, the throughput to electric current, and the gas flow resistance to electrical resistance. The resistance to gas flow is often referred to as *impedance*. The symbols P and W are often used for gas pressure and gas resistance (or impedance). Sometimes the symbol Z is used for impedance. Suppose the pressures at the ends of some portion of a vacuum system, say part of the piping, are P_1 and P_2 respectively, where P_1 is a higher pressure than P_2. Then the flow of gas will be from the region where the pressure is P_1 to the region at pressure P_2. According to the above relationship:

Pressure difference = resistance (or impedance) × throughput

or

$$\text{Resistance (or impedance)} = \frac{\text{pressure difference}}{\text{throughput}}$$

In symbols:

$$W = \frac{P_1 - P_2}{Q} \tag{2.1}$$

A comparison between the electric and vacuum circuits is shown in Fig. 2.3.

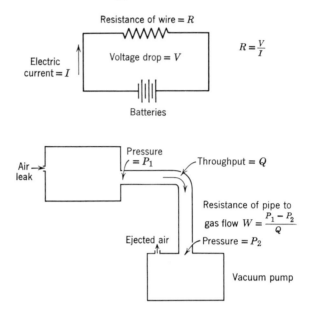

Fig. 2.3   Comparison between electric and vacuum circuits.

2.5   Comparison with Water Flow

Sometimes it is more useful to compare gas flow with water flow. Consider a closed water system where water is circulated through a system of pipes. An example would be the cooling system in an automobile. Then the water pump used would correspond, in a sense, to the vacuum pump(s) in the vacuum circuit. Also, just as the pressure continually drops throughout the water system, so does the pressure in the vacuum system. The flow of water, usually measured in units such as gallons per minute, corresponds to throughput in the vacuum case. Clearly, the smaller the pipe carrying water (for a given water pump), the smaller the rate of flow of water. Also the rate of water flow can be increased by adding more pipes to carry part of the water (pipes in parallel). Naturally, inserting additional pipes between existing pipes (in series) will increase the resistance to the flow of water so the rate of flow will decrease. These ideas can be carried over to the case of a vacuum system. However, it must be kept in mind that the nature of gas flow changes as the pressure changes and it is necessary to treat the resistance offered to gas flow differently for different pressure regions.

Regardless of the pressure region, the resistance to gas flow can be decreased by increasing the sizes of vacuum connections, such as the tubing or piping. Putting constrictions in the system has a drastic effect on the rate of gas flow and must be avoided, consistent with the purpose of the system. Particular attention must be given to the design of traps and valves so that they will perform adequately but will not unduly restrict the flow of gas. Long piping or tubing connections should be avoided, as should the use of pipes or tubes which are too small in diameter.

2.6   Resistances Connected in Series

If two pipes are connected together in line they are said to be in series. Each of the pipes will offer its own resistance to the gas flow and when put together the total resistance will be the sum of the resistances of the individual pipes. One can state for pipes A and B connected in series that

Resistance (or impedance) of [pipe A + pipe B]

= resistance of pipe A + resistance of pipe B

Naturally, this can be extended to any number of pipes or any combination of vacuum components when connected in series. For example, a diffusion pump might be connected to a vacuum chamber by a pipe, a liquid nitrogen trap, a water-cooled baffle, and a valve. Then the total resistance between the pump and the chamber will equal the sum of the resistances of the individual components or total resistance = resistance of pipe + resistance of liquid nitrogen trap + resistance of water-cooled baffle + resistance of valve. This case is illustrated in Fig. 2.4. A general formulation for any number of vacuum components 1, 2, 3, 4, etc., is

Total resistance (or impedance) = resistance of 1

+ resistance of 2 + resistance of 3, etc.

If the individual resistances are labeled W_1, W_2, W_3, etc., then the total resistance, W, is given by

$$W = W_1 + W_2 + W_3 + \cdots \qquad (2.2)$$

This is the same way that electrical resistances in series add together.

Thinking in terms of resistance or impedance gives a reasonable physical picture of the flow of gases through various vacuum com-

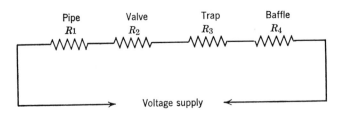

Corresponding electric resistances
total resistance $= R = R_1 + R_2 + R_3 + R_4$

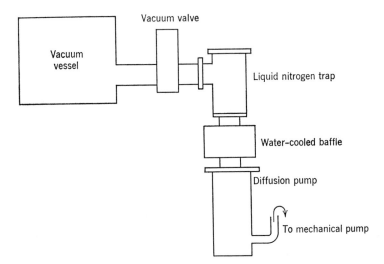

Fig. 2.4 Example of resistances (or conductances) connected in series.

Resistance (impedance) $= W$
Conductance $= U$

Values
  For all piping: W_1
  U_1
  Valve: W_2
  U_2
  Liquid nitrogen trap: W_3
  U_3
  Baffle: W_4
  U_4

Totals
  Impedance: $W = W_1 + W_2 + W_3 + W_4$
  Conductance: $\dfrac{1}{U} = \dfrac{1}{U_1} + \dfrac{1}{U_2} + \dfrac{1}{U_3} + \dfrac{1}{U_4}$
  $U = \dfrac{1}{W}$

32

ponents. However, it is customary to use the concept of *conductance* in vacuum practice since it can be combined with the speed of a pump to get the true pumping speed at the vacuum vessel. The conductance is simply the reciprocal of the resistance or impedance, i.e., conductance = 1/resistance. Consequently, the greater the resistance, the smaller will be the conductance and vice versa. Therefore, as much as possible is done to keep the conductance high so as to minimize interference with gas flow. If vacuum components 1, 2, 3, etc., are connected in series, the total conductance is given by

$$\frac{1}{\text{total conductance}} = \frac{1}{\text{conductance of 1}} + \frac{1}{\text{conductance of 2}} + \cdots$$

If the symbols U_1, U_2, U_3, etc., are used for the individual conductances, then the total conductance, U, is given by

$$\frac{1}{U} = \frac{1}{U_1} + \frac{1}{U_2} + \frac{1}{U_3} + \cdots \tag{2.3}$$

Also the total resistance and impedance are related in the following manner:

$$U = \frac{1}{W} \quad \text{or} \quad W = \frac{1}{U} \tag{2.4}$$

The units of conductance are those of volume per unit time. Common units are liters per second (l/sec), liters per minute (l/min), cubic feet per second (ft³/sec), and cubic feet per minute (cfm). Because of the relationship between resistance and conductance, the resistance units are those of time per unit volume, such as seconds per liter (sec/l), minutes per liter (min/l), seconds per cubic foot (sec/ft³), and minutes per cubic foot (min/ft³).

EXAMPLE

Consider a length of pipe and a water-cooled baffle connecting a pump to a vacuum chamber. If the conductances of pipe and baffle are calculated to be 200 l/sec and 50 l/sec respectively, what is the total conductance of pipe and baffle?

$$\frac{1}{\text{total conductance}} = \frac{1}{\text{conductance of pipe}} + \frac{1}{\text{conductance of baffle}}$$

or

$$\frac{1}{U} = \frac{1}{U_1} + \frac{1}{U_2}$$

and

$$\frac{1}{U} = \frac{1}{200} + \frac{1}{50} = \frac{1+4}{200} = \frac{5}{200}$$

Therefore, total conductance $U = 200/5 = 40$ l/sec. The total resistance will be $W = 1/U = 1/40 = 0.025$ sec/l. It will be noted that the total conductance is less than the conductance of either pipe or baffle and is nearest the smallest value of conductance (50 l/sec for the baffle).

2.7   Resistances Connected in Parallel

Often there will be more than one path available for the flow of gas. Gas will then flow through all available paths, which are said to be in parallel, and the rate of flow through each path will depend on its resistance. Clearly, the total resistance of several paths should be less than the resistance of any one path since there is more space available for the gas. Conversely, the combined conductance of all the paths will be larger than the conductance of any individual path. The paths available might be pipes going from a vacuum pump to a vessel or separate openings in a vacuum component. The total resistance of several vacuum components 1, 2, 3, etc., in parallel is given by

$$\frac{1}{\text{total resistance}} = \frac{1}{\text{resistance of 1}} + \frac{1}{\text{resistance of 2}} + \cdots$$

Using the symbol W again for resistance:

$$\frac{1}{W} = \frac{1}{W_1} + \frac{1}{W_2} + \frac{1}{W_3} + \cdots \tag{2.5}$$

Using   conductances:

Total conductance = conductance of 1 + conductance of 2 + $\cdots$ or using U for conductance:

$$U = U_1 + U_2 + U_3 + \cdots \tag{2.6}$$

An example of parallel paths is shown in Fig. 2.5. The path including the foreline, the diffusion pump, the water-cooled baffle, the valve, and the pipe would be treated as a series of resistances connected in parallel with the resistances of the roughing line and the roughing valve. To find the total resistance (or conductance), first find the total re-

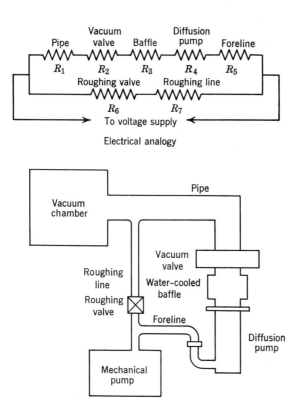

Fig. 2.5   Resistances (impedances) in parallel.

Resistance (impedance) $= W$; Conductance $= U$

Values

Pipe: W_1; U_1

Vacuum valve: W_2; U_2

Baffle: W_3; U_3

Diffusion pump: W_4; U_4

Foreline: W_5; U_5

Roughing line: W_6; U_6

Roughing valve: W_7; U_7

Totals

Roughing section: $W_R = W_6 + W_7$

$$\frac{1}{U_R} = \frac{1}{U_6} + \frac{1}{U_7}$$

Diffusion pump section: $W_D = W_1 + W_2 + W_3 + W_4 + W_5$

$$\frac{1}{U_D} = \frac{1}{U_1} + \frac{1}{U_2} + \frac{1}{U_3} + \frac{1}{U_4} + \frac{1}{U_5}$$

$$\frac{1}{W_T} = \frac{1}{W_R} + \frac{1}{W_D} \qquad U_T = U_R + U_D$$

35

sistance (or conductance) of the path including the diffusion pump and then combine this with the resistance (or conductance) of the roughing line and roughing valve.

EXAMPLE

Consider two pipes in parallel. If the conductances of the individual pipes are 200 l/sec and 1000 l/sec respectively, what is the total conductance of the two pipes?

Total conductance = conductance of pipe 1 + conductance of pipe 2

or

$$U = U_1 + U_2$$

and

$$U = 1000 + 200 = 1200 \text{ l/sec}$$

The total resistance is $W = 1/U = 1/1200 = 0.00083$ sec/l.

In the discussion above conductances for pipes, liquid nitrogen traps, etc., have been used without regard for the effect of the connections between the various parts. Actually this has to be taken into consideration. Various formulas for vacuum components are included in Appendix C.

2.8   The Use of Logarithms

In Chapter 1 some attention was given to the use of decimals and powers of ten. In pumping down a system from atmospheric pressure to the operating pressure there is generally a very large change in pressure. It is much more convenient to express the pressures as powers of ten rather than writing them out in decimal or fractional forms. Suppose the final pressure expressed as a power of ten is 10^{-5} mm Hg. This is a much more convenient form than either the decimal (0.00001) or the fractional ($\frac{1}{100,000}$) form. However, in many cases it is desirable to present data in a graphical form. If the vacuum quantities involved (such as the pressure) change by large amounts then it becomes impossible to present all values on the usual size of rectangular coordinate graph paper. An example of a graph using this type of paper is shown in Fig. 1.4. It is said that the vertical scale (ordinate) and the horizontal scale (abscissa) are linear, i.e., each unit of length represents the same amount of the quantity being plotted. In order to present more data on a graph, a compressed type of scale, called a logarithmic scale, is used.

A number such as 10,000 can be written as 10^4. Instead of writing the number in these ways it is possible to express it by the number 4 as long as it is understood that the number of concern is given by 10 raised to the 4th power. It is said that 4 is the logarithm of 10,000 and this makes it possible to use small numbers to represent very large numbers. In shorthand notation:

$$\log_{10} 10,000 = 4$$

The subscript 10 is the *base* and is the number which must be raised to the power 4 (the *logarithm*) to equal the number 10,000. Of course, the logarithm need not be a whole number. For example, the logarithm of 760 is between 2 and 3 since $10^2 = 100$ and $10^3 = 1000$. The actual logarithm of 760 is 2.88081, or $\log_{10} 760 = 2.88081$. Therefore, $760 = 10^{2.88081}$. It should be noted that when the logarithm changes by 1 the number involved changes by a factor of 10, e.g., $\log_{10} 100 = 2$ and $\log_{10} 1000 = 3$ (a difference in logarithms of 1) and the number 1000 is ten times the number 100. Similarly the numbers 10,000 and 100,000 $(10 \times 10,000)$ have logarithms of 4 and 5 (a difference of 1).

In multiplying or dividing numbers it is possible to express the result as a logarithm by using the logarithms of the individual numbers. If 100 (10^2) is multiplied by 1000 (10^3) the result is 100,000, or 10^5. In logarithms the product is expressed as 5, or $2 + 3$, i.e., the logarithm of the product equals the sum of the logarithms of the individual numbers. Of course, the logarithms need not be whole numbers and can be found from appropriate tables of logarithms. Suppose 10,000 (10^4) is to be divided by 100 (10^2). The resulting number is 100 (10^2), i.e., $10,000/100 = 100$. In terms of logarithms the result is 2, which is the logarithm of the numerator (4) minus the logarithm of the denominator (2). Consequently, in dividing two numbers simply subtract the logarithm of the denominator from the logarithm of the numerator. Again the logarithms need not be whole numbers. Also these rules for multiplication and division can be extended to any number of operations.

2.9 Logarithmic Scales

In vacuum practice logarithms are encountered mostly through reading graphs plotted on logarithmic scales. Suppose pressures ranging from 1 mm Hg to 100 mm Hg are plotted on a linear scale (rectangular coordinates) with 1 cm representing 1 mm Hg. Then 100 cm would be required to represent the pressure of 100 mm Hg. Clearly this would be an impractical method since 100 cm is over a yard. Instead of plotting on a linear scale it is possible to use a logarithmic scale, i.e., a scale based on the logarithms of the numbers involved. In the case of this scale equally spaced divisions on the graph paper represent units of logarithms, i.e., if one division were to represent the distance between the logarithm 1 and the logarithm 2 the next higher division would represent the logarithm 3. Consequently, whole logarithms form a linear scale. Each division on this scale represents a change of a factor of 10 in the actual numbers represented.

Figure 2.6 compares linear and logarithmic scales. In this figure a unit on the logarithmic scale has been taken arbitrarily to equal a distance equal to 0.9 mm Hg (0.1 mm Hg to 1 mm Hg) on the linear scale. The zero point on the logarithmic scale represents 0.1 mm Hg. Actually this zero point is $\log_{10} 0.1 = \log_{10} \frac{1}{10}$ or $\log_{10} 10^{-1} = -1$. The next division on the logarithmic scale would be 0 or $\log_{10} 1 = 0$. Therefore two successive logarithms have been arbitrarily chosen, viz., -1 and 0, to correspond on the linear scale to 0.1 mm Hg and 1 mm Hg. The next division on the logarithmic scale is $+1$ or $\log_{10} 10$. Therefore, it is possible to represent on the logarithmic scale a pressure change of 0.1 to 10 mm Hg whereas the same distance on a linear scale would only represent a pressure change of 0.1 to 1.9 mm Hg. The compression in the logarithmic scale is clearly evident, e.g., the distance between 0.1 mm Hg and 0.2 mm Hg is much greater than the distance from 0.2 to 0.3, etc. It should be noted that the origin on the logarithmic scale always represents finite numbers for the scales. The principal thing to remember is that a whole logarithmic unit repre-

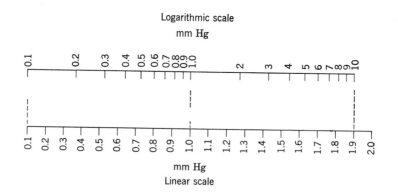

Fig. 2.6   Comparison of logarithmic and linear scales.

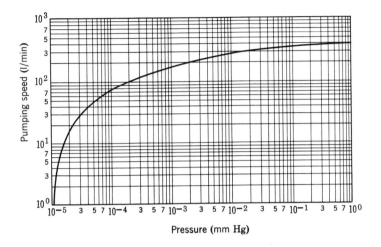

Fig. 2.7 Typical log-log plot. (Courtesy W. M. Welch Scientific Company.)

sents a change of a factor of 10 in the number being plotted. As in the case of Fig. 2.6, the actual numbers being plotted rather than the logarithms are shown on the scales. This is illustrated again in Fig. 2.7, which shows a typical pumping speed versus pressure curve for a commercial rotary mechanical pump. Pumping speed is discussed in Section 2.10. Referring to Fig. 2.7, the compression of the scales is quite clear. In this particular example, the ordinate (vertical scale) represents a change in pumping speed of 1000 ($10 \times 10 \times 10$) through the use of three logarithmic units. Each logarithmic unit represents a factor of 10 and is called a decade. The abscissa (horizontal scale) shows that the pressure changes by a factor of 100,000 (10^{-5} or $\frac{1}{100,000}$ to 10^0 or 1 mm Hg). This change is represented by five logarithmic units or five decades. This particular graph is plotted on three decade by five decade (3 by 5) logarithmic paper. The number of decades used for the scales depends on the ranges of the quantities being plotted. For example, if the pumping speed of the pump did not exceed 100 l/min then the ordinate would only involve two decades. Similarly, in the case of the abscissa, if there were an interest only in the pumping speed over a pressure range of 10^{-5} mm Hg to 10^{-2} mm Hg then three decades ($10^{-2}/10^{-5} = 1000$ or 10^3) would be required.

It should be pointed out that in many cases one vacuum quantity may change much more slowly than another. The slowly changing quantity is then represented on a linear scale while the more rapidly changing quantity is represented on a logarithmic scale. The graph

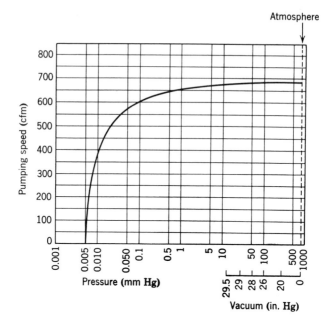

Fig. 2.8 Typical semi-log plot. (Courtesy Kinney Vacuum Division, The New York Air Brake Company.)

paper used for this purpose is called semi-log paper. An example of this type of plot is shown in Fig. 2.8. If the ordinate had been plotted on a logarithmic scale then the origin for this scale would have had to be a finite number and the number used would depend on what detail was desired regarding the pumping speed. Using 0.1 cubic foot per minute (cfm) for the origin would show considerable detail for small pumping speeds but would compress the range 100 to 700 cfm into one decade. Starting with 10 cfm would show great detail between this value and 100 cfm (one decade) and again would compress 100 cfm to 700 cfm into one decade. In this particular example the pumping speed drops very rapidly from 100 cfm to 0 with essentially no change in pressure. Therefore, there is little point in showing this range of pumping speed in detail. The linear scale shows the pumping speeds between 100 and 700 cfm on an expanded scale and this is the range which shows strong dependence on pressure.

Finally it should be pointed out that it is not necessary to use 10 as the base of logarithms although this is most common in graphical representation. For various reasons the other base which is used is $e = 2.71828$. Logarithms using this base are called natural logarithms and are written $\log_e$ or ln. As an example, in this system $\log_e 10 = 2.30259$ or $10 = e^{2.30259} = 2.71828^{2.30259}$ while $\log_{10} 10 = 1$. In practical vacuum work natural logarithms are commonly encountered in connection with the time required to pump down a system to a particular pressure.

2.10   Pumping Speed

In operating a vacuum system there is an interest in how fast gases and vapors can be removed from the system. The proportions of permanent gases and vapors present in any system will depend on the nature and condition of the system. The most common offender in the way of vapor is water vapor and its presence can only be minimized through proper design, choice of materials, and attention to accepted standards of cleanliness. Of course, the very nature of certain vacuum processes, such as vacuum distillation, freeze-drying, etc., involves the handling of large quantities of vapor. Although the usual types of vacuum pumps such as mechanical or diffusion pumps will pump both permanent gases and vapors, cold traps are the most effective for pumping vapors. Consequently, a cold trap can be considered to be a pump for vapors. Regardless of the types of pumps involved it is possible to measure the rate of removal of gases and vapors by pumping speed.

The pumping speed at any point in a vacuum system is defined as the ratio of the throughput to the pressure at that point. Therefore, it can be written

$$\text{Pumping speed} = \frac{\text{throughput}}{\text{pressure}}$$

In symbols, this becomes

$$S = \frac{Q}{P} \tag{2.7}$$

where S is the pumping speed.

Since throughput is given by pressure times volume per unit time, the pumping speed has the units of volume per unit time. The most common units are cubic feet per minute, liters per minute, and liters per second. The pumping speed used by manufacturers is actually a *measured speed*. Depending on its nature and design, any given pump will remove gases and/or vapors only over a particular range of pressure. Furthermore, in general the speed will be a function of the pressure. The speed will drop off rapidly as the lowest pressure attainable by the pump (*the ultimate pressure*) is approached. Finally the ultimate pressure will be reached and at this point the manufacturer's speed curves will show zero pumping speed. This is because the manufacturer usually measures the speed of a pump at various pressures by introducing a variable leak. When this leak is shut off completely the ultimate pressure will be reached and the pumping speed is then indicated as being zero. Actually, the pump is still pumping gases and vapors being released from inside the vacuum system or from natural leaks in the system. An equilibrium condition has been reached where the rate at which gases and vapors accumulate in the systems equals their rate of removal by the pump. Naturally if the system is "dirty," i.e., large quantities of gas and vapor are being released, then it will take much longer to reach the ultimate pressure. Also, if there are large leaks in the system, the ultimate pressure reached will be considerably higher than in the case of no leaks. Again the measured speed of the pump is zero although it is actually handling much larger quantities of gas and vapor. To obtain the pumping speed it is necessary to know the throughput at the ultimate pressure.

Figures 2.7 and 2.8 show speed curves for typical mechanical pumps of the rotary oil-sealed type (see Chapter 3). The curve of Fig. 2.7 is plotted on a log-log scale so zero pumping speed cannot be shown. However, it is evident that the speed is dropping very rapidly at about 10^{-5} mm Hg so this could be taken as the ultimate pressure. This curve was obtained under idealized conditions and the manufacturer quotes a *guaranteed vacuum* of 10^{-4} mm Hg so as to take into consideration various factors occurring in actual vacuum practice. Figure 2.8 shows a speed curve on a semi-log scale so that the measured pumping speed is shown to be zero at 0.005 mm Hg (5 μ). For this curve it would be stated that the ultimate pressure is 0.005 mm Hg. Actually the manufacturer quotes a more conservative value of 10 μ ($10^{-2} = 0.010$ mm Hg) for this class of pump.

2.11   Losses in Pumping Speed

If it were possible to connect the inlet of a pump directly to the vacuum vessel then pumping speeds at the vessel would be given directly by the speed curve of the pump. In practice this is not possible, simply due to space considerations. As a rule some kind of pumping line is used to connect the pump to the vacuum system. The pumping line could be simply a length of piping or tubing with a valve in it (as for roughing lines and for forelines connecting a mechanical

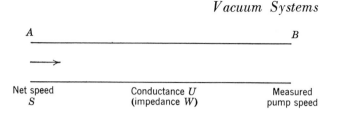

Fig. 2.9 Effect of an impedance (resistance) on pumping speed.

pump to a diffusion pump) or it could be a more complicated arrangement including piping or tubing, a liquid nitrogen trap, a water-cooled baffle, a valve, and perhaps other components. Regardless of the particular type of connection used, this connection will offer resistance to the flow of gases and vapors. As a consequence the pumping speed at the vessel will be less than the pumping speed at the inlet to the pump.

In setting up a vacuum system one of the most important considerations is the choice of connections between pump and vessel so as to keep pumping losses to a minimum, consistent with proper performance of the system for the intended purpose. As has been indicated, the resistance of the connection will depend on the pressure. Rules for estimating the resistance (actually conductance) of various vacuum components for the different flow regions are given in Chapters 3, 4, and 11. At this point the concern will be only with the effect of a conductance on pumping speed without regard to the exact nature of the components involved.

Consider a connection between two points A and B of a vacuum system as is shown in Fig. 2.9. Although the connection is shown as a cylinder (pipe or tube) it could be a combination of any number of vacuum components. The flow of gas is from A to B. The pumping speed at A (net speed) is related to the resistance (impedance) of the connection and the pumping speed at B (measured pumping speed) by the following relationship:

$$\text{Net speed} = \frac{\text{measured pumping speed}}{1 + \text{impedance} \times \text{measured pumping speed}}$$

In symbols this becomes

$$S_n = \frac{S_m}{1 + WS_m} \tag{2.8}$$

where S_n = net speed
S_m = measured pumping speed
W = impedance of connection

In practice it is more common to use conductance rather than impedance. In this case the relationship becomes

$$\text{Net speed} = \frac{\text{measured pumping speed} \times \text{conductance}}{\text{measured pumping speed} + \text{conductance}}$$

In symbols this becomes

$$S_n = \frac{S_m U}{S_m + U} \qquad (2.9)$$

Here S_m and S_n have the same meaning as for eq. 2.8 and U stands for conductance. This is the basic design formula used in high vacuum practice.

Rather than use eq. 2.9 it is possible to use simple graphs to represent the relationship between S_n, S_m, and U. Such a graph is shown in Fig. 2.10. This graph shows the relationship between U and S_n for values of S_m between 1 and 10. It will be noted that no units are shown on this graph. In order to cover the complete range of commercial pumps all scales can be multiplied by any desired factor. For example, suppose the net speed (S_n) required at a vacuum vessel is 1200 cfm and the conductance (U) of the connecting lines is 5000 cfm. Then both ordinate and abscissa should be multiplied by 1000. Choosing 1200 on the ordinate and 5000 on the abscissa brings us to a point approximately two-thirds of the distance between the curves $S_m = 1$ and $S_m = 2$. Using this approximation for the curves $S_m = 1$ and $S_m = 2$ where they level off at high conductances, a value of S_m of about 1600 cubic feet per minute is obtained for the measured pumping speed. If the formula is used instead of the graph, then

$$S_n = \frac{S_m U}{S_m + U}$$

or

$$S_n S_m + S_n U = S_m U$$

Therefore,

$$S_m (U - S_n) = S_n U$$

or

$$S_m = \frac{S_n U}{U - S_n}$$

and

$$S_m = \frac{1200 \times 5000}{5000 - 1200} = \frac{1200 \times 5000}{3800}$$

$$= 1580 \text{ cfm}$$

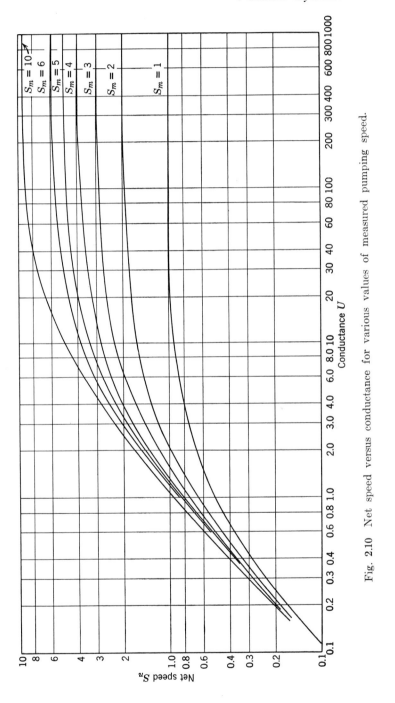

Fig. 2.10 Net speed versus conductance for various values of measured pumping speed.

This compares very well with the value of 1600 cfm obtained from the graph. It must be kept in mind that whatever factor is used to multiply one scale must also be used to multiply all other scales. In the example above, the factor 1000 is used for all three scales, viz., S_n, S_m, and U. The manner in which these graphs are used can be illustrated as follows.

EXAMPLE 1

A mechanical pump with measured pumping speed (S_m) of 15 cfm is connected by a pipe with a valve to a vacuum vessel for roughing down purposes. The conductance (U) of the pipe and valve is found to be 23.6 l/sec. What is the net speed at the vacuum vessel?

Since the same units must be used, convert the conductance (U) into cubic feet per minute. Then

$$23.6 \text{ l/sec} = 23.6 \times 60 = 1416 \text{ l/min}$$

$$= \frac{1416}{28.3} = 50.0 \text{ cfm}$$

From Fig. 2.10, when a factor of 10 is used it is seen that the value $S_m = 15$ falls between $S_m = 1$ (10) and $S_m = 2$ (20). At large values of the conductance (right side of graph) the curves for $S_m = 1$ (10) and $S_m = 2$ (20) approach the values 1 (10) and 2 (20) respectively for S_n. A curve for $S_m = 15$ for these large values of conductance would fall at $S_n = 15$ (ten times the ordinate) which is somewhat more than one-half the distance between $S_n = 1$ (10) and $S_n = 2$ (20). By picking out the value of the abscissa for $U = 50$ (ten times the abscissa) and following the ordinate to a point somewhat over one-half the distance between the curves for $S_m = 1$ (10) and $S_m = 2$ (20) it is seen that S_n can be estimated to be about 12 or perhaps a little less. Therefore, the net speed at the vessel is about 12 cfm as against a measured pumping speed of 15 cfm. Consequently the loss in pumping speed is about $(15 - 12)/15 \times 100 = 3/15 \times 100 = 20\%$. This is about the usual allowed pumping loss for a roughing line. It is desirable to keep down the pumping loss for such a case to 20% or less.

If the formula (eq. 2.9) is used instead of the graph to obtain the net speed, the following result is obtained:

$$S_n = \frac{US_m}{U + S_m} = \frac{50 \times 15}{50 + 15} = \frac{50 \times 15}{65}$$

$$= 11.5 \text{ cfm}$$

This compares quite well with the value of 12 cfm estimated from the graph.

EXAMPLE 2

A net speed of 100 l/sec is required at a vacuum vessel and an available diffusion pump has a measured pumping speed of 250 l/sec. What must be the conductance of the connecting lines between pump and vessel?

From Fig. 2.10, using a factor of 100, the measured pumping speed $S_m = 250$ falls between the curves $S_m = 2$ (200) and $S_m = 3$ (300). Checking at large values of conductances it is seen that $S_m = 250$ falls at a point somewhat over one-half the distance between $S_m = 200$ and $S_m = 300$. Picking out the value $S_n = 100$ and moving along the abscissa to a point somewhat over one-half the distance between the curves $S_m = 2$ (200) and $S_m = 3$ (300) on the ordinate, the conductance required is found to be about 175 l/sec.

In this case it is also possible to use the formula

$$S_n = \frac{US_m}{U + S_m}$$

Then

$$S_n(U + S_m) = US_m$$

$$S_nU + S_nS_m = US_m$$

Collecting terms in U,

$$U(S_m - S_n) = S_nS_m$$

or

$$U = \frac{S_nS_m}{S_m - S_n}$$

Therefore

$$U = \frac{100 \times 250}{250 - 100} = \frac{100 \times 250}{150}$$

$$= 167 \quad \text{l/sec}$$

This agrees quite well with the value from the graph.

EXAMPLE 3

If the net speed at a vacuum vessel is 400 cfm and the conductance of the connecting lines is 1200 cfm, what is the measured pumping speed of the pump which must be used?

Multiply the scales in Fig. 2.10 by a factor of 100. Then the abscissa $U = 1200$ and the ordinate $S_n = 400$ intersect almost exactly on the curve $S_m = 6$ (600). Therefore, the value of the measured pumping speed is just about 600 cfm.

Again the measured pumping speed can be obtained by means of the formula

$$S_n = \frac{US_m}{U + S_m}$$

or

$$US_n + S_nS_m = US_m$$

Collecting terms in S_m,

$$US_m - S_nS_m = US_n$$

and

$$S_m(U - S_n) = US_n$$

or

$$S_m = \frac{US_n}{U - S_n}$$

Therefore,

$$S_m = \frac{1200 \times 400}{1200 - 400} = \frac{1200 \times 400}{800}$$

or

$$S_m = 600 \text{ cfm}$$

This is the same value that was obtained directly from the graph.

THREE

Rotary Oil-Sealed Pumps

3.1 The Nature of Such Pumps

Rotary oil-sealed pumps (or simply rotary pumps) are the most common type of mechanical pump used in vacuum systems. As the name implies, this type of pump involves the use of a rotary member (the *rotor*) with sealing against air leakage effected by the use of a vacuum oil.

Although the various commercial pumps may differ considerably in detail, they all operate on the same basic principle. The three most common arrangements are illustrated in Figs. 3.1a, 3.1b, and 3.2. In each case air enters the pump from the vacuum system through the *inlet port* ("intake" in the case of Fig. 3.2). This air is trapped, compressed, and ejected to the atmosphere through the *discharge* (or *exhaust*) valve by means of the rotor arrangement. In Fig. 3.1a, sealing is done by means of two sliding (or scraping) *vanes* and the point of contact between the rotor and the *stator* or *housing*. In Fig. 3.1b, seals are provided by a fixed, spring-loaded vane riding on the rotor and by the point of contact between the rotor and the inside of the housing. The arrangement of Fig. 3.2 makes use of a sliding valve and the point of contact between the rotor (plunger) and the inside of the housing. In all cases oil is used as the sealant. However, close tolerances must be used to prevent leaks and by-passing of gases. Consequently, care

49

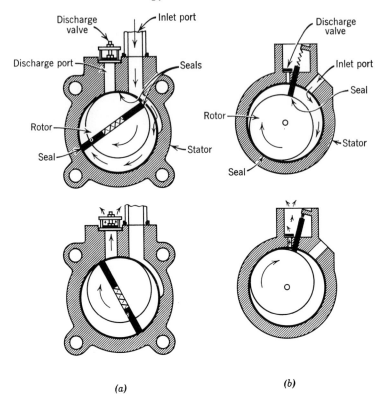

Fig. 3.1 (*a*) Rotary vane type of pump. (*b*) Fixed spring-loaded vane type
of pump.

must be taken to prevent foreign particles from entering the pumping
chamber.

The discharge or exhaust valve is usually a *feather* or *poppet* valve.
A feather valve consists of a thin strip of metal, which in some cases is
held in place by springs. A poppet valve is constructed from a metal
disk in a suitable housing with a spring to retain the disk and limit its
displacement. Some pumps are provided with *ball valves*, which are
constructed from a metal ball that is usually restrained by a spring.

3.2   Pump Oils and the Oil System

The operating fluid used in any type of pump is called the *pump
fluid, working medium, working fluid,* or *pump oil.* With rotary pumps,

normally a good quality light petroleum oil, with the high vapor pressure fractions removed, is used to provide sealing and lubrication. A list of some suitable pump oils is given in Appendix B. These oils are dehydrated and, therefore, should not be left exposed to air.

In all rotary pumps, the rotating element inside the pump housing is driven by a shaft that must enter the vacuum region and so requires a rotating seal. Tight packing (a *packing seal*) is used by some manufacturers to effect the seal (usually for smaller pumps) while others prevent leakage by supplying oil at atmospheric pressure on the outside of a close-fitting bearing (*oil-bleed seal*), thereby permitting oil rather than air to leak into the pump. In this case a certain amount of air dissolved in the oil enters the system but this is usually a negligible leakage.

The oil for lubricating and sealing is contained in an *oil reservoir*. The arrangement of the reservoir differs from manufacturer to manufacturer. In some small pumps the pump chamber is actually immersed in the reservoir. For the larger pumps the reservoir is usually separated from the pump chamber, often being mounted above the pump itself. The oil entering the pump is forced out with the gas and is returned to the reservoir from a suitable trap. The actual quantity of oil which

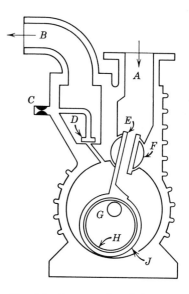

Fig. 3.2 Rotary oil-sealed pump using principle of rotating plunger with slide valve. (Courtesy Consolidated Vacuum Corp., Rochester, N.Y.)

A intake
B discharge
C gas ballast valve
D discharge valve
E slide valve
F slide pin
G drive shaft
H eccentric cam
J plunger—with integrally cast slide valve

is lost is very small. With larger pumps an *oil separator* is used which is simply a reservoir provided with baffles to reduce the loss of oil. Some manufacturers provide heated baffle systems so as to drive off various vapors. Heating is done electrically or with steam. Another method used to remove absorbed water vapor, solvents, etc., is the use of an *oil purifier*. The purifier is normally attached to the oil reservoir and operates by filtering, heating, etc.

Suitable valving is normally provided to control the flow of oil or shut off the flow completely. The amount of oil allowed to circulate through a pump is sometimes determined by the particular use which is to be made of the pump. If the oil is adjusted for the production of the lowest possible pressure then the amount of oil will usually be insufficient to seal the moving parts effectively when large amounts of gas are actually being pumped. If valving were not provided to shut off the oil flow, the pump chamber could fill up with oil when it was not operating. Ejection of the oil in the pump chamber under power can damage the valve mechanism, shear shaft keys, or crack castings. For pumps with oil-bleed seals, each seal often has its own metering valve. The supply of oil to all metering valves is normally controlled by two shut-off valves in series, usually mounted close to each other. One of these valves is hand operated, for normal starting and stopping, and the other is solenoid operated for cutting off the oil supply in case of power failure.

3.3   Ultimate Pressure

The *limiting pressure* approached in a vacuum system after sufficient pumping time to establish that further reductions in pressure will be negligible is called the *ultimate pressure* or sometimes *ultimate vacuum* (not recommended). In the case of a pump under test, the terms *blank-off pressure* or *base pressure* are sometimes used. The ultimate pressures listed by manufacturers are actually base pressures. It must be kept in mind that these values are obtained under the best possible conditions, with a clean, new, blanked-off pump and with clean, dehydrated oil. In practice, the manufacturer's value is often not obtained, due to wear of pump parts, contamination of the oil, or leaks in the vacuum system. Base pressures of commercial rotary pumps range from about 2×10^{-5} mm Hg to 5 mm Hg. The lowest base pressures can only be obtained with compound pumps, i.e., pumps using several pumping stages (usually two) in series. Of course, single-stage pumps could also be connected in series to achieve similar results.

As will be seen shortly, one of the main difficulties with rotary pumps arises from contamination of the oil. As the gases and vapors are compressed, the vapors will tend to condense and contaminate the oil. The most common vapor encountered in vacuum practice is water vapor. Of course, the problem will be minimized if recommended cleaning techniques, such as those discussed in Chapter 9, are followed. However, many vacuum processes such as vacuum distillation, evaporation, drying, and concentration involve the production of large amounts of vapors. The use of oil purifiers for separating condensed material from the oil has already been mentioned. Also, pumps can be purchased in which a feature is incorporated which minimizes the condensation of vapors in the oil. These are called *gas ballast* or *vented exhaust* pumps. These pumps work on the basis of admitting air (sometimes dry air) into the compression chamber of the pump so as to dilute the vapors to the point where they do not condense during compression. The air is admitted through a special valve. Because of the higher pressure in the chamber during compression there is more leakage back into the vacuum system. Consequently, gas ballast pumps will not achieve as low a pressure as pumps which do not have this feature.

3.4 Pumping Speed

Apart from the ultimate pressure which can be achieved by any particular pump, there is an interest in how fast the pump can reduce the pressure in a vacuum system to the operating value. Manufacturers normally specify the pumping speeds of their rotary pumps at atmospheric pressure; this speed is called *free air displacement* or *free air capacity*. Specifically, this refers to the volume of air passed per unit time through the pump when the pressures on the intake and exhaust sides are both equal to atmospheric pressure. Sometimes the term *displacement* is used, which simply means the geometric volume being swept out per unit time by the working mechanism of the pump at normal speeds. In general, rotary pumps start pumping at atmospheric pressure and as the pressure is reduced the pump becomes less efficient. Eventually the pumping speed becomes zero at the ultimate pressure. The free air displacement of a pump depends on the speed with which it is operated. The manufacturer normally specifies both the free air displacement and the speed at which it is to be run. Pump speeds generally range from about 300 rpm to 650 rpm. Higher free air displacements can be obtained by increasing the pump speed. However, this is not recommended because of accelerated pump wear. The most common units used for pumping speeds are liters per minute and cubic feet per minute. To convert from liters per minute to cubic feet per minute simply divide the number of liters per minute by 28.3 or multiply by 0.353. Conversely, to convert cubic feet per minute to

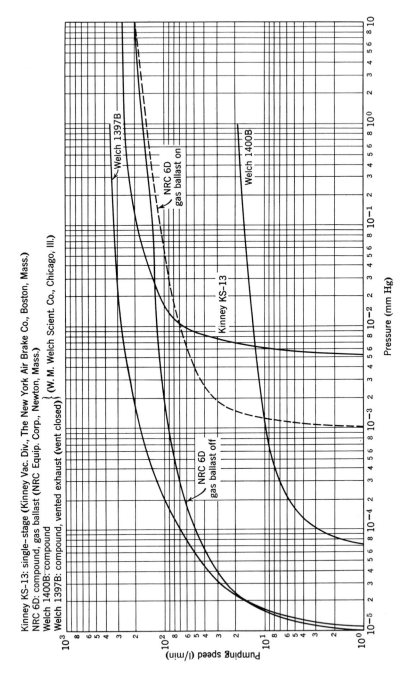

Kinney KS-13: single-stage (Kinney Vac. Div., The New York Air Brake Co., Boston, Mass.)
NRC 6D: compound, gas ballast (NRC Equip. Corp., Newton, Mass.)
Welch 1400B: compound
Welch 1397B: compound, vented exhaust (vent closed) } (W. M. Welch Scient. Co., Chicago, Ill.)

Fig. 3.3   Pumping speed curves for some small pumps (manufacturers' data).

liters per minute, multiply by 28.3. Sometimes liters per second are used for pumping speed.

Some typical pumping speed curves are shown in Figs. 3.3 and 3.4. Single-stage, compound, and rotary ballast pumps are included. It will be noted that the speed drops off rapidly as the ultimate pressure is approached. Consequently, to accurately determine the time required to pump a vacuum system down to a required pressure the reduction in pumping speed as the pressure is reduced must be considered. It is often possible to use some simple, general rules which give sufficiently accurate results for most purposes. Several of these rules are discussed in succeeding sections of this chapter. Included in Figs. 3.3 and 3.4 are curves for the pumps shown in Figs. 3.5 and 3.6. The pump of Fig. 3.5 is a typical small compound pump which is very popular in laboratories. Its free air displacement is 21 l/min and its base pressure is 10^{-4} mm Hg. The pump of Fig. 3.6 is a large single-stage pump with a free air displacement of 850 cfm (about 24,000 l/min) and a base pressure of 0.010 mm Hg. The free air displacements of rotary

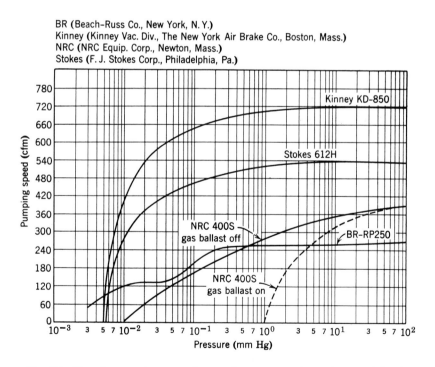

Fig. 3.4 Pumping speed curves of some large pumps (manufacturers' data).

Fig. 3.5   Small compound vane pump. (Courtesy W. M. Welch Scientific Co.)

pumps range from about 0.7 to 850 cfm. To obtain adequate pumping speed, choose the proper pump from manufacturers' data or connect several pumps in parallel.

3.5   Effect of Connecting Lines

Factors determining the type of vacuum pump required for a particular process include: (1) size of vacuum system, (2) operating pressure required, (3) nature and quantity of vapors and gases to be handled, (4) frequency of pumping the system down to the operating pressure, (5) time required to pump the system down to the operating pressure, and (6) effect of connecting lines on pumping speed. Estimates of factor 1 can be made from the external dimensions of the system while factors 2, 4, and 5 are generally specified as part of the vacuum process. It is difficult to estimate factor 3, with experience usually being the important determinant. The connecting lines will result in the pumping speed at the vacuum vessel being less than that at the pump. Of course, the interest is in the pumping speed at the vessel. Most connecting lines are round pipes or tubes so these will be considered here. The effect of bends and valves will be considered briefly.

As has been mentioned previously, the resistance of a pipe to gas flow depends on the pressure. The flow regions are generally divided into: *turbulent, viscous, intermediate,* and *free molecule* or *molecular*. Turbulent flow is only important for large systems operating in the mm Hg region. In the viscous flow region molecules make many collisions with each other before striking the walls of the container, i.e., the mean free path is much less than the dimensions of the container.

Fig. 3.6 Large single-stage rotary piston pump. (Courtesy Kinney Vac. Div., The New York Air Brake Co.)

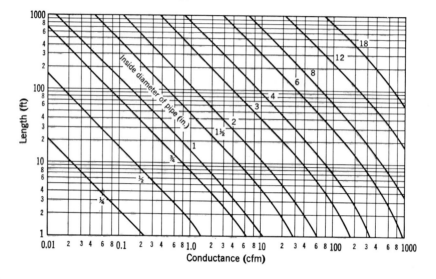

Fig. 3.7   Molecular conductance versus length for several diameters of pipe.

In molecular flow the molecules cross the container with little chance of striking other molecules, i.e., the mean free path is greater than the dimensions of the vacuum container. The intermediate flow region occurs between viscous and molecular flow. The flow region will depend on the pressure and on the dimensions of the vacuum component. Rotary piston pumps will be treated as operating in the viscous flow region. This is not strictly accurate since some pumps, particularly compound pumps, will pump into the intermediate and molecular flow regions. However, the treatment considered here is adequate for most practical situations.

Figure 3.7 shows the conductances of pipes plotted against pipe length for various inside diameters of pipe. These curves are for the molecular flow region and for air at 20°C. For another gas the values from these curves must be multiplied by the square root of the molecular weight of air (about 29) divided by the molecular weight of the other gas. The conductance of air for viscous flow can readily be found from the curves of Fig. 3.7 by the following relationship:

Viscous conductance = 0.04 × conductance from Fig. 3.7

(molecular conductance)

× diameter of pipe in inches

× pressure in microns

or

$$U_v = 0.04 \, U_m \, dP \qquad (3.1)$$

The pressure, P, is the average pressure in the pipe. However, the pipe length should be chosen so the loss in pumping speed is no more than 20%. Consequently, the pressure drop in the pipe is relatively small and the pressure at the process vessel (inlet to pipe) can be used. By using eq. 3.1 and Fig. 3.7, together with Fig. 2.10, it is possible to find any one of the three quantities viscous conductance, pumping speed at the inlet to the pipe (process vessel), and pumping speed at the outlet of the pipe (at the pump), knowing the other two quantities.

EXAMPLE 1

What is the viscous conductance of a pipe 1 in. in diameter and 20 ft long at a pressure of 1 mm Hg (at pipe inlet)?

From Fig. 3.7 the molecular conductance is about 0.7 cfm. From eq. 3.1 the viscous conductance is given by

Viscous conductance $= 0.04 \times 0.7 \times 1 \times 1000 = 28$ cfm

EXAMPLE 2

A Kinney KS-13 pump is to be used to pump down a vessel through the pipe of Example 1. From the speed-pressure curve for this pump, the pumping speed at 1 mm Hg is 9 cfm. Using a conductance of 28 cfm and a pump speed (S_m) of 9, the net pumping speed from Fig. 2.10 is estimated to be 7.2 cfm. This is the pumping speed at the inlet to the pipe (or the process vessel) at 1 mm Hg. The loss of pumping speed is

$$\frac{9 - 7.2}{9} \times 100 = \frac{1.8}{9} \times 100 = \text{approx } 20\%$$

This is just the allowable limit of 20%. If a higher speed were required at the pipe inlet then it would be necessary to use a larger pump if the same pipe connection were used.

EXAMPLE 3

The pumping speed required at the above pipe inlet is 20 cfm. From Fig. 2.10 the pump speed (S_m) required is 65 cfm. In this case the abscissa, ordinate, and S_m values should all be multiplied by 10, i.e., the abscissa value is 2.8×10 and the ordinate value is 2×10, giving at the point of intersection $S_m = 6.5 \times 10$ or 65 cfm. In this case the pumping speed loss is

$$\frac{65 - 20}{65} \times 100 = \frac{45 \times 100}{65} = 70\%$$

The only way the 20 cfm pumping speed can be obtained at the pipe inlet with 20% loss in pumping speed or less is to use a shorter pipe or a pipe of larger diameter.

EXAMPLE 4

A vacuum process requires a net pumping speed (S_n) of 50 cfm at a pressure of 50 μ. If a pumping speed loss of 20% is allowed and a pipe length of 10 ft must be used, what diameter of pipe should be used?

$$80\% \text{ of pump speed} = \text{net pumping speed}$$

$$= 50 \text{ cfm}$$

Therefore,

$$\text{Pump speed} = \frac{100}{80} \times 50 = 62.5 \text{ cfm}$$

From Fig. 2.10, the conductance equals about 250 cfm, using a factor of 10 on the scales. This is assumed to be viscous conductance.

From eq. 3.1

$$250 = 0.04 \times U_m \times d \times 50$$

or

$$U_m \times d = \frac{250}{0.04 \times 50} = \frac{5}{0.04}$$

$$= 125$$

The pipe diameter must be found by trial and error. From Fig. 3.7, for a pipe length of 10 ft try a pipe diameter of 3 in. This gives a conductance of about 35 cfm. Now $35 \times 3 = 105$, which is less than the required value of 125. Consequently, a larger size of pipe must be used. Since the next commercial size of pipe has an inside diameter (ID) of 4 in., try this value. From Fig. 3.7, the pipe of length 10 ft and diameter 4 in. has a conductance of about 76 cfm.

Then $76 \times 4 = 304$, which is considerably greater than the requirement of 125. Consequently, a pipe with diameter 4 in. is more than adequate.

EXAMPLE 5

In the preceding example, what commercial pump should be used?

A first estimate of pump speed has been given as 62.5 cfm. Using a connecting line which is larger than necessary (due to sizes available) will reduce this value to some extent. However, using the value of 62.5 cfm will give a safety factor. From the manufacturers' literature, the following pumps should be adequate for the job:

	Pumping Speed at 50μ
Kinney KDH-130	93 cfm
Consolidated E-225	75
Consolidated DK-180	78
Stokes Microvac 212-H	95
Beach-Russ 135RP	78

It will be noted in the above examples that a pumping speed curve is used to find the pumping speed at any given pressure. Clearly the pumping speeds at the inlet and outlet of a connecting pipe will depend on the operating pressure. Furthermore, the viscous conductance depends directly on the pressure (eq. 3.1) while the molecular conductance is independent of the pressure (Fig. 3.7). Some manufacturers quote *volumetric efficiency* as well as or instead of pumping speed. Volumetric efficiency is the pumping speed at any pressure divided by the free air displacement and multiplied by 100, i.e.,

$$\text{Volumetric efficiency} = \frac{\text{pumping speed at any pressure}}{\text{free air displacement}} \times 100$$

Volumetric efficiency is expressed as a percentage.

No exact method is available for taking into consideration the effects of elbows and foreline or roughing valves on pumping speed. However, some rough rules can be given which are adequate for most situations. For short radius elbows, such as right-angled elbows, simply add the length of the elbow measured along its center line plus the inside diameter (ID) of the elbow to the straight length of pipe. For more gradual bends of large radius only the length of the elbow measured along the center line need be added. Globe-type vacuum valves are often used in the forevacuum and roughing lines. The effect of such a valve can be taken into account by assigning a nominal value of 5 straight pipe sections to the valve. For example, a 2 in. globe valve has a dimension of $5\frac{9}{16}$ in. between flanges. This valve is equivalent to $5\frac{9}{16} \times 5 =$ about 28 in. or $2\frac{1}{3}$ ft so $2\frac{1}{3}$ ft of 2 in. piping should be added to the rest of the connecting line to take care of the globe valve. If a gate valve is used it is only necessary to add a straight pipe section equal to the flange-to-flange dimension of the valve. These rules apply to the micron pressure range. Valves will offer considerably greater opposition to gas flow in the mm Hg range. However, as will be seen in Section 3.6, the pump-down time is of little concern in this pressure range.

3.6   Pump Size and Pump-Down Time—Method 1

The time required to pump down a vacuum system from atmospheric pressure to a specified pressure is called the *time of evacuation* or *pump-down time*. Since the pumping speed drops as the pressure is reduced, some procedure must be adopted to take this into consideration. The method described here is only one of several commonly used methods. In view of the fact that the condition of the vacuum system will have a considerable effect on the evacuation time, some assumptions must be made regarding the system. Assume that the vacuum system has been constructed using reasonable standards of cleanliness (Chapter 9). Any large amount of water will have a drastic effect on the pump-down time since the vapor pressure of water is 17 mm Hg at 20°C. Consequently, under these circumstances the pressure will not drop below this value until the bulk of the water has been pumped out.

The method to be described here involves simple rules for making use of pumping speed curves. The expression giving the time to pump down a volume V to any particular pressure at a pumping speed S is given by

$$\text{Time} = \frac{2.3 \times \text{volume}}{\text{pumping speed}}$$

or

$$T = \frac{2.3V}{S} \tag{3.2}$$

Here, V and S must be in the same units, i.e., if V is in cubic feet then S must be in cubic feet per unit time. Whatever time units are used for S will be the units of T. Furthermore, S is the pumping speed at the volume V. In order to find the required size of pump, the connecting pipe must be taken into account according to the method of Section 3.5. Finally, eq. 3.2 must be modified according to the pressure which is involved. The rules to be used are:

Time to Pump from Atmospheric Pressure to	Multiply Right Side of eq. 3.2 by
100 μ	4
10	5
1	6

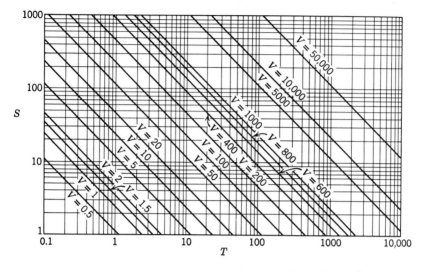

Fig. 3.8 Pumping speed versus pump-down time for various volumes.

The next question is the value of S that should be used. In general, an average value of pumping speed between that at atmospheric pressure and that at the final pressure can be used. If a larger safety factor is desired then S can be chosen as the value at the end pressure.

Instead of using eq. 3.2, it is possible to find the pump-down time from a graph. Such a graph is shown in Fig. 3.8. This graph shows curves of time of evacuation versus pumping speed for various volumes. Here, again, the pumping speed is the average value between that at atmospheric pressure and that at the final pressure. The factors for various final pressures must be used, e.g., if the final pressure is 50 μ the factor to be used is between 4 and 5. As a safety factor use 5. Then the time read off the graph must be multiplied by 5.

EXAMPLE 1

A Kinney KS-13 is used to pump a volume of 100 ft³ from atmospheric pressure is 50 μ. What is the pump-down time?

Assume the connecting line is short and large enough to have negligible effect on the pumping speed. The pumping speed curve for the KS-13 shows a speed of 10 cfm at atmospheric pressure and a speed of about 6 cfm at 50 μ. Therefore, the average speed, S, is $(10 + 6)/2 = 16/2 = 8$ cfm. Using eq. 3.2,

$$T = \frac{2.3 \times 100 \times 5}{8} = \text{about 144 min}$$

The factor 5 is used since the final pressure is 50 μ. The time involved here is probably too long and in practice a larger capacity pump would be used.

Instead of using eq. 3.2, Fig. 3.8 could be used. In this case, the value of the ordinate, $S = 8$, intersects the curve for $V = 100$ at $T = 28$ min. Multiplying by the factor of 5 gives a pump-down time of 140 min. This agrees quite well with the above value. It will be noted that S, T, and V are given number units in Fig. 3.8. Any units can be used as long as they are consistent. For example, if S is in liters per minute then V must be in liters and T will come out in minutes.

EXAMPLE 2

A 200-l vessel is to be pumped down to 5 μ in 10 min. What size of pump should be used?

From eq. 3.2

$$10 = \frac{2.3 \times 200}{S}$$

or

$$S = \frac{2.3 \times 200}{10} = 46 \text{ l/min}$$

Since the end pressure is 5 μ, a factor of 6 should be used. Therefore, the pump used should have a speed of $6 \times 46 = 276$ l/min. Instead of using the formula, use Fig. 3.8. Then for $T = 10$ and $V = 200$, $S = 46$. Again using the factor 6, the required pumping speed is $6 \times 46 = 276$ l/min, which is the same as obtained from eq. 3.2.

Suppose a pumping line is used which allows a pumping speed loss of 20%. Then the pumping speed required is obtained as follows:

$$\text{Pump speed} - \frac{20}{100} \times \text{pump speed} = 276 \text{ l/min}$$

or

$$\text{pump speed} \times \frac{80}{100} = 276$$

therefore

$$\text{Pump speed} = \frac{100}{80} \times 276 = \text{about } 345 \text{ l/min}$$

Some pumps which would be adequate for the purpose are:

Pump	Free Air Displacement	Ultimate Pressure
Welch 1397	375 l/min	0.0001 mm Hg
Kinney KS-13	15 cfm = 424 l/min	0.0001 mm Hg

Pumps of higher capacity (and sufficiently low ultimate pressure) could be used.

3.7 Pump Size and Pump-Down Time—Method 2

The method described in Section 3.6 for finding the pump size or pump-down time for a system makes use of the pumping speed curve of a pump together with certain rules of thumb. The method to be described here involves the use of an empirical curve employing what is called a *pump-down factor*. The pump-down factor is the product of the time to pump down to a given pressure and the free air displacement (for a service factor of 1) divided by the volume of the system. In symbols:

$$F = T\frac{D}{V} \tag{3.3}$$

or

$$T = \frac{FV}{D}$$

Here T is the pump-down time, F is the pump-down factor, V is the volume of the system, and D is the free air displacement. The term *service factor* is an empirical factor equal to or greater than 1, which is specified by the engineer for given pressure ranges and which is multiplied by the displacement as calculated from formulas for a service factor of 1 to obtain the *equivalent pump displacement* required by a mechanical pump to meet unusual demands due to outgassing and other service conditions in average industrial systems.

Any one of the four factors in eq. 3.3 can be found if the other three are known. The pump-down factor, F, will depend on the pump being used and is a function of the pressure. In spite of differences between commercial pumps, average values of the pump-down factor can be plotted against pressure. This has been done in Fig. 3.9. Two curves are shown here—one for an "average" single-stage pump and one for an "average" compound pump. Admittedly many commercial pumps will not have the specific pump-down factors shown in these curves. However, using these curves will give results that are adequate for most practical situations. The free air displacement, D, used in eq. 3.3 is the pumping speed at the vessel. When there is a loss of pumping speed due to connecting lines, this must be taken into consideration; an approximate method for doing this was described in Section 3.5.

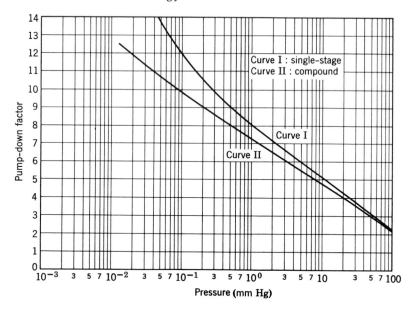

Fig. 3.9   Pump-down factor versus pressure.

EXAMPLE 1

What size of pump should be used to evacuate a 300 ft³ vacuum dryer to a pressure of 5 mm Hg in 10 min?

For the final pressure required, a single-stage pump will be adequate. Curve I in Fig. 3.9 gives a pump-down factor of 6 at a pressure of 5 mm Hg. Using eq. 3.3, the free air displacement is

$$D = \frac{FV}{T} = \frac{6 \times 300}{10} = 180 \text{ cfm}$$

A pump with at least this displacement is then required.

EXAMPLE 2

Consider Example 1 of Section 3.6.   The KS-13 pump is a single-stage pump with a free air displacement of 10 cfm.   Using curve I in Fig. 3.9 gives a pump-down factor of 13.8.   The pump-down time is then given by eq. 3.3 as follows:

$$T = \frac{FV}{D}$$

or

$$T = \frac{13.8 \times 100}{10} = 138 \text{ min}$$

This compares quite well with the value of 144 min found by the method of Section 3.6.

EXAMPLE 3

A test chamber with a volume of 12 ft^3 is to be evacuated to 100 μ, raised to 100 mm Hg, and then re-evacuated to 100 μ. The cycling from 100 mm Hg to 100 μ must be done 10 times per hour. What size of pump is needed?

Assume it takes 2 min to raise the pressure from 100 μ to 100 mm Hg. A total of 20 min/hr is then used for increasing the pressure. This leaves 40 min/hr for re-evacuation or 4 min per cycle. From Curve I of Fig. 3.9, the pump-down factors at 100 μ and at 100 mm Hg are 12.0 and 2.2 respectively. Subtracting the latter from the former gives an effective pump-down factor of 12.0 − 2.2 = 9.8, to reduce the pressure from 100 mm Hg to 100 μ. Using eq. 3.3, the pump displacement is

$$D = \frac{FV}{T} = \frac{9.8 \times 12}{4} = 29.4 \text{ cfm}$$

A single-stage pump of this capacity will be required for this cycling operation.

If a compound pump is employed then use curve II in Fig. 3.9. In this case the pump-down factors at pressures of 100 μ and 100 mm Hg are 9.9 and 2.2 respectively. The effective pump-down factor to reduce the pressure from 100 mm Hg to 100 μ is then 9.9 − 2.2 = 7.7. Again applying eq. 3.3, the pump displacement is

$$D = \frac{FV}{T} = \frac{7.7 \times 12}{4} = 23.1 \text{ cfm}$$

A compound pump of this capacity, or somewhat greater, should be adequate.

3.8 Pump Size and Pump-Down Time—Method 3

The methods described in Sections 3.6 and 3.7 can be used to obtain pump-down times and pump sizes with reasonable accuracy for tight, clean vacuum systems down to pressures of a few microns or less.

However, many average industrial systems may have various sources of gas and vapor and leaks which will require larger pump sizes for any given pump-down time. In order to take care of this type of vacuum, certain modifications to the method of Section 3.7 can be made. *Absorption* refers to the binding of gas in the interior of a solid (or liquid) while *adsorption* refers to the condensing of gas (vapor) on the surface of a solid. As the pressure is reduced there is a spontaneous evolution of gas (and vapor) from materials in the vacuum, which is referred to as *outgassing*. In spite of efforts to adhere to certain standards of cleanliness there will be variations in outgassing rates from system to system. Also, the number and sizes of leaks will vary from system to system. An approximate method for taking care of these factors, together with the effects of average types of connecting lines, can be set up by using the procedure of Section 3.7.

The method involves the use of the pump-down curves of Fig. 3.9 together with the assignment of various service factors to different pressure regions. The service factors to be used for average industrial systems are as follows:

Pressure Region	Service Factor
760–100 mm Hg	1.0
100–10	1.25
10–0.5	1.5
0.5–0.05	2.0
0.05–0.0002	4.0

EXAMPLE 1

Consider again Example 1 in Section 3.6. This problem was treated as Example 2 in Section 3.7 by using the pump-down curve of Fig. 3.9. The value of pump-down time obtained was 138 min. The method to be used here introduces a service factor. At a pressure of 50 μ a service factor of 2.0 can be used. This gives a pump-down time of $2 \times 138 = 276$ min, which is a more conservative figure than the 138 min noted above or the 144 min found in Section 3.6.

EXAMPLE 2

Consider Example 3 of Section 3.7. The treatment shown gave a pump displacement of 29.4 cfm. At a pressure of 100 μ the service factor can be taken as 2.0. This then gives a pump displacement of $2 \times 29.4 = 58.8$ cfm, which is considerably larger than the 29.4 cfm and therefore more conservative.

3.9 Gas Handling Capacity of a Pump

Up to this point the treatment has been concerned with the choice of pump and with pump-down time. The prime factors involved were the operating pressure and the pump-down time. The amount of gas and vapor due to leaks and outgassing was considered in an average sense in Section 3.8. In many cases the pump-down time is not the important factor and then the choice of pump is dictated by the quantity of gas and vapor to be handled and by the operating pressure. In general, pumps chosen on the basis of pump-down time and operating pressure are more than adequate for handling the gases and vapors. However, it is uneconomical to use pumps with higher capacity than needed. For example, a roughing pump used to reduce the pressure in a system to the operating value is generally of considerably higher capacity than a forepump which is used to handle gases and vapors at the operating value.

According to eq. 2.7, the pumping speed is given by $S = Q/P$, where Q is the throughput and P is the pressure at the point where the throughput is measured. As was pointed out in Chapter 2, the pressure P is due to the throughput Q. Suppose Q is measured in cubic feet per minute at 1 mm Hg. Then S is given in cubic feet per minute. Also, if Q is measured in liters per second at 1 mm Hg, then S is given in liters per second. Q is then 760 times the volume rate of evolution and flow of gas and vapor measured at atmospheric pressure.

EXAMPLE 1

Consider a vacuum system in which a flow of 1 cc of air per minute at atmospheric pressure is expected. What pumping speed is needed to maintain a pressure of 10 μ or less?

The throughput is

$$Q = \frac{1.0 \times 760}{1000} = 0.76 \text{ l/min at 1 mm Hg}$$

The required pumping speed at 10 μ is then

$$S = \frac{Q}{P} = \frac{0.76}{0.010} = 76 \text{ l/m}$$

Now this is a minimum pumping speed which will barely fulfill requirements. To take care of bursts of gas and losses in connecting lines, use a service factor. At a pressure of 10 μ, the service factor as given in Section 3.8 is 4.0. Therefore, a pump with a pumping speed of $4 \times 76 = 304$ l/min at a pressure of 10 μ is required.

3.10   Some General Rules

No hard and fast rules can be given for treating any particular vacuum system. Vacuum systems vary considerably in their characteristics, particularly with regard to the quantity of gas and vapor involved. However, some general rules can be summarized as follows:

1. Where the vacuum process necessarily involves large quantities of vapors, refrigerated traps should be used to reduce the burden on the rotary piston pumps.

2. Such connecting lines should be chosen as to keep the pumping speed loss to 20% or less.

3. The methods considered in Sections 3.6, 3.7, and 3.8 for treating pump size and pump-down time become progressively less reliable as the pressure is reduced. In general these methods should be used only down to pressures of a few microns.

4. In the cases of small laboratory-type vacuum systems, where particular attention is given to cleaning techniques and materials of construction, the methods of Sections 3.6 and 3.7 can be used.

5. With larger, industrial-type installations it is best to use the method of Section 3.8. This method should also be used for any "dirty" system involving fair amounts of gas and vapor.

6. Compound pumps should be used for operating pressures of a few microns or less.

3.11   Setting up Rotary Oil-Sealed Pumps

In dealing with a brand-new pump it is advisable to turn it by hand to ensure freedom of motion. In spite of the most careful precautions taken by the manufacturer, occasions will occur when foreign material, such as machining chips, etc., are found inside new pumps. In the rare case where the pump cannot be rotated freely it will be necessary to take it apart and investigate the difficulty. With the pump turning smoothly the next step is to fill it to the prescribed level with a good quality vacuum oil which has not been left open to air. It is a good idea to flush out the pump a couple of times with new oil before filling it to the proper level.

In dealing with a pump which has been in use it is advisable to thoroughly inspect and clean the pump before putting it back in service. This again involves checking for freedom of action, flushing with clean vacuum oil, and filling with clean oil. With any pump, new or used, it is important to check both the belt tension (as well as the condition of the belt) and the pulley ratio to ensure that the pump is being operated at the proper speed.

Many pumps include various valving arrangements such as oil flow control valves, metering valves for oil-bleed seals, and air release valves. Also many larger pumps are water-cooled and may include control valves. Some of these valves may be manual while others are solenoid operated. In any case the operation of all valves should be checked before starting the pump. Some of the valves, such as the metering valves, are supposed to be set at particular values as specified by the manufacturer. The proper values should be checked. In the case of pumps operated with three-phase motors it is important to check for correct rotation of the pump. Incorrect rotation will result in the pump getting flooded with oil, which could mean damage to pump parts or, at least, a difficult cleaning job.

In mechanically setting up the pump-motor system certain precautions should be taken. Some of these precautions are:

1. The pulley-belt system should be lined up properly.

2. The pump should be mounted in a reasonably level position.

3. Where noise and vibration are objectionable, the pumps should be placed at some distance from the vacuum process chamber and other types of pumps, say at a different level. Of course, the loss of pumping speed in pumping lines must then be taken into consideration.

4. Attention must be given to venting pumps so as to avoid a health hazard from oil vapor.

5. A flexible section should be included in the line connecting the pump to the vacuum chamber or other pumps to avoid the transmission of vibrations. Sylphon bellows, rubber tubing, plastic tubing, and corrugated metal tubing can be used. It must be kept in mind that this connection must withstand atmospheric pressure. Fairly thin plastic tubing can often be used by inserting a steel spring so that the tubing will not collapse under atmospheric pressure.

3.12   Starting Rotary Oil-Sealed Pumps

The precautions to be taken before starting a pump will depend to a large degree on the type of pump involved. When a pump has been shut off for a very short period of time (a matter of minutes) it can usually be turned on immediately. However, in the cases of new pumps, pumps being put into new service, and pumps which have been idle for a considerable length of time, two precautionary steps usually should be taken: (*a*) check the oil level and (*b*) turn the pump over by hand to ensure freedom of operation.

In the case of larger pumps other checks should be made prior to starting these pumps. Some precautions which should be taken can be summarized as follows:

1. Do not start the pump under load. This means having the inlet and outlet both at atmospheric pressure when the pump is first turned on. In many cases this is done by closing the roughing and forevacuum valves and opening the release valve.

2. Open oil valves just before or at the time the pump is started. In the case of a pump with metering valves, both a manually-operated and a solenoid-operated valve are connected in series. The manually-operated valve should be opened.

3. After the pump has reached full speed close the release valve.

4. Some water-cooled pumps are interlocked so that when the pump is turned on a valve is opened to permit water flow. When the control is manual, the control valve should be turned on before starting the pump. In any case a check should be made to ensure that cooling water is flowing.

3.13   Stopping Rotary Oil-Sealed Pumps

When a pump is shut down for a very short time (a matter of minutes) usually no particular precautions are necessary. However, any extended shut-down requires that certain safety measures be taken. Some general precautionary steps can be summarized as follows:

1. Let the pump down to air when it is stopped. This can be done through the use of a release valve or some auxiliary valve. Some pumps incorporate a solenoid valve which opens automatically when the power is turned off. However, the correct operation of this valve

should be checked. If the pump is not let down to air, oil will get into the pump chamber. This may cause damage to parts when the pump is started or at least oil contamination of pumping lines and perhaps the vacuum system.

2. If the pump is equipped with oil valves, these valves should be closed at the time the pump is stopped. For pumps equipped with manual and solenoid valves for metering purposes, the manual valve should be closed. This is simply a matter of insurance in case the solenoid valve fails to operate. Failure to close the oil valves may again result in oil getting into the pump chamber.

3. With water-cooled pumps, the water flow should be shut off at the time the pump is stopped. If this is not done, water vapor will condense inside the pump and contaminate the oil. In the case of inter-locked systems an actual check on water flow should be made rather than relying completely on the interlock.

3.14 Maintenance Problems

Rotary oil-sealed pumps are extremely reliable as long as the recommended precautions are taken. For example, small pumps can be operated continuously for months without attention when used on "tight" systems which are rarely let down to air. On the other hand pumps used in systems which are cycled regularly, that is, let down to air and then pumped down again, or systems which handle large amounts of vapors (cycled or otherwise) may require a fair amount of routine maintenance. This usually involves cleaning by flushing out with new vacuum oil and then filling with new or reclaimed vacuum oil. Of course, problems stemming from contamination of oil can be minimized by use of a refrigerating system (cold traps, etc.) to freeze out the vapors or by purification of the oils through various types of filters or heating systems. However, in spite of such precautions any pump used in a system which is handling large quantities of vapors will eventually have contaminated oil. This matter of contaminated oil is by far the major maintenance problem experienced with rotary oil-sealed pumps.

The common indication of a malfunctioning pump is an abnormally high ultimate or base pressure. However, other possibilities must be considered before immediately attributing such a high pressure to the pump. Real leaks in the system should first be eliminated. The prob-lem is simplified when the pump can be isolated from the system, perhaps by roughing and forevacuum valves. An appropriate vacuum gauge near the pump can then be used to good effect. When it is clear that the high pressure is not due to real leaks, it can then be attributed to the pump itself. Possible causes of the trouble are:

1. Faulty oil valves.
2. Faulty exhaust valves.
3. Slipping belts.
4. Contamination of the oil.

Proper inspection procedures should have detected items 1, 2, and 3. However, these points should still be checked as a matter of insurance. Item 4 is the usual cause of high base pressure. One way of showing that the high pressure is due to contaminating vapors is to use trapped and untrapped gauges. The untrapped gauge will show a considerably higher pressure than the trapped gauge.

Other difficulties with rotary oil-sealed pumps may not show up immediately in terms of a high base pressure. These difficulties usually stem from corrosion and gumming of working parts of the pump due to contaminating vapors. Such corrosion and gumming will be indicated by slipping belts or heating of the pump. The correction is to try cleaning the pump and if this doesn't work to take the pump apart and remove gums or other foreign material by scraping. Excessive corrosion may require replacement of pump parts. Of course, such a condition should have been avoided by observing pump performance. Inadequate attention to gumming and corrosion can result in a frozen pump.

An adequate inspection and maintenance schedule should include some or all of the following steps:

1. Inspect oil periodically.
2. Check for slipping belts.
3. Check for overheating of the pump.
4. Check setting of exhaust valve.
5. Check setting of oil valves.
6. Check for proper water flow.
7. Examine a sample of oil periodically.

Actually step 1 is not adequate in checking the condition of the oil. A sample of oil should be examined, as in the case of step 7. A whitish, emulsified appearance indicates the presence of condensed vapors. Foreign particles can also appear in the oil. In either case the oil should be purified or replaced with new or reclaimed oil.

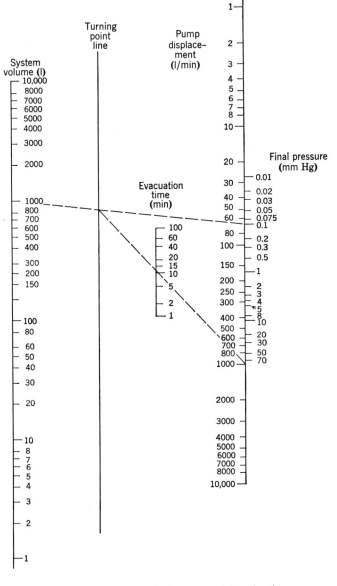

Fig. 3.10 Nomograph for pump determination.

3.15   Use of a Nomograph

Instead of using the methods of Sections 3.6, 3.7, and 3.8, it is some-times possible to use a nomograph such as that shown in Fig. 3.10. This method can be used for systems which are reasonably clean. Although the pumping speeds given in this figure refer to those at the vacuum vessel, they can still be used for the pumps when losses in the pumping lines are nominal. Appropriate service factors have been used in making up the nomograph.

As an example of the use of Fig. 3.10, consider a system volume of 1000 l, an operating pressure of 0.1 mm Hg, and a desired pumping time of 12 min. The pump speed is required.

The solution is:

1. Draw a line from the volume line to 0.1 on the pressure line.
2. Draw a line from the intercept of this line with the turning point line through the 12 point on the time line and extend it to the pump displacement line.
3. The required pump speed is given by the intercept on the pump displacement line (1000 l/min).

It should be kept in mind that some manufacturers have slide rules available which relate various vacuum quantities such as system volume, pump speed, pump-down time, etc.

FOUR

Vapor Pumps

4.1 General Nature of Vapor Pumps

Up to this point, the most common type of mechanical pump used in vacuum work has been discussed, viz., the rotary oil-sealed pump. Although many of these pumps have base pressures well below 1 μ Hg, they are generally used in the mm Hg or μ Hg ranges. To obtain pressures well below 1 μ Hg other types of pumps are used (see Chapter 5). However, the most common pump used to achieve pressures down to around 10^{-7} mm Hg or lower is a vapor pump. A *vapor pump* can be defined as any pump employing a vapor as the pumping means. The principle of operation is entirely different from that of a rotary oil-sealed pump, where the gases and vapors are compressed by a rotating mechanical member and expelled to the atmosphere.

Vapor pumps can be divided into two general classes which are determined by the pressure range in which they are designed to operate. All of these pumps must include a component for producing vapor from the pump fluid, usually referred to as the *boiler*. The first class of vapor pump, a diffusion pump, operates in the molecular flow region and the basic principle involved is shown in Fig. 4.1a. Here vapor from a boiler passes through a narrow opening (the *nozzle*) with high speed, being directed at an angle downwards. Molecules of

77

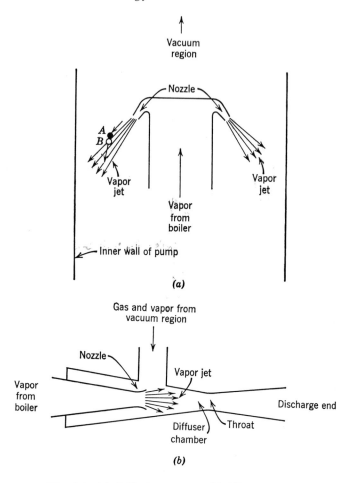

Fig. 4.1   (*a*) Diffusion pump.   (*b*) Ejector pump.

gas or vapor which wander or diffuse from the vacuum region toward the jet stream will be struck by vapor molecules.   The gas or vapor molecules will first encounter the low density scattered vapor and, on the average, will be forced by collisions into the denser forward moving core of the freely expanding vapor jet.   In Fig. 4.1*a*, the gas or vapor molecule *B* has diffused from the vacuum region into the outer part of the jet stream, where it is struck by the vapor molecule *A*. Molecule *B* is given a generally downward motion into the dense part of the vapor jet.   The net effect is to compress gases and vapors to the point where they can be removed by some other type of pump.

Of course, the compression can be carried out in stages, i.e., by using several nozzles in line. It is desirable to give the gas or vapor molecules as rapid a motion as possible downwards. This is done by using heavy vapor molecules and having them move as rapidly as possible. The optimum condition for this process to occur is for the mean free path of the molecules being pumped through vapor molecules in the region of the nozzle to be greater than the nozzle clearance. The nozzle clearance is the perpendicular distance between the outer rim of the nozzle exit and the inner wall of the pump. Clearly it would be desirable to minimize intermolecular collisions between gas (or vapor) molecules since this would result in many of these molecules going back to the vacuum region.

At higher pressures vapor pumps depend for their pumping action on entrainment of gas by viscous drag and by diffusion of gas into the vapor at the boundary of a dense vapor stream. Such pumps are called *ejector pumps* and their basic principle is illustrated in Fig. 4.1*b*. The dense vapor stream from the boiler passes through the nozzle into a converging chamber called the *diffuser*. The vapor stream entrains the gases entering the diffuser near the end of the nozzle and carries them at supersonic speeds through the *throat*. For the best pumping conditions the mean free path of gas molecules through the vapor issuing from the nozzle should be considerably less than the nozzle clearance. Ejector pumps are generally used between diffusion pumps and mechanical pumps. Some vapor pumps combine diffusion stages with ejector stages and are called *diffusion-ejector pumps*.

4.2   Some Features of Diffusion Pumps

Historically the earliest work with diffusion pumps was carried out in 1915 by Gaede, who used mercury as a pump fluid. This work was followed in 1916 by the studies of Langmuir, who introduced the concept of condensation of the pump fluid on the walls of the pump. The words *condensation pump* and *diffusion pump* are often used interchangeably. Such pumps are also sometimes called high vacuum, high speed pumps since they are designed to have high pumping speeds at low pressures. Until 1928 the only pump fluid in general use was mercury. At that time Burch distilled certain low vapor pressure fractions of natural hydrocarbon oils. This work was followed by the studies of Hickman, who developed certain synthesized and purified esters. The most commonly used pump oils now are refined hydro-

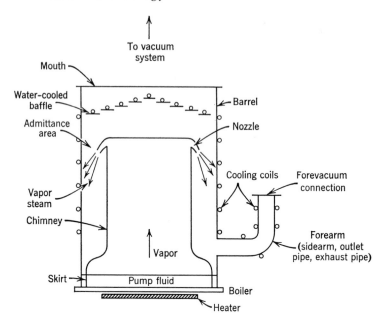

Fig. 4.2   Elements of a diffusion pump.

carbon oils, esters, chlorinated biphenyls, and the semiorganic silicone oils. The properties of certain pump oils are given in Appendix B.

Figure 4.2 shows the elements of a practical diffusion pump. This is a single-stage metal pump, i.e., it uses only one nozzle. Some small pumps of this type are also constructed from glass. It is not indicated in Fig. 4.2 whether this pump is to be used with mercury or oil. Actually mercury and oil diffusion pumps look very much the same. However, they cannot be used interchangeably, i.e., a pump designed for use with an oil should never be used with mercury, and vice versa. This is because the nozzle opening, heat input, and cooling requirements are different for oil and mercury. Most of the designations shown in Fig. 4.2 are self-explanatory. It will be noted that the nozzle is oriented downward toward the pump *barrel* (or *casing*), which is water cooled. The water used normally comes from the water mains and, as a rule, is a few degrees cooler than room temperature. Vapor striking the barrel is condensed and flows back to the boiler. The *forevacuum connection* (also called *forearm, outlet flange,* and *exhaust fitting*) is the flange or coupling on the forevacuum side (discharge port) of the pump. It connects to some form of backing pump, usually a rotary oil-sealed pump or an ejector pump backed by a rotary oil-sealed pump. The section of pipe between the forevacuum connection and the pump body is called the *forearm*. It is normally directed upwards and is water cooled so as to avoid loss of pump fluid. The *mouth* of the pump is also called the *intake port, entrance,* or *inlet.* The combination of nozzle and chimney (with its associated skirt) is called the *nozzle* or *jet* assembly and is removable from the casing. Most commercial pumps use a cooling coil (usually copper) which is brazed or soldered to the casing (and forearm). However, some pumps are designed with a chamber around the casing through which cooling water is circulated. Figure 4.2 shows a water-cooled baffle near the mouth of the pump. This is to reduce the amount of pump oil getting into the vacuum system. Such baffles are usually mounted above the pump as a separate component. Some type of gravity drain should be provided on the baffle to return oil to the boiler and to prevent oil from striking hot surfaces. Often wires attached to the baffle and touching the pump wall can be used.

Most commercial diffusion pumps use more than one nozzle and are called multi-stage pumps. The features of two types of three-stage diffusion pumps are shown in Figs. 4.3a and 4.3b. The main difference between these pumps is in the arrangement of the nozzles and chimney. In the case of Fig. 4.3a the vapor to all nozzles passes through the same chimney, whereas in the case of Fig. 4.3b the chimney is divided into parts corresponding to the various nozzles. In the latter case the vapor to the various nozzles originates at different parts of the boiler. The net effect is to have the more volatile constituents of the oil issue through the nozzles nearest the forepump connection. The high speed jet, i.e., the one nearest the vacuum system, will operate on the least volatile oil constituents, which results in a lower ultimate pressure than that available with the pump of Fig. 4.3a. Pumps incorporating this feature are falled *fractionating, self-fractionating,* or *purifying* pumps. Naturally such pumps are not used with mercury since mercury is a pure element and therefore cannot be broken down into other constituents. However, multi-stage mercury pumps are of the same general design as the pump of Fig. 4.3a. In all these pumps the nozzles are designed so that the pressure rise across the various jet streams increases progressively as one goes from the vacuum system side to the forepump side. The smallest pressure rise is across the high speed jet stream (nearest the vacuum system). Regardless of the number of stages, the speed of a pump will be determined by what is called the *admittance area, nozzle clearance area, aperture,* or *annular gap.* This is the area of the annular cross section between the outer rim of the

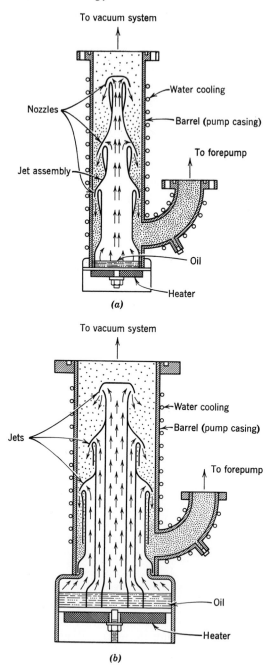

Fig. 4.3 (*a*) Three-stage nonfractionating pump. (*b*) Three-stage fractionating pump. (Courtesy Consolidated Vacuum Corp., Rochester, N.Y.)

nozzle exit and the wall of the pump casing. As this area is increased the pumping speed also will increase.

The examples shown in Figs. 4.2, 4.3*a*, and 4.3*b* all indicate the use of water cooling. This is generally true for pumps with casings of diameter greater than 2 in. However, some small pumps are manufactured which are air cooled. In this case the pump casing is generally provided with fins to radiate away heat. The fins are cooled by an air blower, which is normally mounted on the pump. Clearly it is an advantage with portable vacuum systems to have an air-cooled pump since the only service then required is electrical. The heaters used to vaporize the pump fluid are either mounted directly in the boiler (immersion heaters) or are attached to the boiler.

Mercury and oil pumps differ somewhat in the matter of construc-

Fig. 4.4 Four-stage nonfractionating oil diffusion pump (Consolidated PMC-1440). (Courtesy Consolidated Vacuum Corp., Rochester, N.Y.)

tional materials. This is because mercury tends to amalgamate with various nonferrous metals and alloys such as copper and brass, whereas diffusion pump oils are inert. The pump casings for both mercury and oil pumps are generally made of low carbon steel or stainless steel. Some small pump casings are made of aluminum. Many users of vacuum equipment prefer stainless steel because it takes a high polish and forms an impervious oxide layer rather than a spongy scale (as in the case of carbon steel), thus leading to less outgassing and to ease of cleaning. The nozzle assemblies in oil pumps are generally constructed of aluminum or carbon steel (often nickel plated). Untreated aluminum cannot be used in mercury pumps. Nickel plating nonferrous metals makes them suitable for use with mercury. Most diffusion pumps have flanged inlets and outlets to connect to other vacuum components. In the cases of some small pumps, usually 1 and 2 in. sizes, tubing connections are provided. Metal pumps are often designated by the inside diameter of the barrel or casing, e.g., a 4 in. pump has a nominal 4 in. inside barrel diameter. Glass pumps are not commonly used in industrial processes because they are so fragile. However, for various laboratory scale operations glass pumps offer several advantages over metal pumps, including ease of cleaning and outgassing, corrosion resistance to many gases, and the possibility of continuous inspection of pump operation. Both mercury and oil glass pumps are available commercially. A typical metal oil diffusion pump is shown in Fig. 4.4. This is a four-stage, nonfractionating, water-cooled pump with a nominal pumping speed of 1440 l/sec at 2×10^{-5} mm Hg.

4.3   Some Features of Ejector Pumps

The basic element of all types of ejector pumps is the nozzle arrangement and the basic features have already been shown in Fig. 4.1b. Naturally the exact details of any particular nozzle arrangement will be dependent on the function of the pump. Ejector pumps are designed to operate with steam, oils, or mercury and may be single-stage or multi-stage. In all cases the fluid vapor expands through the nozzle at supersonic speeds and entrains gases and vapors in the process.

Steam ejector pumps are generally used in connection with rather large scale vacuum installations and in the rough or low vacuum regions, i.e., down to a few millimeters of mercury. However, such ejectors can be designed to go into the micron range. In general, each stage in a steam ejector gives a compression ratio of around 10 to 1. Approximate values of pressures which can be achieved with various numbers of stages are as follows:

Number of Stages	Absolute Pressure
1	3 in. Hg
2	⅓ in. Hg
3	1 mm Hg
4	150 μ Hg
5	20 μ Hg

The value of 3 in. Hg corresponds to a gauge reading of 27 in. Hg. Gauges of the Bourdon and diaphragm types (see Chapter 6) are generally calibrated so that 0 reading corresponds to atmospheric pressure and 30 in. Hg corresponds to a very good vacuum. The steam supplied to the nozzle must be at high pressure, usually more than 125 psig. The unit psig (pounds per square inch gauge) refers to gauges calibrated in pounds per square inch, with the 0 corresponding to atmospheric pressure (15 psi). Consequently, 15 psi must be added to the gauge reading to get an absolute pressure value. The steam plant required represents a considerable part of the initial financial outlay for any steam ejector installation. The steam must be liquefied by the use of cooled condensers and the temperature of the cooling water is usually less than 75°F. Steam ejectors discharge directly to the atmosphere. An increase or decrease in the quantity of gas or vapor being handled under constant suction and discharge conditions cannot be accomplished without changing the dimensions of the nozzle or diffuser chamber (see Fig. 4.1*b*). Consequently, commercial ejectors are designed for certain conditions and flexibility is achieved by operation in parallel as required. Most steam ejector failures result from the steam pressure dropping below a certain minimum value. Too high a steam pressure results in higher steam consumption but no increase in capacity (sometimes a decrease). Dry steam should be used since wet steam causes rapid corrosion of parts.

Oil or mercury ejector pumps are often used to bridge the gap between diffusion pumps and rotary oil-sealed pumps. This allows both the diffusion pump and the rotary pump to operate at maximum efficiency. Sometimes ejector pumps are called booster pumps. This is not strictly true since a *booster pump* is any vapor pump or specially designed mechanical pump which is used between a vapor pump and

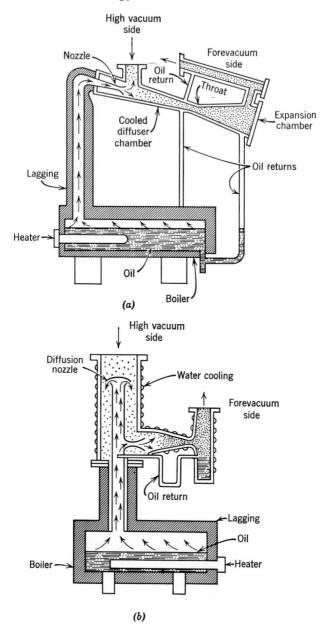

Fig. 4.5 (a) Single-stage ejector pump. (b) Diffusion-ejector pump. (Courtesy Consolidated Vacuum Corp., Rochester, N.Y.)

a forepump (usually a rotary oil-sealed pump) to increase the maximum gas throughput which can be handled. Oil and mercury ejector pumps will give base pressures from around 1 μ to several hundred microns depending on the number of stages. Oil pumps are far more commonly used than mercury pumps. Multi-stage oil ejectors are often used directly (backed by a mechanical pump) in various vacuum processes such as dehydration and metallurgical applications. Oil and mercury ejectors have relatively small admittance areas with nozzles designed to produce as high a forepressure tolerance as possible and a large volume of vapor. The density of the jet stream is much higher than in the case of diffusion pumps. The primary function of an ejector pump is to handle large quantities of gas or vapor. Ejector pumps can usually be recognized by the large boilers which are used to produce the large amounts of vapor needed. Figure 4.5a shows a typical single-stage oil ejector pump. The principal components are the boiler with associated heater (in this case an immersion heater), the nozzle, the diffuser chamber, the throat, and the expansion chamber, together with associated oil returns and connections. The most common constructional material is low carbon steel.

In some cases diffusion and ejector nozzles are combined in a single pump which is called a diffusion-ejector pump. Again such pumps can use oil or mercury, with oil the most common. Commercial diffusion-ejector pumps have base pressures around 10^{-4} to 10^{-5} mm Hg. Figure 4.5b shows a typical diffusion-ejector pump. This particular unit has a single-stage diffusion unit followed by a single-stage ejector unit. The ejector unit is normally backed by a rotary pump. These pumps can be constructed with several diffusion and/or ejector stages. The necessity for a dense jet stream through each ejector stage makes it necessary to have a large boiler. This puts a limit on the lowest pressure which can be obtained. Lowering the heat input would give lower pressures but this would defeat the purpose of a diffusion-ejector pump, which is designed to handle large quantities of gas or vapor. These pumps are commonly used in various metallurgical processes, such as melting and casting furnaces. Carbon steel is a commonly used construction material with glass confined to small pumps.

4.4 Mercury versus Oil

The oils listed in Appendix B have been tested in vacuum practice over a period of years. The treatment to follow is concerned largely with diffusion pumps. However, much of the material presented is applicable to ejector and diffusion-ejector pumps. Although oil vapor pumps have largely replaced mercury vapor pumps in most high vacuum processing operations, there are some applications where mercury presents some advantages. Examples where mercury may have advantages over oil include the processing of special x-ray tubes and the operation of nuclear particle accelerators. Oils tend to dissociate on hot surfaces and in electrical discharges, which may give rise to certain detrimental effects in certain special processes.

ADVANTAGES OF MERCURY

1. It cannot decompose no matter how hot the boiler.
2. It is unlikely to react with most gases and vapors.
3. It is a known vapor (easy to recognize).
4. It cannot decompose on hot filaments or in an electrical discharge.
5. It is readily trapped by liquid nitrogen.
6. It returns easily to the boiler due to its high density.
7. It can be used in pumps which are designed to discharge into a mechanical pump at relatively high pressures.

DISADVANTAGES OF MERCURY

1. It has a high vapor pressure—approximately 1 μ at room temperature.
2. Refrigerated traps must be used to obtain pressures comparable to those available with oil diffusion pumps not using traps.
3. It amalgamates with various metals and alloys.
4. It is sensitive to oil and grease contamination (resulting in lowered pump efficiency).
5. It will oxidize if exposed to air when hot.
6. It is toxic and must be trapped or vented so as not to get into the working area.

ADVANTAGES OF OILS

1. They are available with low vapor pressures (less than 10^{-7} mm Hg at room temperature).
2. In many systems they can be used to obtain operating pressures down to their vapor pressures without the use of refrigerated traps.
3. They consist of heavy molecules which are effective in giving high speeds to gas and vapor molecules.

4. They are not toxic.

5. They are inert with respect to most materials used in vacuum system construction (particularly pumps).

6. Pumps using oils can be designed to be self-purifying, i.e., the effect of deterioration of the oil can be minimized.

DISADVANTAGES OF OILS

1. They consist of complex chemical compounds which are subject to decomposition even under normal operating conditions. Higher temperatures such as occur in boilers or on hot filaments accelerate the decomposition.

2. They can decompose in electrical discharges. The products of decomposition (whether from heat or an electrical discharge) can be harmful to electrical instrumentation through the production of undesired conducting paths. Also, the composition of these products is largely unknown.

3. They decompose when exposed to air at their working temperature. Even silicone oils are subject to decomposition due to prolonged exposure to air.

4.5   Pump Performance

In order to discuss the performance of any vapor pump certain vacuum terms have come into general use. Many of these terms are applicable to other pumps besides vapor pumps. The outlet and inlet sides of a vapor pump were mentioned in Section 4.2. Although used in connection with diffusion pumps these terms are also applicable to ejector pumps. Instead of outlet, terms such as *discharge side, forevacuum side,* or *forepressure side* are sometimes used. The terms, *ultimate pressure, throughput,* and *pumping speed,* which are important factors determining vapor pump performance, were defined in Sections 3.3, 2.3, and 2.10, respectively. Vapor pumps are most commonly used with another pump, the *backing pump* or *forepump,* connected to the outlet. The total pressure on the outlet side of a pump, measured near the outlet port, is called the *forepressure.* This is also known as *outlet pressure, exhaust pressure, discharge pressure,* or *backing pressure.*

Commercial vapor pumps are designed to operate with a certain heat input to the boiler and specified cooling requirements, as well as a certain forepressure range. Departing from recommended heat input and cooling requirements will generally result in poor performance. The *normal forepressure* is the forepressure produced by a given forepump as measured near the outlet port of a given vapor pump at a given load without leaks or other (large) sources of gas in the forevacuum line. This is also called the *dynamic forepressure* or *operating forepressure*. Load refers to the throughput, which is the same in the vapor pump and the mechanical pump, i.e., the throughput being handled by the vapor pump must also be handled by the mechanical pump. The *limiting forepressure* is the forepressure which corresponds to an inlet pressure about 0.2 μ (millitorr) greater than the inlet pressure corresponding to the normal forepressure for a vapor pump at a given load. This is sometimes called *static breakdown forepressure*. The *cross-over forepressure* can be defined as the forepressure at which the inlet pressure of a vapor pump becomes equal to the forepressure. It is sometimes called the *take-hold* pressure. Another term often used in discussing pump performance is *tolerable forepressure*. This is the forepressure which corresponds to an inlet pressure 10% greater than the inlet pressure corresponding to the normal forepressure for a vapor pump at a given throughput. This is sometimes called *forepressure tolerance*. Generally a pump is considered to fail when the forepressure exceeds this value.

Apart from listing pertinent design features of vapor pumps such as dimensions, construction materials, and power and cooling water requirements, manufacturers usually publish curves of pumping speed versus pressure or curves of throughput versus pressure, or both. The pumping speed curve is often referred to as a *speed-pressure curve* or simply a *speed curve*. Some manufacturers also include curves of the limiting forepressure versus throughput. Reference to these various curves will point out the significance of various terms which have been defined above. Figure 4.6 shows the typical form of a pumping speed–pressure curve for a diffusion pump. Many such pumps will not show a flat region in their pumping speed curves (P_a to P_b), the exact shape being dependent on the design. Ejector pumps will often show a peak in their pumping speed curves at some particular pressure, which is determined by the design. Curve 1 in Fig. 4.6 represents an unbaffled pump while curve 2 represents a pump equipped with a refrigerated baffle. It will be noted that the effect of the baffle is to reduce the maximum pumping speed from S_b to S_b' and the ultimate pressure from P_0 to P_0'. The actual changes effected will depend on the designs of the pump and the baffle as well as the condition of the vacuum system. Curve 3 is a typical pumping speed curve for a rotary backing pump (see Chapter 3). It will be noted that the pumping speed curves

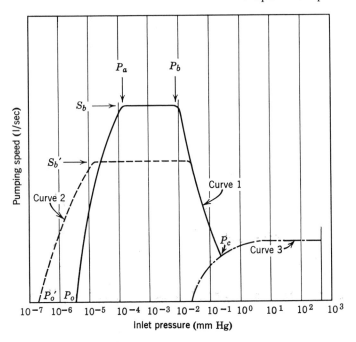

Fig. 4.6 Form of pumping speed–pressure curve for a diffusion pump. (Courtesy Consolidated Vacuum Corp., Rochester, N.Y.)

for the diffusion pump and the backing pump cross at a pressure P_c, which is the cross-over forepressure. As the forepressure drops below this value, the pressure at the inlet of the diffusion pump drops rapidly, corresponding to an increase in speed of the diffusion pump (P_c to P_b on curve 1). The ultimate pressure obtained in a system free of leaks and containing no sources of gas or vapor depends primarily on the vapor pressure of the pump fluid.

Figure 4.7 shows typical curves of throughput versus pressure. Curve 1 is the throughput versus pressure at the inlet of a vapor pump while curve 2 is a similar curve at the inlet of the forepump. Curve 3 represents the limiting forepressure of the vapor pump as a function of throughput. The curves in Fig. 4.7 are carried to larger values of throughput than is normal practice. Clearly at point P_c the pressures at the inlets of the vapor pump and the forepump are equal and therefore the vapor pump is not pumping. Manufacturers will show curves going up to the maximum values of throughput for which measurements have been made. Such curves usually look like the solid parts of the curves in Fig. 4.7.

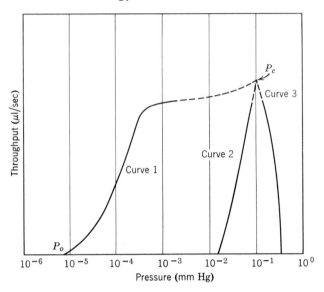

Fig. 4.7 Throughput versus pressure curves. (Courtesy Consolidated Vacuum Corp., Rochester, N.Y.)

4.6   Performance Characteristics of Diffusion Pumps

The principal factors affecting the performance of any vapor pump are the type of pump fluid used and the heat input to the boiler. Diffusion pumps are generally used only when pressures below 10^{-5} mm Hg are needed. Since the vapor pressure of mercury is about 1 μ at room temperature, pressures below this value can only be obtained by the use of refrigerated baffles or traps. With oil diffusion pumps on a clean, tight vacuum system, i.e., one with no large leaks or sources of gas and vapor, the ultimate pressure will be essentially the vapor pressure of the oil when no refrigerated baffles or traps are used. Consequently, the choice of oil is important when low ultimate pressures are desired. By using refrigerated baffles or traps ultimate pressures well below the vapor pressure of the pump oil can be obtained. Commercial oil diffusion pumps are generally designed to be used with a particular oil specified by the manufacturer. Other oils can be used but the performance characteristics will then change. Also, it may be necessary to change the heat input and cooling water requirements.

The gas-handling capacity of a diffusion pump depends on the heat input and the speed of the backing pump. As the gas flow rate increases the pressure below each jet increases. The pump will not fail as long as a certain maximum forepressure is not exceeded. The required forepressure for any given throughput can be obtained from curves of the type shown in Fig. 4.7. The maximum value of forepressure which can be tolerated is determined by the nozzle design and the pump fluid. Heavy molecules are more effective than light ones. The backing pump is usually the limiting factor in handling larger quantities of gas and vapor. Although the heat input can be increased so as to increase gas handling capacity this will result in a higher ultimate pressure. This is because the higher pressure increase across the high speed jets near the inlet results in more pump fluid and gases and vapors being scattered back into the vacuum system. A low ultimate pressure means a small throughput. The power consumption by the heater depends on the general design of the pump and, in particular, on the design of heater, boiler, and nozzle assembly. There should be good thermal contact between the heater and boiler (for non-immersion-type heaters) and the use of lagging (asbestos, etc.) will minimize heat losses. Mounting the pump in an area free of drafts will also help. Once a pump has been brought up to operating temperature, the power required is needed primarily to replace heat lost through radiation, convection, and conduction. Most of the heat is lost to the cooling water. Too much cooling water serves no useful purpose and results in excessive power consumption. Too little cooling water causes poorer pump performance due to the pump fluid not being condensed properly, resulting in pump fluid getting into the vacuum system. Oil diffusion pumps are usually cooled with water near room temperature while mercury pumps are sometimes operated with water cooled to a few degrees centigrade. Commercial diffusion pumps have cooling water requirements ranging from a few hundredths of a gallon per minute for a 1 in. pump to several gallons per minute for a 32 in. pump. Corresponding heater power requirements are about 100 watts (w) and 25 kilowatts (kw) respectively.

Manufacturers will often tabulate the maximum speeds together with the pressures at which these speeds occur. The speeds at other pressures can be obtained from the approximate pumping speed–pressure curves. Maximum pumping speeds range from about 15 l/sec for a 1 in. pump to 30,000 l/sec for a 32 in. pump. The speeds listed by manufacturers are generally measured values for dry air or nitrogen. The pumping speeds for other gases will be different, being greater for gases or vapors that are lighter than air (smaller molecular weight) and less for heavier gases or vapors (larger molecular weight). It must be kept in mind that the pumping speeds quoted by manufacturers are upper limits since they are determined at the inlet of the pump. In practice there will be pumping speed losses in connecting lines, which often include baffles and traps, and these losses can be quite high. Methods for treating such losses are considered in Section 4.10 and in Chapter 11. The pumping speed of a diffusion pump is often described in terms of a *figure of merit* or *speed factor* called the *Ho coefficient*. This is the ratio of the measured pumping speed to, basically, a perfect pumping speed. The perfect pumping speed is obtained by replacing the diffusion pump by a perfect vacuum. All gas and vapor molecules with motions toward the perfect vacuum will enter it with no rejections. Such a situation will result in pumping speeds equal

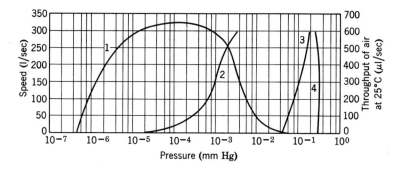

Fig. 4.8*a*   Performance curves for a 4 in. fractionating metal oil diffusion pump (Consolidated Vacuum Corp. Type MCF-300).

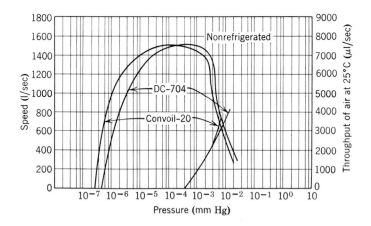

Fig. 4.8*b*   Speed curve for nominal 7 in. ID metal fractionating pump (Consolidated Vacuum Corp. Model PMC-1440).

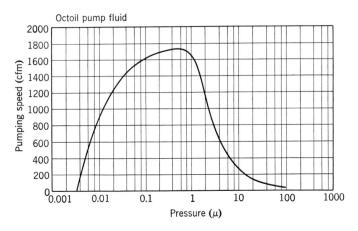

Octoil pump fluid

Fig. 4.8c Speed curve for a 6 in. metal oil diffusion pump (F. J. Stokes Machine Co., Philadelphia, Pa., ring jet pump, Series 160, 6 in. size).

to 11.6 l/sec per cm² or about 75 l/sec per in.² Commercial diffusion pumps have Ho coefficients which are usually considerably less than 50% (0.50). An average value might be around 0.30 to 0.40. Naturally a higher value would indicate a more efficient pump.

The limiting forepressures of diffusion pumps are sometimes shown by manufacturers in the form of graphs (as in Fig. 4.7) and in other cases they are listed for full-load and no-load conditions. Limiting forepressures range from around 100 μ to over 600 μ (full load). The quantity of pump fluid required ranges from a few cc for a 1 in. pump to around 3 gal for a 32 in. pump.

Typical curves showing the performance of several commercial diffusion pumps are shown in Figs. 4.8 through 4.11. Figures 4.8a, b, and c show curves for several metal oil diffusion pumps. These curves were obtained without refrigerated baffles or traps. The labeling of the curves is as follows: 1 = speed − pressure, 2 = throughput − pressure at inlet, 3 = throughput − pressure at outlet, 4 = throughput − limiting forepressure. Curves for the pump of Fig. 4.4 are shown in Fig. 4.8b. Figure 4.9 shows the effect of pump oil on pump performance. The heat inputs used are shown on these curves. It is evident that the pump oil has a considerable effect on the ultimate pressure. The maximum pumping speed is not affected too much. Figure 4.10 shows performance curves for a commercial metal mercury pump. These curves do not show the ultimate pressure since this is determined by the type of trapping that is used. With suitable traps these pumps can be

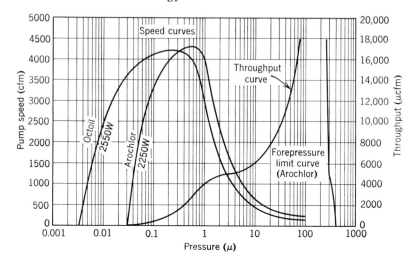

Fig. 4.9   Effect of pump oil on speed (F. J. Stokes Machine Co., 10-in. diffusion pump with 115 cfm forepump, Series 160).

used to obtain pressures down to 10^{-6} mm Hg or less.   Speed-pressure curves for mercury pumps are usually obtained by using a McLeod gauge which does not measure the pressure of mercury vapor.   Figure 4.11 shows performance curves for a glass oil pump.   The curves for glass mercury pumps are similar to those in Fig. 4.10.   Figure 4.12 shows the effects of refrigerated baffle, pump oil, and type of sealing O-ring on pump performance.   It must be kept in mind that the design of pump, test system, and refrigerated baffle, as well as the amount and nature of O-ring material, will have considerable effect on the actual

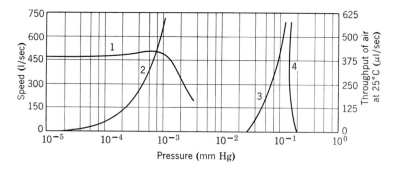

Fig. 4.10   Performance curves for 6 in. metal mercury diffusion pump (Consolidated Vacuum Corp. Type MHG-300).

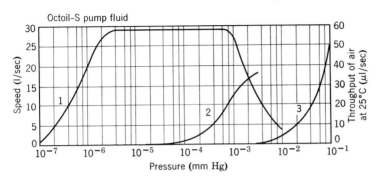

Fig. 4.11 Performance curves of glass oil diffusion pump (Consolidated Vacuum Corp. Type GF-25).

curves obtained. It is evident from Fig. 4.12 that diffusion pumps can be used to obtain pressures in the ultra-high vacuum region, i.e., 10^{-9} mm Hg or below as long as certain special procedures are followed and a careful choice of construction materials is made. The method of using curves of the type shown in Figs. 4.8 through 4.11 is discussed in Section 4.9.

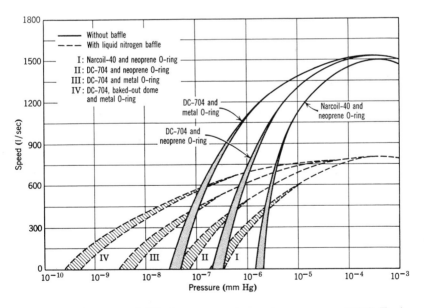

FIG. 4.12 Performance of diffusion pumps at very low pressures (NRC Equipment Corp., Newton Highlands, Mass., pump HG-1500).

4.7   Performance Characteristics of Ejector Pumps

Many of the comments made in Section 4.6 concerning diffusion pumps are also applicable to ejector pumps. The concern here is with oil and mercury pumps. The main function of an ejector pump is to handle large quantities of gas and vapor at pressures which are normally in the micron range. Pump fluids with vapor pressures as high as 1 μ are commonly used. Ejectors are designed to produce fairly high forepressures, usually a few hundred microns to several millimeters of mercury. Many of these pumps have limiting forepressure curves which are different from those for diffusion pumps in that the full-load value is often greater than the no-load value. Pump fluid charges are much larger than for diffusion pumps because of the requirement for large quantities of vapor. This leads to the need for high power requirements. A 5000 l/sec oil ejector may require over 500 gal of oil and a power requirement of 25 kw or more. Cooling water requirements are also much greater for ejector pumps than for diffusion pumps. Typical performance curves for oil ejector pumps show a peak speed at some particular pressure. For example, the Consolidated metal oil ejector, type KB-300, rated at 300 l/sec, has a maximum speed of about 350 l/sec at 250 μ, using Convoil-20. This is a single-stage pump. The Consolidated two-stage metal oil ejector, rated at 5000 l/sec, has a peak speed of 5500 l/sec at 90 μ, using Convachlor-8.

Diffusion-ejector pumps are also designed to handle large quantities of gas and vapor and therefore require large quantities of pump fluid vapor. In general they operate at pressures between those produced by ejector pumps and diffusion pumps, say down to 10^{-5} mm Hg. The large quantity of vapor needed leads to requirements similar to those in the case of ejector pumps—large quantities of oil and cooling water and high power consumption. Typical performance curves for oil and mercury diffusion-ejector pumps are very similar to the corresponding curves for diffusion pumps. Small glass oil and mercury diffusion-ejector pumps are available.

4.8   Migration of Pump Fluid

Pump fluid getting into the high vacuum system or into the forepump represents a loss of fluid from the boiler and may also cause damage to constructional materials or interference with the vacuum process. The direct flight of vapor molecules by scattering from the hot vapor jet or evaporation from hot nozzle parts in the direction of the mouth or inlet of a vapor pump is called *backstreaming*. Backstreaming can be reduced by careful design of the nozzles and by cooling the pump barrel so that there is no re-evaporation of fluid droplets. In spite of such precautions some vapor will get into the vacuum system. Many systems incorporate a baffle at the top of the vapor pump. Such baffles are normally water cooled so as to condense the pump fluid vapor and return it to the boiler. Freon cooling is often used with mercury pumps because of the high vapor pressure of mercury. The baffle should not be cooled to the point of freezing the pump fluid since this then means a progressive loss of fluid. Sometimes a water-cooled valve-baffle system is used. The passage of vapor molecules into a vacuum system by re-evaporation from baffles, the inside wall of the pump casing, and connections to the system is called *back migration*. A common type of baffle is called an *optical baffle*. In the molecular flow region molecules heading toward the baffle will generally reach it without colliding with other molecules. In an optical baffle, the design is such that molecules cannot "see" through the baffle regardless of the angle of approach (see Chapter 11).

No baffle, regardless of design, is 100% efficient in stopping all vapor molecules. Where traces of pump fluid vapor cannot be tolerated, such as in systems to be operated at pressures well below the vapor pressure of the fluid, the use of a baffle is augmented by the use of refrigerated traps. These traps are usually cooled by liquid nitrogen or Freon and also act as very effective pumps for other condensable vapors (see Chapter 5). The more effective baffles and traps are in trapping pump fluid, then the greater will be the reduction in pumping speed.

The usual method for preventing pump oil from getting into the forepump (and hence into the atmosphere) is by use of a vertical, water-cooled discharge tube. Sometimes a dry ice trap or Freon refrigeration is used, particularly to remove toxic vapors, including mercury vapor. Where several pumps are connected in series, such as a diffusion pump, followed by an ejector pump and then a rotary pump, there is a tendency for one pump to lose oil to another. Where the same pump fluid can be used, then boilers of the pumps can be connected by a small tube so as to equalize the fluid levels. Where different pump fluids must be used the best solution is to use suitable baffling between pumps.

4.9   Backing Pump Requirements

The attention here will be confined to the case of a diffusion pump backed by a rotary pump. However, the method used is applicable to any pump backing another pump. The choice of backing pump is determined by the required speed at the operating forepressure. In addition, if the backing pump is used to pump down the system to the pressure where the diffusion pump can start pumping, then its ultimate pressure must be considered. In most cases a separate roughing pump is used to pump the system down to the pressure where the diffusion pump can take over. The capacity of the roughing pump must be considered in terms of the required pump-down time according to the methods of Chapter 3. The choice of backing pump is based on the use of pump performance curves.

EXAMPLE 1

A Consolidated MCF-300 oil diffusion pump is used on a vacuum system. What size backing pump should be used?

Assume the backing pump is used to bring the system down to the point where the diffusion pump can start pumping and to then back it. The backing pump must handle the maximum throughput which can be passed by the diffusion pump. Referring to Fig. 4.8a, the maximum throughput for an MCF-300 pump is 600 μl/sec. At the inlet to the forepump this occurs at a pressure of about 190 μ. Since pumping speed is throughput divided by the pressure, the required speed of the backing pump must be $600/190 = 3.16$ 1/sec (at 190 μ pressure). No allowance has been made for losses in the foreline or bursts of gas and vapor. For an average clean industrial system, a service factor of 2 should be used at a pressure of 190 μ. This gives a required pump speed of 2×3.16 or about 6.3 1/sec. For the MCF-300 pump the manufacturer recommends a backing pump with a speed of 6 1/sec. At the maximum throughput the limiting forepressure from curve 4 of Fig. 4.8a is about 250 μ. The diffusion pump does not start pumping until the pressure reaches about 6×10^{-2} mm Hg. The backing pump must pump down to at least this pressure. Practically the backing pump should have an ultimate pressure of 10^{-2} mm Hg or better.

In the above example the pump-down time has not been considered. This is usually the factor determining the size of backing pump. When such a pump is used to pump down a vacuum system through a direct pipe or tube connection then the methods of Chapter 3 can be used to determine pump-down time. However, if the pump-down is carried out through the diffusion pump and associated piping, traps, baffles, etc., then allowance must be made for the increased resistance to the flow of gas. Methods for doing this are considered in Chapter 11. If pump-down is carried out through the diffusion pump then this pump cannot be heated up until the pressure is reduced to a sufficiently low value so as to avoid damage to the pump fluid. Usually a pressure of a few microns is adequate. The best procedure is to carry out the initial pump-down through a separate line, using either the backing pump or a separate roughing pump.

EXAMPLE 2

A vacuum system is expected to have a continuous gas flow of 5 cm³/min at atmospheric pressure. An ultimate pressure of 10^{-4} mm Hg is to be maintained by diffusion pumps with an operating fore-pressure of 100 μ. What capacity backing pump is required?

A throughput of 5 cm³/min at atmospheric pressure (5 std cc/min) equals $5 \times 760 = 3800$ mm Hg cc/min $= 3800/1000 = 3.8$ mm Hg l/min $= 3800$ μl/min. This could have been obtained directly from Table 2.1. Again the pumping speed of the backing pump is given by throughput divided by pressure or $3800/100 = 38$ l/sec. At a pressure of 100 μ the service factor is 2 for a clean industrial system, which gives a required pump speed of $2 \times 38 = 76$ l/min. However, this pump size may not be adequate to handle the throughput as the diffusion pump pumps down through its range or to pump down in a short enough time. These factors must be considered if the pump is to be used both for pump-down and backing purposes. Also, in this case the ultimate pressure of the pump must be low enough to permit the diffusion pump to start pumping.

4.10 Effect of Connecting Lines

Consideration has been given to the effect of round pipe on the pumping speed of rotary pumps where the flow is in the viscous region. In dealing with vapor pumps the concern is largely with molecular flow. Vacuum components such as refrigerated traps and baffles will reduce the pumping speed by a considerable amount. These components are also involved in the viscous region during pump-down. Some common designs of traps, baffles, and valves are considered in Chapter 11.

Emphasis here is on the effect of round pipe on pumping speed. The curves of Fig. 3.7 apply to the molecular flow region and include the effect on gas flow of the pipe end where it is considered to enter a vessel of considerably larger cross section (the usual case). Figure 3.7 together with Fig. 2.10 can be used to immediately find the pump speed required to obtain a certain speed at a vessel (or vice versa). A bend is considered to be an equivalent length of straight pipe by measuring its length along the center line.

EXAMPLE 1

A Stokes 6 in. oil diffusion pump (using Octoil) is connected to a vacuum vessel by 6 in. pipe (ID) which is 4 ft long. If the pump has an inlet pressure of 10^{-5} mm Hg, what is the pumping speed at the vessel?

Figure 3.7 shows that the conductance of a 4 ft length of 6 in. pipe is about 460 cfm. Figure 4.8c shows the pumping speed curve for the pump being used. At a pressure of 10^{-5} mm Hg the pumping speed is about 900 cfm. From Fig. 2.10 a conductance of 460 cfm and a measured pumping speed, S_m, of 900 cfm gives a net speed, S_n, of about 310 cfm. It will be noted that 4 ft of 6 in. pipe reduces the pumping speed to one-third of the value at the pump inlet. It is also possible to find the pressure at the vessel near the pipe inlet. The throughputs at both ends of the pipe must be equal, i.e.,

$$\text{Pressure at vessel} \times 310 = 10^{-5} \times 900$$

or

$$\text{Pressure at vessel} = \frac{900}{310} \times 10^{-5} = \text{about } 3 \times 10^{-5} \text{ mm Hg}$$

EXAMPLE 2

A 10 in. diffusion pump is connected by 10 in. pipe (ID) to a vacuum process chamber and it is required that there be no more than 50% loss in pumping speed. A pumping speed of 1000 l/sec is required at the vessel. What is the maximum length of pipe that can be used?

$S_n = 1000$ l, which is 50% less than the measured speed, S_m. Therefore, $S_m - 0.50 S_m = 1000$, or $S_m (1 - 0.50) = 0.50 S_m = 1000$ and

$$S_m = \frac{1000}{0.50} = 2000 \text{ l/sec}$$

Using these values of S_m and S_n, the conductance required is estimated from Fig. 2.10 to be about 2000 l/sec. Liters per second can be converted to cubic feet per minute by multiplying by 0.47. Therefore,

2000 l/sec is equal to 2000 × 0.47 = 940 cfm. Using this value of conductance and the curve for a 10 in. pipe, Fig. 2.10 shows the pipe length should be no more than about 10 ft.

EXAMPLE 3

What diameter of pipe should be used to connect a diffusion pump operating at a speed of 100 cfm to a vacuum vessel where a pumping speed of 70 cfm is required? Because of the physical setup the pipe must be 3 ft long.

From Fig. 2.10, the curve for $S_m = 1$ (corresponding to 100 cfm) intersects the abscissa $S_n = 0.7$ (corresponding to 70 cfm) at a conductance of about 2.2 (220 cfm). From Fig. 3.7 the abscissa at 3 ft intersects the ordinate for a conductance of 220 cfm at a point between the curves for 4 in. and 6 in. pipe. To be on the safe side 6 in. pipe should be used.

4.11 Pump-Down Time for Vapor Pumps

Vapor pumps are designed to operate at relatively low pressures and in their operating ranges they have high pumping speeds. As a rule the time involved in pumping down a system to its operating value is used in getting down to the pressure value where the vapor pump takes hold, i.e., while the forepump is doing the pumping. Once the vapor pump starts pumping, the pressure drops rapidly due to the high speed of the pump. Therefore, with many systems the time involved in going from the time of initial operation of the vapor pump to the final operating pressure is negligible. However, with systems involving considerable release of gases and vapors the pump-down time with a vapor pump operating may be quite extended. The pressure will remain fairly constant until most of the gases and vapors are removed. Refrigerated traps are very effective in reducing pressure due to condensable vapors. Even with traps the time involved will depend on the condition of the system and on the final pressure desired.

The method for determining pump-down time for rotary piston pumps as described in Section 3.6 is also applicable to vapor pumps for average, clean industrial vacuum systems. Equation 3.2 is only applicable to the constituents of the atmosphere in a vacuum vessel and not to gases and vapors being released from surfaces. Under these circumstances it can also be applied to the pump-down time for vapor pumps. Conservative estimates of pump-down time are obtained by using the minimum pumping speed for the pressure range being considered.

EXAMPLE 1

A Consolidated MCF-300 pump, backed by a rotary pump, is used to pump down a vessel of volume 200 l from 10 μ (where it begins to pump effectively) to 2×10^{-6} mm Hg (where the pumping speed is dropping rapidly). How long will this take?

An average value of pumping speed between 10 μ and 2×10^{-6} mm Hg can be estimated from the pumping speed curve for an MCF-300 pump (Fig. 4.8a) to be about 200 l/sec. The time required is then $2.3 \times (200/200) = 2.3$ sec. The minimum speed involved is about 50 l/sec. Using this value, the time is $2.3 \times (200/500) = 9.2$ sec. Instead of using eq. 3.2, Fig. 3.8 could have been used to obtain values very similar to these. In connection with rotary pumps, a factor was used in estimating pump-down times which increased to 6 at 1 μ. Even using a factor of 20 in the case above would give a maximum pump-down time of only $20 \times 9.2 = 184$ sec or about 3 min. In practice the loss of pumping speed in pumping lines must be taken into consideration.

4.12   Operating a Vapor Pump

The attention here is confined primarily to diffusion pumps, although many of the following remarks apply to both ejector and diffusion-ejector pumps. Some steps to follow in setting up and operating a diffusion pump are as follows:

1. Choose an installation area with suitable power outlets, water supply, and drain (for water-cooled pumps). The area should be free of strong drafts.

2. Check the pump to be sure that the nozzle assembly is aligned properly.

3. Fill the pump with the correct amount of recommended oil.

4. Mount in a vertical position so as to have a reasonably equal depth of oil throughout the boiler.

5. Make tight vacuum connections on the high vacuum and forevacuum sides. These will generally be tubing or flange connections involving the use of gaskets. As has already been noted, the connection to a mechanical pump should have some flexibility so as to avoid detrimental effects from vibration.

6. Check the flow of water (for water-cooled pumps) to make sure there are no obstructions. The flow should conform to manufacturer's specifications. In the case of air-cooled pumps the blower should be checked for correct operation.

7. Check the operation of any protective devices such as might be used in the cooling water and heater circuits.

8. Turn on the heater and water cooling when the pressure has reached a point where no damage will result to pump fluid. A pressure of a few microns is usually satisfactory. This implies the incorporation of a suitable vacuum gauge in the system.

9. Check the operation of the heater. This can usually be done by feeling the boiler.

10. The pressure should drop rapidly when the pump has reached operating temperature. The boiler should then be too hot to touch comfortably.

In shutting down a diffusion pump the important thing to remember is to avoid letting air into the pump while it is still hot so as to prevent damage to the pump oils. Even silicone oils can be decomposed to some extent by exposure to air when hot. Some specific steps to be followed in shutting down a diffusion pump are as follows:

1. Isolate the diffusion pump and its backing pump by suitable valving from the rest of the system, which can then be let down to air.

2. Shut off the heater but not the water cooling or air blower.

3. After the boiler has cooled sufficiently, shut off the heater. The boiler can be considered to be cooled sufficiently when the hand can be held against it without undue discomfort.

4. Let the pump down to atmosphere through suitable valving. Normally the backing pump is shut off at the same time or is valved off.

5. Shut off the cooling water (or air blower). Failure to do this will result in water vapor condensing inside the pump. This will cause large bursts of gas and vapor when the pump is turned on again, i.e., a large gas load.

4.13   Poor Operation or Failure of Vapor Pumps

The following discussion is again concerned primarily with diffusion pumps but is applicable to a large extent to ejector and diffusion-ejector pumps. A vapor pump may show the following unusual types of behavior:

1. Failure to pump at all.
2. Low pumping speed.
3. Low gas handling capacity.
4. High ultimate (or base) pressure.

Causes of the above defects and possible cures are:

1.(a) No oil in pump. *Cure:* Add proper amount of recommended oil. Sometimes it may be necessary to clean the pump.

(b) Insufficient or no heat to boiler. *Cure:* Check heater for damage (burned out, worn, etc.)—replace as necessary. Otherwise, check power circuit for heater.

(c) Forepressure above the tolerable value due to poor operation of backing pump or blockage of nozzles. *Cure:* Check performance of backing pump and correct as necessary. Check nozzle assembly.

2. The general cause of this condition is not enough vapor being supplied to the first stage.

(a) Insufficient oil in pump. *Cure:* As for 1(a).

(b) Insufficient heat to boiler. *Cure:* As for 1(b).

(c) Partial blockage of nozzles. *Cure:* Clean nozzles and align nozzle assembly correctly.

3.(a) Insufficient heat to boiler. *Cure:* As for 1(b).

(b) Too high a foreline pressure, probably due to poor operation of the backing pump. *Cure:* As for 1(c).

(c) A leak in the pump itself. *Cure:* Check for leaks in pump and correct as necessary.

4.(a) Insufficient or warm cooling water. *Cure:* Provide additional cooling or increase rate of flow.

(b) Partial blockage of nozzles. *Cure:* As for 2(c).

(c) Too much heat being supplied to boiler. *Cure:* Check condition of heater, particularly for shorts, and replace or repair as necessary. Check heater circuit.

(d) Insufficient oil in pump. *Cure:* As for 1(a).

(e) High vapor pressure contaminants in oil. *Cure:* Clean or replace oil.

The causes of some of the problems discussed above can often be attributed to:

1. Exposure of pump fluid to the atmosphere. This is the most common problem in pump operation.

Results:

a. Loss of pump fluid which could lead to pump stoppage.

b. Contamination of working volume with resulting high ultimate pressure.

c. Damage to vacuum components (mercury) or interference with various vacuum processes (oil).

d. Possibility of blocking nozzles with gummy residues (oil) with resultant drop in pumping speed or complete pump stoppage.

Prevention:

a. Set up a standard, manual operating procedure so as to minimize the possibility of letting the vapor pump down to air while hot.

b. Interlock the operating mechanism of the valve above the diffusion pump so this valve closes automatically when the pressure rises above a certain value. Some type of vacuum gauge is used to control the valve.

2. Interruption of coolant flow.

Results:

a. Extensive backstreaming and fluid migration—oil and mercury.

b. Decomposition of oil—interference with vacuum processes, formation of gummy residues on nozzle assembly.

Prevention:

a. Interlock coolant flow with heater power so power is shut off if coolant is interrupted.

b. Provide flow indicator so coolant flow can be readily checked. The coolant flow can also be matched to heater input so as to get best pump performance.

c. Where feasible, periodically check rate of flow of coolant.

3. Interruption of heater power.

Results:

a. Mechanical pumps stop immediately—start when power comes on again. Vapor pumps stop pumping due to cooling and lack of a backing pump.

b. Mechanical pumps do not have completely tight seals and air gets into them. This air gets into the vapor pumps when power comes on again, getting same results as for 1.

c. Mechanical pump oil gets into the vapor pump.

Prevention:

a. Install foreline valve which closes automatically when power goes off. At the same time the valve above the vapor pump should be closed since an unbacked vapor pump will backstream into the vacuum system.

Problems with vapor pump operation can be minimized by following standard operating procedures and setting up a standard maintenance schedule. Actually, properly designed vapor pumps are quite reliable and the maintenance schedule is generally simple. The schedule should include:

1. Check heater operation. Too high or too low a boiler temperature sometimes is used as an indication of heater troubles.

2. Check the coolant flow. If a flow indicator is not provided simply measure the flow by collecting water in a graduated vessel for a given length of time.

3. Change the oil periodically. How often the oil has to be replaced depends on the nature of the vacuum process.

4. Check the condition and level of the oil when feasible.

5. Set up a cleaning procedure. Again the nature of the vacuum process determines how often the pump should be cleaned. In many cases the pump is cleaned only when it shows signs of poor performance.

FIVE

Some Other Types of Pumps

5.1   Molecular Drag Pumps

A *molecular drag pump* is a mechanical vacuum pump which creates
a gas flow toward a suitable forepump by imparting a component of
motion (momentum) to gas molecules by means of a rapidly rotating
body. The sealing between individual stages is effected by the use of
narrow air gaps. This type of pump is often called simply a *molecular
pump*. The earliest pump of this type was developed by Gaede (*Z.
Physik*, **13**, 864–870, 1912). Constructional difficulties in making this
form of pump led to various other designs. The Holweck design (*J.
Phys.*, **3**, 645, 1922) will be considered here in order to illustrate the
principle of a molecular pump. Figure 5.1 shows the main elements of
a Holweck pump. The inner cylinder, *A*, is made of duralumin and
is smooth. The housing, *B*, is made of bronze and has a spiral groove
cut into its inside surface. The clearance between the cylinder and
the housing is made not more than 0.05 mm (about 2 mils). Two
spiral grooves are used, one right handed and the other left handed,
and they meet at the inlet port, *C*, which is at the center and connects
with the high vacuum system. The depth of the groove increases from
the ends where it is about 0.5 mm (about 20 mils) to 5 mm depth (about
200 mils or ⅕ in.) at the center where the grooves join and connect to

109

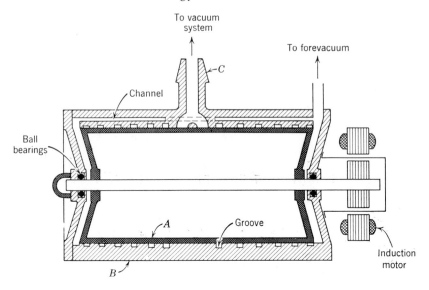

Fig. 5.1   Holweck molecular pump.

the inlet port, C.   In operation the vacuum system is first pumped by a suitable forepump down to a pressure of a few microns.   The molecular pump is then turned on, being backed by a forepump.   Molecules wandering through the inlet port will strike the groove wall and be given a motion down the groove.   Clearly the dimensions of the groove at the inlet port must be such that the molecules have a good chance of hitting the walls of the groove without making a lot of collisions with other molecules.   This means that the dimensions of the groove at the inlet port must be comparable to the mean free path of the molecules at the pressure at which the pump is started.   As the gas is compressed down the groove it is necessary to decrease the dimensions of the groove so as to keep them comparable to the mean free path of the molecules.   The ends of the grooves are connected together by a channel in the housing.   The Holweck pump operates at speeds of about 5000 rpm and produces pressures down to around 10^{-6} mm Hg.

Pumps of the Gaede, Holweck, and Siegbahn design all make use of some type of groove and a high speed rotor.   Getting high rotational speeds presents no real problem since modern ground spindles can be used at speeds up to 100,000 rpm.   However, pumps using some form of groove require close tolerances.   Current designs of pumps use air gaps of $\frac{2}{100}$ to $\frac{5}{100}$ mm maximum (1 or 2 mils).   Nonuniform thermal expansion, the presence of small foreign particles of the order of tenths of a millimeter (a few mils) in diameter, or a sudden air shock can result in seizure of the rotor. A recent design intended to overcome this type of difficulty is shown in Fig. 5.2. In Fig. 5.2a the housing, 1, contains stator disks, 2, and the rotor, 3. The stator disks are fixed in the housing while the rotor disks are attached to the shaft and rotate

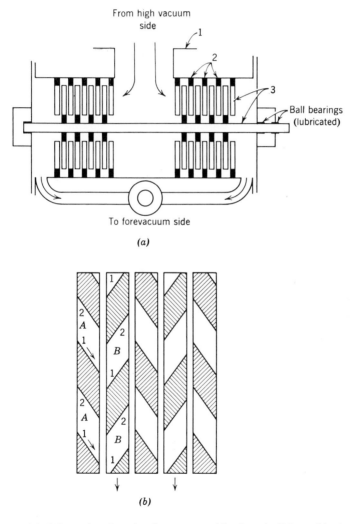

Fig. 5.2 (a) Schematic of molecular pump with slotted disks. (b) Arrangement of slotted disks. (From W. Becker, *Vakuum-Tech.*, Oct. 1958, pp. 149–152.)

at high speed with it. All disks contain inclined slots such that those in the stator disks are arranged as mirror images to those in the rotor disks. Some details of the disk arrangement are shown in Fig. 5.2*b*. Consider a slot *A* in the first stator disk. The wall indicated by 1 forms a wedge shaped channel with the general surface of the second rotor disk. This causes the gas to be accelerated in the direction of the arrow. The wall of slot *B* in the rotor disk also forms a wedge shaped channel with the surface of the stator disk so that again an accelerating effect occurs. Simultaneously, the wall, 2, of the same disk forms a wedge shaped channel with the surface of the next stator disk so that again the gas is speeded up in the direction of the arrow. This process is repeated in all disks. With thin disks, say around $\frac{1}{8}$ in., only short channels are obtained in the disks and, therefore, there are small pressure differences. However, many pairs of disks can be used so that a high total pressure can be obtained. Because of the small pressure difference across a pair of disks it is possible to use relatively large clearances without significantly affecting the pump characteristics. Many channels can be used simultaneously and the clearance between disks can be 40 mils or more. The gap between rotor and housing can also be about this value. The smaller the angle the slot makes with the surface of the disk, the greater will be the pressure difference with a subsequently low pumping speed. Generally the disks toward the center (high vacuum side) show a large angle for high pumping speed whereas the outer disks (forevacuum side) show small angles. It is claimed that a pump of this type with a rotor diameter of about $6\frac{1}{2}$ in., an overall length of about 26 in., a speed of rotation of 16,000 rpm, and using 40 channels with clearances of 40 mils can achieve pressures down to 10^{-9} mm Hg at a speed of 70 l/min (for air). At 10^{-8} mm Hg the speed is about 140 l/min.

The principal advantage of a molecular pump is that it will provide an essentially vapor-free vacuum without using cold traps. The objection to early designs involving the use of close tolerances was the the possibility of seizure of the rotor due to thermal expansion, foreign particles, or air shock. Designs such as that shown in Fig. 5.2*a*, allowing larger clearances, could well lead to wider use of molecular pumps. Some possible special areas of application of these pumps are:

Manufacturing large transmitting tubes for high frequencies.
Accelerators in nuclear physics.
Mass spectrometer high vacuum equipment.
Vacuum furnaces for producing materials of high purity.

5.2 Roots Blower Pumps

A *rotary blower pump* is a pump without a discharge valve which moves the gas by the propelling action of one or more rapidly rotating members provided with lobes, blades, or vanes. It is sometimes called a *mechanical booster pump* when used in series with a mechanical forepump. A special form of this type of pump is called a *Roots blower pump;* it has a pair of two-lobe interengaging impellers of special design.

Figure 5.3*a* shows a sectional view and the mode of operation of a typical Roots blower pump. It consists primarily of two figure-eight intermeshing rotors revolving in a close-fitting housing, and a motor. The rotors revolve in opposite senses, as can be seen in Fig. 5.3*b*, and are synchronized by some appropriate method such as timing gears. The gas (and vapor) being pumped enters the pump on the inlet side and fills the area shown by the shaded area in view *A* of Fig. 5.3*b*. Referring again to Fig. 5.3*b*, as the rotors advance, the gas flows into the area between the rotors and the chamber wall (view *B*) until a point is reached where a definite volume is trapped (view *C*). With a further advance of the rotors, the trapped volume is discharged into the outlet of the pump (view *D*), which normally is the inlet of a

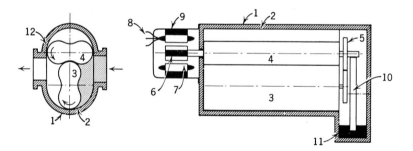

Fig. 5.3*a* Sectional view of Roots pump. (Heraeus Pumps. Courtesy Consolidated Vacuum Corp., Rochester, N.Y.)

1. Outer, vacuum-tight housing
2. Inner, cast iron housing
3 and 4. Rotors
5. Timing gears
6. Cage rotor of motor
7. Motor coil
8. Electrical leads for motor
9. Water cooling for motor
10. Oil pump for gear and bearing lubrication
11. Oil reservoir
12. Opening between inner and outer casting

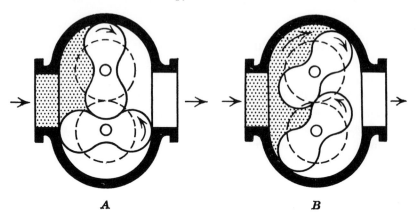

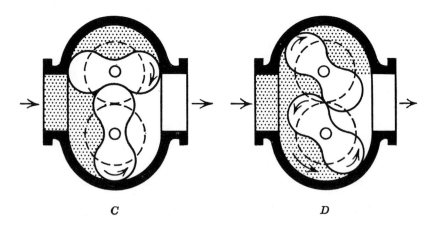

Fig. 5.3*b*   Operation of Roots pump.   (Heraeus Pumps.   Courtesy Consolidated Vacuum Corp., Rochester, N.Y.)

forepump.   Although Roots pumps can discharge directly to the atmosphere in order to produce a rough vacuum, this is not recommended since it results in more heating and higher power consumption.   It must be kept in mind that these pumps involve close tolerances between the rotors and chamber walls with no lubrication.   Prolonged operation against atmospheric pressure may produce enough heat to cause seizure of parts due to expansion.   The speed and operating pressure of these pumps depend in part on the performance characteristics of the backing pump.   In general, the lower the ultimate pressure of the backing pump the lower will be the ultimate pressure of the combination. Also, the higher the speed of the backing pump, the higher will be the speed of the combination.

Figure 5.4 shows speed-pressure curves for several Roots blower pumps. It will be noted that pressures down to almost 10^{-6} mm Hg can be achieved. As noted above, the choice of backing pump is a factor that must be considered. The maximum pumping speed shown in Fig. 5.4 is 1600 l/sec (3392 cfm). Actually commercial pumps are available which have speeds greater than 10,000 cfm. Many Roots pumps are designed to operate against forepressures of a few mm Hg.

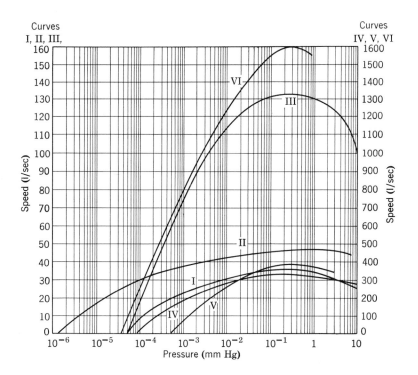

Fig. 5.4 Speed curves of some typical Roots pumps. (Courtesy Consolidated Vacuum Corp., Rochester, N.Y.)

Curve I:	Heraeus VPR-150 A forepump—15 cfm	Curve IV:	Heraeus VPRG-1000 A forepump—130 cfm
Curve II:	Heraeus VPR-152 A forepump—15 cfm	Curve V:	Heraeus VPR-1600 A forepump—130 cfm
Curve III:	Heraeus VPRG-350 A forepump—47 cfm	Curve VI:	Heraeus VPR-6000 A forepump—VPR 1600 and 130 cfm

Reference gauge: McLeod gauge trapped with liquid nitrogen

The high rotational speeds of these pumps results in high pumping speeds per unit weight.

Roots pumps have come into fairly general usage in various vacuum processes, particularly in the pressure range which lies below the efficient operating range of oil-sealed mechanical pumps and above the efficient range of diffusion pumps. The most generally useful range is from around 1 mm Hg to about 10^{-4} mm Hg, although this range can be extended considerably depending on the design of pump. The operating costs for these pumps are low and they are capable of handling large bursts of gas. The design of any particular pump is such that it should only be turned on at the pressure at which it starts pumping efficiently. This can readily be handled automatically by using an appropriate vacuum gauge. Turning on the pump prematurely usually will not harm it unless it is operated at fairly high pressures for a long enough time to cause it to overheat. Because of these various advantages Roots pumps are used in metallurgical processes, such as sintering, annealing, melting, and outgassing, in electronic tube evacuation, in studies with environmental chambers, in vacuum metallizing, and in certain supplementary applications such as vacuum drying, temperature reduction of liquefied gases, and house vacuum systems for drying ovens and dessicators.

In spite of the many advantages of Roots pumps, they still suffer from several disadvantages. The initial cost is high compared to other types of pumps which could be used for the same purpose. Because they are operated at high speeds and involve close tolerances, foreign material entering the pump can cause damage to parts. This problem can be minimized by the use of screens with a loss in pumping speed. As was noted above, operating at pressures above the recommended operating range results in overheating, which could lead to seizure of parts. Although the rotors themselves are not lubricated, oil must be provided for the various gears and bearings required in the operation. A diffusion or oil ejector pump oil is normally used so that any oil getting into the system will contribute a low vapor pressure. The smaller pumps are often air cooled while the larger ones are generally water cooled. A typical commercial Roots pump is shown in Fig. 5.5. This pump has a maximum speed of 840 cfm at a pressure of about 300 μ and an ultimate pressure of 5×10^{-4} mm Hg. In Fig. 5.4 it should be noted that the curves shown were obtained with a McLeod gauge trapped with liquid nitrogen. Such an arrangement does not measure the pressures of vapors (see Chapter 6). The forepump recommended for the pump shown in Fig. 5.5 has a speed of 130 cfm and the forepressure required to start the pump is 10 mm Hg.

Fig. 5.5 Typical Roots pump (Consolidated Vacuum Corp. Heraeus VPR-1600).

5.3 The Use of Getters as Pumps

Getters are materials included in a vacuum system or device for removing gas by sorption. Sorption refers to the taking up of gas (or vapor) by absorption, adsorption, chemisorption, or any combination of these processes. Absorption and adsorption are physical processes which were discussed briefly in Chapter 4. *Chemisorption* refers to the binding of gas (or vapor) on the surface or in the interior of a solid (or liquid) by chemical action. Getters have been widely used in the vacuum industry for many years, particularly in the manufacture of vacuum (electronic) tubes. These tubes are pumped down to a low pressure (say 10^{-5} mm Hg or less), usually by means of mechanical and diffusion pumps together with cold traps, although other types of pumps are used. After being pumped down the tubes are degassed by baking and then sealed off. Getters in the tubes remove residual gas, part of which is released in the sealing-off process, as well as gas evolved in the tubes at later times. Electronic tubes are examples of static vacuum systems. It is often remarked that getters are used for the clean-up of gases. Although they do not remove gases and vapors from inside the vacuum system, these gases and vapors are held so they do not contribute to the pressure and, therefore, getters act as pumps.

A getter pumps while in the evaporated state ("dispersal" gettering) and also after condensing on surfaces ("contact" gettering). The pumping speed of the condensed film depends on the area it presents to the gas molecules. The rate at which the getter becomes saturated with gas and ceases to pump depends on the pressure at which it is used. Getters are generally not used at pressures above a few microns. In dynamic systems it is necessary to keep replacing the getter surface so as to avoid saturation. Besides using evaporation to obtain a gettering action it is also possible to use a process called sputtering. In this process electrically charged particles strike a surface in vacuum and remove particles of material from the surface which deposit throughout the vacuum system. In the case of evaporation the pressure may rise if the getter has not been degassed or contains gases only liberated at the evaporation temperature. Slow evaporation produces the highest gas absorption in the growing deposit, so that the gas pressure may fall even though gas is liberated from the getter.

Getters are normally used in the form of thin films or metal ribbons and powders (solid getters). Metal films prepared by vacuum evaporation generally clean up gases rapidly at normal temperatures because of the freshness of their surfaces. Evaporated films also possess large surface areas because of their granular structure. On the other hand the pumping speed of solid getters decreases rapidly because of their limited area which becomes covered with gas molecules. Solid getters are usually operated at high temperatures to aid diffusion of the adsorbed gas into the solid. Thin film getters are usually chosen from the alkaline earth metals (barium, strontium, calcium, and magnesium) because of their chemical activity and the relative ease with which they can be volatilized in a vacuum. Solid getters are normally selected from elements with high melting points and include titanium, zirconium, hafnium, tantalum, molybdenum, tungsten, thorium, and uranium. Several of these solid getters are also used as thin film getters. Table 5.1 shows the characteristics of some metals which are useful as getters. From this table it is evident that the only metals which clean up a range of gases and have sufficient volatility to be readily prepared in thin film form are barium, calcium, and titanium. However, metals which are difficult to evaporate can be sputtered. The properties of other getters are discussed in the article, "Getter Materials for Electron Tubes," by W. Espe, M. Knoll, and M. P. Wilder (*Electronics*, 80–86, Oct. 1950).

Table 5.1 Characteristics of Some Metals Useful as Getters *

Getter Film	Gas	Initial Adsorption Rate ($\sim 20°C$), l or μl/sec/cm² †	Film Temperature for Continuous Sorption, °C	Sorptive Capacity, μl/mg (°C) ‡	Remarks
Aluminum mp = 660°C 1291°C (vp = 100 μ)	O_2	C	500	7.5–38.5 (20)	—
	H_2	0.0	—	0	
	N_2	0.0	—	0	
	CO_2	<0.005 l	—	0	
	CO	<0.005 l	—	—	
Barium mp = 717°C 730°C (vp = 100 μ)	Air	Ch	—	56 (400)	H_2 gettering increased by presence of heated cathode. Often requires extensive degassing or distillation. Obtainable in alloy form with Al.
	O_2	0.3 l	>40	57 (300)	
	H_2	0.05 l	200	100 (400)	
	H_2O	C	>100	72 (300)	
	N_2	0.003 l	—	{ 3–25 (<100), 43–51 (>100) }	
	CO_2	5.0 l	—	66 (400)	
	CO	3.5 l	>80	100 (400)	
	C_2H_2, C_2H_4	Ch	—	—	
Calcium mp = 810°C 700°C (vp = 100 μ)	O_2	C	425	—	Electrolytically prepared metal may contain CaH_2, which dissociates at the evaporation temperature.
	H_2, N_2, CO, C_2H_2	Ch	—	—	
	CO_2, H_2O, SO_2, NH_3		—	—	
	C_2H_4	Ch	—	—	
Magnesium mp = 651°C 515°C (vp = 100 μ)	O_2	C	>450	20–200 (20)	H_2 gettering increased by presence of heated cathode.
	H_2	0.0	—	—	
	N_2	0.0	—	—	
	CO_2	<0.005 l	—	—	
	CO	<0.005 l	—	—	
Titanium mp = 1660°C 1742°C (vp = 100 μ)	O_2	C	—	—	Vacuum-melted metal in form of wire and sheet is commercially available.
	H_2	Ch	—	$\sim$0.0 (30)	
	N_2	3.0 l	—	1.9–2.5 (30–300)	
	CO_2	4.3 l	—	4.3 (20)	
	CO	12.0 l	—	3.4–4.2 (30–200)	

Table 5.1 (Continued)

Getter Film	Gas	Initial Adsorption Rate (~20°C), l or μl/sec/cm² †	Film Temperature for Continuous Sorption, °C	Sorptive Capacity, μl/mg (°C) ‡	Remarks
	SF_6	Ch	—	—	—
	C_2H_2, CH_4,	Ch			
	C_2H_4,		—	—	
	CCl_2F_2, NH_3	Ch			
Titanium (solid)	O_2	2.01 μl (800°)	>650	—	CH_4 dissociates when striking Ti at 1200°C. C is sorbed and H_2 taken up at lower temperatures.
	H_2	D	20–400	90 (800)	
	N_2	0.08 μl (1000°)	>700	160 (1000)	
	CO_2	0.81 μl (1100°)	>700	50 (1100)	
	H_2O	D	300–400	—	
Zirconium mp = 2127°C 2212°C (vp = 100 μ)	O_2	C	—	—	—
	H_2, C_2H_2, C_2H_4	Ch		—	
	N_2, CO	>2.5 l	—	—	
Zirconium (ribbon)	O_2	C	885	1.99 (400)	—
	H_2	D	300–400	13.3 (350)	
	N_2	C	1527	1.46 (800)	
	CO_2	C	—	3.04 (800)	
	CO	C	—	3.65 (800)	
Tantalum mp = 2996°C 2820°C (vp = 1 μ)	O_2	C	—	—	Difficult to evaporate. Low vapor pressure.
	H_2, C_2H_2, C_2H_4	Ch	—	—	
	N_2	>2.5 l	—	—	
	CO	>2.5 l	—	—	
Molybdenum mp = 2622°C 2533°C (vp = 10 μ)	O_2	C	—	—	Difficult to evaporate thick films. Low vapor pressure. May be sublimated.
	H_2, C_2H_2, C_2H_4	Ch	—	—	
	N_2	2.7 l	—	1.0 (30)	
	CO	3.5 l	—	3.0 (30–200)	
Tungsten mp = 3382°C 3309°C (vp = 10 μ)	O_2	Ch	—	—	Difficult to evaporate thick films. Low vapor pressure. May be sublimated.
	H_2, C_2H_2, C_2H_4	Ch	—	—	
	N_2, CO	>2.5 l	—	—	

Thorium mp = 1827°C 2431°C (vp = 100 μ)	O_2 H_2 CO_2	C Ch Ch	450 — 650	7.5–33.1 (20) 19.5–53.7 (20) —	Difficult to evaporate. Low vapor pressure. —
Uranium mp = 1132°C 2098°C (vp = 100 μ)	O_2 H_2	C Ch	240 —	10.6–9.3 (20) 8.9–21.5 (20)	
Misch metal	O_2	C	—	21.7–51 (20)	Chiefly cerium and lanthanum.
Cerium mp = 785°C 1439°C (vp = 100 μ)	H_2 N_2 CO_2	Ch Ch Ch	— — —	46.1–64 (20) 3.2–16 (20) 2.2–45 (20)	

* Based on the article "Theory and Design of Getter-Ion Pumps" by L. Holland (*J. Sci. Instr*, **36**, 105, March 1959).
† C = chemical reaction, Ch = chemisorption, D = diffusion.
‡ Generally values are for bright films. Where two values are quoted the second figure is for a black deposit.

mp = melting point
vp = vapor pressure

O_2 = oxygen
H_2 = hydrogen
N_2 = nitrogen
H_2O = water (vapor)
CO = carbon monoxide
CO_2 = carbon dioxide

C_2H_2 = acetylene
C_2H_4 = ethylene
SO_2 = sulfur dioxide
SF_6 = sulfur hexafluoride
CH_4 = methane
NH_3 = ammonia

5.4 Gettering and Ionizing

Getters will not pump the inert gases which occur in the atmosphere. Table 1.1 shows that argon is the most abundant inert gas, with helium and neon present to a smaller extent. To obtain low pressures it is necessary to pump these inert gases. Of course, conventional pumps, such as diffusion pumps, will pump the inert gases, although not at the same speeds as oxygen or nitrogen. In order to obtain pressures less than or comparable to those obtainable with diffusion pumps, it is possible to combine gettering with what is called ion pumping. A true *ion pump* depends for its action on the breaking apart (ionizing) of gas and vapor molecules into electrically charged particles (positive and negative ions). Both chemically active and inert gases can be broken apart by this means. If the ions thus formed could be physically removed from the vacuum system this would constitute a form of pumping. There have been various attempts to make true ion pumps, in which the molecules are ionized and the ions are removed directly by some type of auxiliary pump or trap. However, such a method, to date, has involved the use of large amounts of electrical power. Commercial pumps use both ionizing and gettering actions to remove gases and vapors. The ionization is carried out by bombarding molecules with high speed electrons (negatively charged particles). The electrons are speeded up by a voltage (potential) difference between two electrodes. Any gas always contains some electrons and additional electrons can be provided by heating various materials, particularly metals such as tungsten, to a high temperature. When an electrical voltage is applied between two electrodes, the electrons will be speeded up. However, at higher pressures, such as atmospheric pressure, the electrons will make many collisions with gas molecules and will not gain enough speed to break apart the molecules. As the pressure is reduced, the electrons can go farther before they collide with molecules. Eventually the pressure will be low enough for the electrons to ionize molecules.

Figure 5.6*a* shows an arrangement consisting of a glass tube which can be pumped out and has two electrodes sealed into it. When a voltage is applied between the electrodes and the pressure is reduced, eventually ionization occurs and this will show up as a glow discharge when the right values are reached. This glow discharge will have a color determined by the gases in the tube. For example, if the tube

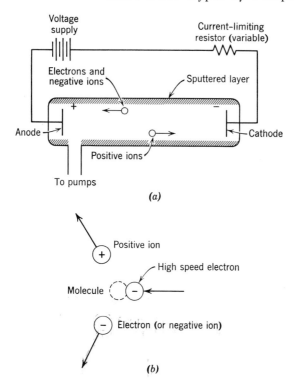

Fig. 5.6 (*a*) Glow discharge tube. (*b*) Collision between an electron and a molecule.

is filled with neon gas the gas discharge is blood red in color ("neon tube"). The voltages used will depend on the dimensions of the tube, and usually range from hundreds of volts to thousands of volts. Pressures of a few microns or less are common. When electrons reach a great enough speed (energy) to ionize a molecule, the molecule is generally broken up into a positive ion and an electron (or electrons). The electrons will be speeded up and can produce more ionization, or they may be "captured" by molecules which produce negative ions. A cumulative ionization process occurs which soon results in a glow discharge. The ions are attracted to the electrodes and are neutralized. The ions striking the electrodes may break off small particles which will deposit on the walls of the tube. This process is called sputtering and the deposited layer provides pumping by gettering, the effectiveness depending on the types of electrodes being used. Additional gettering action can be provided by evaporating a getter material in the tube. The collision of a high speed electron with a molecule is shown in Fig. 5.6*b*.

The pumping action of various devices involving ionization and gettering has been observed for many years. In particular, it has been known for a long time that certain types of vacuum gauges act as pumps. Ionization gauges and Philips (Penning or cold cathode) gauges will act as effective pumps for small volumes. Observation of the performance of such gauges (and other electrical devices) led to the development of commercial pumps combining ionization and gettering. Such pumps provide for continuous gettering and ionizing. Some pumps use evaporation of a suitable material for the gettering while others make use of sputtering.

5.5   Pumps   Using   Evaporation   and   Ionization

An example of a commercial pump of this type is the Evapor-ion pump which is marketed by the Consolidated Vacuum Corporation, Rochester, New York. The principal features of this pump are shown in Fig. 5.7. A titanium wire is fed onto a post which is heated by a nearby filament. The post is made of some refractory material such

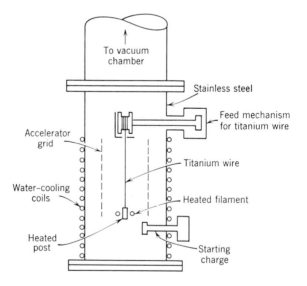

Fig. 5.7 Principle of operation of the Evapor-ion pump (Consolidated Vacuum Corp. Model EL-2000).

as tantaloy. When the titanium wire strikes the heated post some metal is evaporated and vapor deposits on the cooled walls of the pump. Molecules of the chemically active gases which wander into the pump from the vacuum chamber will combine chemically with the titanium on the walls or with the titanium vapor before it deposits. To remove the inert gas molecules, they are ionized by electrons from the heated filament. The electrons are speeded up by the accelerator grid and they then ionize molecules between the grid and the pump wall. The voltage on the grid is such as to repel the positive ions to the pump wall, where they strike. Also the depositing layer of titanium prevents these molecules from escaping. The pump is first started by pumping it down to a few microns or μ with a roughing pump (usually a mechanical pump), which is then valved off. A primer charge of titanium is then fired to reduce the pressure to a fraction of a micron. At this point the evaporation and ionizing processes can be started. Instead of a roughing pump and a primer charge it is possible to use a diffusion pump, pumping down to about 10^{-5} mm Hg. The pump is water-cooled so as to prevent the release of gases from the pump walls. With continuous evaporation of titanium an ultimate pressure of 1×10^{-7} mm Hg is claimed. Pressures of 1×10^{-7} to 1×10^{-8} mm Hg and lower are claimed with ion pumping alone. These values are obtained with aluminum gaskets. A typical pumping speed curve for such a pump (for air) with an ID of around 12 in. shows a speed of about 1800 l/sec at a pressure of 10^{-6} mm Hg. The exact curve obtained will depend on the amount of cooling of the walls and the rate of feed of titanium. The pumping speed depends on the gas being pumped, being greatest for hydrogen and decreasing through nitrogen, oxygen, and the rare gases.

5.6 Pumps Using Sputtering and Ionization

Instead of evaporating a suitable material, sputtering can be used to provide the gettering action. Pumps operating on this principle are often referred to as sputter-ion pumps by the manufacturers. They operate in the same manner as a cold cathode vacuum gauge (Chapter 6) except that the overall size is increased and an active metal is used for the cathodes so as to obtain maximum pumping speeds. The principle of operation is illustrated in Fig. 5.8. Two cathodes and one anode are mounted in a suitable enclosure, which is connected to the vacuum system. The anode shown is a hollow, cell structure while the cathodes are flat plates. The actual geometry of

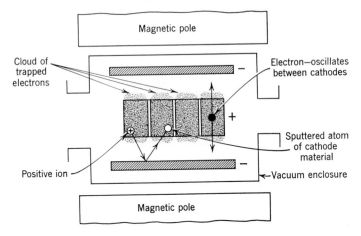

Fig. 5.8   Principle of operation of sputter-ion pump.

the electrodes will vary from manufacturer to manufacturer. The anode is usually made of copper and the cathodes of titanium, which is a good gettering material (see Table 5.1). The electrode assembly and enclosure are mounted between the poles of a magnet. When a high voltage (several thousand volts) is applied between the anode (+) and the cathodes (−) electrons are speeded up toward the anode. However, the magnetic field keeps them from going directly to the anode. Also, they cannot reach the cathodes because of their negative voltage. Consequently, the electrons go back and forth through the anode, finally reaching it by slow migration. In this process the electrons are much more effective in ionizing gases and vapors than in the case of no magnetic field. The positively charged ions are speeded up toward the cathodes. When they hit the cathodes they are buried and at the same time knock out small particles of the cathode material (sputtering). The sputtered material deposits on the anode structure and combines with the chemically active gases (oxygen, nitrogen, carbon dioxide, etc.), thus pumping them. The inert gases are also pumped, probably by being covered with sputtered metal at the cathodes and by entrapment on the anode. It would be expected that eventually the sputtering action would release buried gas atoms, resulting in a pressure build-up. No detailed theory covering this matter is available. In practice, such pumps appear to fail after a considerable length of time due to other causes (see below). The total current produced will be determined by the number of gas molecules present, which is measured by the pressure. Consequently a meter placed in the power supply line to measure the ion current (pump current) gives a reading of the pressure. Figure 5.8 shows the general arrangement of such pump with one type of anode structure.

The permanent magnets used with these pumps generally provide field strengths of between 2000 and 4000 gauss. Increasing the strength of the magnetic field will not increase the pumping speed. However, too low a strength of field will result in poor performance and, finally, failure. Voltages on the electrodes are usually a few kilovolts, say 2 kv to 10 kv, or more. Commercial pumps with speeds of several thousand liters per second are available. There appears to be essentially no limit to the pumping speed that can be obtained. The problem is to obtain large enough magnets and power supplies. Ultimate pressures below 10^{-10} mm Hg can be achieved although special precautions must be taken. These precautions involve the proper choice of construction materials, such as metal instead of rubber gaskets; the use of adequate cleaning techniques; and the application of an appropriate bake-out procedure. To obtain very low pressures it is necessary to bake both the pump and the vacuum system.

To start these pumps it is necessary to reduce the pressure to a few microns so that the electrons can gain enough energy to produce ionization. Mechanical pumps are commonly used for this purpose. Of course, other types of pumps can be used. In some cases, sorption-type pumps are used so as to avoid oil vapor getting into the sputter-ion pump. These types of pumps are discussed in Section 5.8. The use of cold traps between mechanical pump and sputter-ion pump will achieve much the same results. After the pressure has been reduced to a few microns, the roughing pump is valved off and voltage applied to the anode and cathodes. Some erratic behavior may be expected due to bursts of gas unless the pump has been baked out. However, pumps can be started without baking to achieve pressures in the 10^{-8} mm Hg range. To get much lower pressures requires baking the pump and vacuum system (while being pumped). Gases and vapors will be driven off more rapidly at higher temperatures. The highest temperature that can be used will be determined by the construction materials. When the pump and magnet are both baked, the upper temperature is usually limited to around 250°C because of the magnet construction. Of course the magnet can be removed from the pump, although this sometimes presents difficulties. Manufacturers generally rate the life of a pump in hours at a particular pressure. A typical lifetime is 20,000 hr at 10^{-6} mm Hg. The lifetime is a function of the pressure, being longer at lower pressures (e.g., about 200,000 hr at 10^{-7} mm Hg) and shorter at higher pressures (e.g., 2000 hr at

Fig. 5.9   A commercial sputter-ion pump (Ultek Corp., Palo Alto, Calif., Model 20-451).

10^{-5} mm Hg). The lifetime seems to be largely determined by collection of titanium metal (from the cathodes) on the anodes, which, after a sufficient length of time, begins to flake off. This causes erratic operation and ultimately failure. Of course, the electrode assembly could be cleaned and the pump put back into operation or a new assembly could be installed. It is best to consult the manufacturer on the proper cleaning procedure. A typical commercial pump with the arrangement of Fig. 5.8 is shown in Fig. 5.9. A pump using two sets of electrodes, as in Fig. 5.8, is called a diode pump. The pump of Fig. 5.9 is rated at 500 l/sec, which applies over a wide pressure range. This can be seen in Fig. 5.10, which shows a pumping speed curve for air. The pumping speeds for chemically active gases such as hydrogen, oxygen, and nitrogen, are considerably higher than that for an inert gas such as helium. Although a high pumping speed is usually shown for hydrogen, there appears to be some evidence that when hydrogen is bled into a system, as is done with some nuclear accelerators, a sputter-ion pump will simply not pump this gas after a while, although it will continue to pump oxygen and nitrogen. Perhaps hydrogen diffuses out of the sputtered deposit on the anode since it is a light gas. Also, hydrogen ions are too light to produce appreciable sputtering.

In most pumping applications there is little need for high pumping speeds for the inert (noble) gases. However, in some cases the pumping of argon becomes somewhat important because of the ap-

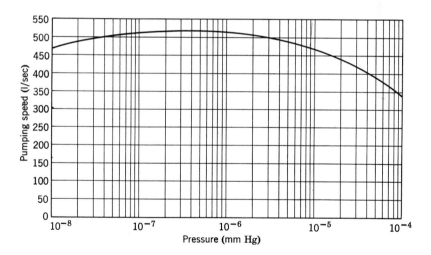

Fig. 5.10 Pumping speed versus pressure for a sputter-ion pump (for air) (Ultek Corp. Model 20-451).

proximately 1% argon content in the atmosphere. In pumping against
a continuous air leak the conventional diode pump of the type shown
in Fig. 5.8 may exhibit periodic pressure fluctuations which are associated with the argon normally present in the air. This phenomenon
is often referred to as "argon instability." This instability arises with
a new or freshly cleaned pump only after many hours of operation.
At a pressure of 1×10^{-5} mm Hg several hundred hours of operation
are required, while at 1×10^{-6} mm Hg several thousand hours will
be required. However, the time involved is a function of the pump.
This instability depends on many factors and many methods have been
tried to eliminate it. Two methods which have been adopted by manufacturers are shown in Figs. 5.11a and b. Figure 5.11a shows an arrangement developed by Varian Associates, Palo Alto, California.
Here the cathodes have been modified so as to more effectively trap
the atoms of the inert gases, in particular those of argon. Figure 5.11b

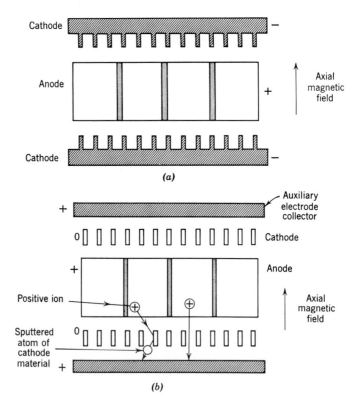

Fig. 5.11 (a) Diode arrangement for pumping inert gases. (b) Triode arrangement for pumping inert gases.

shows a triode arrangement of a sputter-ion pump. This type of pump has been marketed by Consolidated Vacuum Corporation, Rochester, New York, with the claim that the argon instability is minimized. Besides the usual cathodes and anode, a third set of electrodes (the collector) is added and is operated at such a voltage that the positive ions are collected. The theory is that the positive ions reaching the collector have low energies and, therefore, do not knock out gas molecules (such as those of argon) which are already held there. In addition, positive ions striking the cathodes (at high energies) will drive particles from the cathodes toward the collector, thus building up a sputtered layer to trap gas molecules and thereby provide a gettering action.

The sputter-ion type of pump represents a significant advance in the art of high vacuum. Some advantages of these pumps may be listed as follows.

ADVANTAGES

1. They can be used to produce very low pressures (of the order of 10^{-10} mm Hg or less).

2. They provide an essentially constant pumping speed over a wide pressure range (about 10^{-5} mm Hg to 10^{-8} mm Hg).

3. They do not contribute contamination to the vacuum system since they do not involve the use of any type of oil.

4. At low pressures, say less than 10^{-5} mm Hg, they have an advantage in economy of operation over other types of pumps.

5. The fact that the pump current depends on the gas density allows a single meter to read this current and the pressure (with proper calibration).

6. They can operate for extended periods of time without attention since cold traps are not needed. The system remains under vacuum (except for leaks) in the case of power failure.

7. Accidental let down to air does not damage these pumps.

8. They are silent in operation.

In spite of these many advantages of such pumps, they still suffer from certain disadvantages. These disadvantages can be listed as follows.

DISADVANTAGES

1. They are not suitable for routinely handling large bursts of gas.

2. They are not generally suitable for use on systems which are

operated through cycles, being let down to air periodically. This has to do with problems in handling bursts of gas.

3. They become uneconomical at higher pressures, say above 10^{-5} or 10^{-4} mm Hg, in comparison to diffusion pumps. This is because diffusion pumps are essentially constant power devices, i.e., the power required is constant over the pressure range through which the pump operates. On the other hand, the getter-ion pump requires more power as the pressure increases.

4. They are not too satisfactory in handling large quantities of the inert gases, as has been discussed above.

From balancing off advantages against disadvantages it becomes evident that getter-ion pumps (whether of the evaporation or sputter types) will find increasing use in industry. It is doubtful that they will replace the usual diffusion pump–mechanical pump combination in such processes as vacuum melting of metals, freeze drying of pharmaceuticals, dehydrating of various materials, commercial metallizing, etc. However, these types of pumps are very useful in various vacuum applications where "clean" vacuum conditions are desired without the use of cold traps. Typical applications are: processing various types of electronic tubes (receiver tubes, transmitter tubes, microwave tubes, etc.), evacuating nuclear accelerator sections, pumping mass spectrometers and electron microscopes, specialized industrial and laboratory processes. Sometimes it is advantageous to attach a small pump of this type to large sealed-off x-ray tubes and to activate the pump periodically as the tube becomes "gassy."

5.7   Cryopumping

Cryopumping is the reduction of pressure by low temperature condensation of gases in the vacuum system. If a surface is introduced into the system at a temperature low enough to condense the gas present, the resultant pressure in the system will be that of the vapor pressure of the condensate (liquid). As the condensate temperature is lowered, the vapor pressure is further decreased. At even lower temperatures the condensate will freeze and become a solid. A slight decrease in the temperature of the solid will usually result in a sharp reduction in the vapor pressure. Cryopumping is based on this principle. As an example, if the temperature of solid nitrogen is lowered from $30°K$ ($-243°C$) to $20°K$ ($-253°C$) its vapor pressure is reduced from about 10^{-4} mm Hg to 10^{-10} mm Hg.

Cold traps, which are used in connection with mechanical and/or diffusion pumps, can be considered to be cryopumps. In some cases traps are used to keep pump fluid vapors (oil or mercury) out of the system or out of other pumps by condensing or freezing these vapors. In other cases the primary function of a trap is to cut down on the pump-down time by pumping various condensable vapors. The most common condensable vapor which causes a long pump-down time is water vapor. Although diffusion pumps remove this vapor effectively, rotary piston pumps are rather poor in this regard. The water is absorbed by the mechanical pump oil and contributes vapor pressure. Rotary ballast pumps will help to remove water vapor as long as it is not present in large quantities. However, cold traps act as very effective pumps for water vapor and various other condensable vapors such as grease vapors, etc.

The gas and vapors that can be frozen out of a vacuum system will depend on the temperature of the cold surface. For any particular gas, the temperature should be below the freezing (melting) point of the gas so that its vapor pressure is very low. It is possible then to speak of a pumping speed of the cold surface for this particular gas. In the molecular flow region, which is the pressure range in which cold traps are usually used, the pumping speed is given by

$$S = 11.6 \sqrt{\frac{29}{M}} \left(1 - \frac{P_2}{P_1} \right) A \; \text{l/sec} \qquad (5.1)$$

Here, M is the molecular weight of the gas being considered, P_1 is the partial pressure of this gas at the temperature of the vacuum system as a whole (usually room temperature), P_2 is the partial pressure of the gas at the temperature of the cold surface, and A is the area of this surface in square centimeters. When P_2 is very small compared to P_1, then P_2/P_1 can be neglected and the pumping speed then becomes $11.6\sqrt{29/M}$. The number 29 is the molecular weight of air. The molecular weights, together with other characteristics of various gases, are listed in Appendix A.

A surface cooled by liquid air will drop to a temperature of about $-187°C$. At this temperature, carbon dioxide gas and water vapor will be frozen. The freezing point of carbon dioxide (*dry ice*) is about $-78°C$, and it has a vapor pressure of 7×10^{-7} mm Hg at $-187°C$. Suppose a vacuum vessel is pumped down to a pressure of 10^{-5} mm Hg and the residual gas contains 10% carbon dioxide. Then the partial pressure of this gas, P_1, is $\frac{1}{10} \times 10^{-5} = 10^{-6}$ mm Hg. As noted above, the pressure of carbon dioxide gas at the cold surface temperature (P_2) is 7×10^{-7} mm Hg. The molecular weight of carbon dioxide is 44. Therefore, the pumping speed of the cold surface for carbon dioxide, according to eq. 5.1, is

$$11.6 \sqrt{\frac{29}{44}} \left(1 - \frac{7 \times 10^{-7}}{10^{-6}}\right) A = 11.6 \sqrt{0.66} \, (1 - 0.7)A$$
$$= 11.6 \times 0.81 \times 0.3A$$
$$= 2.8A \; \text{l/sec}$$

Water has a vapor pressure of about 10^{-21} mm Hg at $-187°C$. This is extremely small compared to the partial pressure of water vapor at the general temperature of the system. Therefore, P_2/P_1 can be neglected and the pumping speed for water vapor is $11.6\sqrt{29/18}A$, where 18 is the molecular weight of water. The speed is then $11.6\sqrt{1.61}A = 11.6 \times 1.27A = 15A$ l/sec. The exact value of the effective area A can only be determined experimentally. However, when the trap is located so that it is accessible to gas from all sides, then A can be taken as the area of the cooled surface. On the other hand, if part of the trap is obscured by being near some part of the vacuum system, then A will be less than the area of the cooled surface.

As has been noted previously, the above considerations apply only to the molecular flow region. In this case molecules which are moving toward the cold surface have a very good chance of striking this surface without hitting other molecules and being deflected. On the other hand, when the pressure is such that the gas flow is viscous, the pumping speed is considerably less than for the molecular flow region. Now the flow of gas is more like the flow of a river, without turbulence or eddy currents. In most cases much of the flow is due to gases which are not held by the cold surface. Therefore, to get the maximum pumping speed, the cold surface should be placed so that the gas is forced to flow close to it and over as large an area as possible.

Table 5.2 shows the vapor pressures of water and mercury at various temperatures, including the boiling points of liquid air (taken as $-190°C$) and carbon dioxide $(-78°C)$. It is clear that the vapor pressure of water at the temperature of dry ice is still substantial. To pump out the bulk of the water and mercury vapor, liquid air, or preferably liquid nitrogen is used. The use of liquid nitrogen minimizes the problem of oxygen collecting in the system. It must be kept in mind that as frozen vapors collect on the cold surface, they produce an "ice" layer consisting usually of water ice and dry ice. This layer is an insulator, so the temperature at the outside of the layer is higher

Table 5.2 Vapor Pressure of Water and Mercury at Several Temperatures

Temperature		Approx Vapor Pressure (mm Hg)	
°C	°F	Water	Mercury
+100	+212	760	0.27
+20	+68	18.0	1.2×10^{-3}
0	+32	4.58	2×10^{-4}
−78	−108.4	5.6×10^{-4}	3×10^{-9}
−190	−310	10^{-22}	10^{-32}

than the temperature of the refrigerant and the cold trap becomes less effective. The rate of build-up of the ice layer depends on the amount of vapor in the system which has to be pumped. The problem of icing, together with various design and maintenance problems with cold traps, is considered in Chapter 11. As the temperature is further reduced it is possible to freeze out various permanent gases, i.e., gases which cannot be liquefied by pressure at room temperature. Table 5.3 shows the boiling points and melting points of some common gases.

Table 5.3 Melting Points and Boiling Points of Several Gases

Gas	Melting Point (mp), °C	Boiling Point (bp), °C
Argon (A)	−189.3	−186
Carbon dioxide (CO_2)	—	−78
Carbon monoxide (CO)	−207	−192
Bromine (Br)	−7.3	58.8
Chlorine (Cl)	−102	−34
Fluorine (F)	−223	−187
Helium (He)	−272	−268.8
Hydrogen (H_2)	−259	−252.7
Krypton (K)	−157	−153
Neon (Ne)	−249	−245.9
Nitrogen (N_2)	−210	−195.8
Oxygen (O_2)	−219	−183
Xenon (Xe)	−111.8	−109

From this table it is clear that liquid nitrogen (bp $= -195.8°C$) should freeze out bromine, chlorine, krypton, neon, and carbon dioxide, as well as water vapor. Argon will be frozen out but its vapor pressure will be substantial since its melting point ($-189.3°C$) is not much above the boiling point of nitrogen. Fluorine, helium, hydrogen, neon, and nitrogen will not be frozen out, while oxygen will just be liquefied. To freeze out more gases, it is necessary to use a lower temperature. Liquid hydrogen (bp $= -252.7°C$) or liquid helium (bp $= -268.8°C$) is usually used as a refrigerant. The vapor pressures of several substances are shown in Table 5.4 at the temperatures of liquid air and liquid helium. As will be noted, when the substance has a boiling point near that of the refrigerant, its vapor pressure will be quite high, e.g., nitrogen, carbon monoxide, argon, and oxygen. Liquid

Table 5.4   Vapor Pressures of Some Substances at Liquid Air and Liquid Helium Temperatures

	Approx Vapor Pressure (mm Hg)	
Substance	at $-190°C$	at $-268.8°C$
Argon (A)	500	10^{-90}
Carbon dioxide (CO_2)	10^{-7}	—
Carbon monoxide (CO)	760	10^{-94}
Helium (He)	—	760
Hydrogen (H_2)	—	10^{-6}
Mercury (Hg)	10^{-32}	—
Oxygen (O_2)	350	10^{-104}
Neon (Ne)	—	10^{-26}
Nitrogen (N_2)	760	10^{-81}
Pump oils	10^{-35}	—
Water (H_2O)	10^{-22}	—

References:
S. Dushman, *Scientific Foundations of Vacuum Techniques*, 2nd ed., J. M. Lafferty, editor, John Wiley and Sons, New York, 1962.

D. E. Gray, editor, *American Institute of Physics Handbook*, sec. 4, McGraw-Hill, New York, 1937.

L. C. Jackson, *Low Temperature Physics*, 4th ed., Methuen Monographs on Physical Subjects, London, 1955.

K. K. Kelley, *Contributions to the Data on Theoretical Metallurgy III. The Free Energies of Vaporization and Vapor Pressures of Inorganic Substances*, Bull. 383, Bureau of Mines, 1935.

helium will effectively pump all gases met in vacuum practice except helium gas. Table 5.4 also shows that liquid air will effectively pump water vapor, mercury vapor, and pump oil vapors.

In the case of liquid hydrogen the gases which are not frozen are neon, hydrogen, and helium. Under standard atmospheric conditions, the pressures contributed by these gases are: neon, 1.14×10^{-2} mm Hg; hydrogen, 7.60×10^{-2} mm Hg; and helium, 0.38×10^{-2} mm Hg. Consequently, these gases contribute a total pressure of about 0.1 mm Hg. If a volume is pumped down with liquid hydrogen, the final pressure achieved cannot be lower than 0.1 mm Hg. When liquid helium is used, the final pressure cannot be less than about 0.38×10^{-2} mm Hg. In practice, the volume is generally reduced to a pressure of 1 mm Hg or so by mechanical pumps or other means before applying refrigeration. The advantages offered by this procedure are: (1) the pressure contributed by the gases which are not pumped by the refrigerant is reduced in proportion to the reduction in total pressure, (2) less refrigerant is needed to do the final pumping, and (3) there is less icing of the cold surfaces. By bringing down the pressure to around 1 mm Hg and then using the refrigerant, the final pressures obtainable with liquid hydrogen and liquid helium are about 10^{-4} mm Hg and 4×10^{-6} mm Hg, respectively. Before applying the refrigerant, if the system is flushed out with a gas which can readily be frozen, then the amount of gas left after using the refrigerant will be reduced, the result being lower pressures. Gases such as carbon dioxide or nitrogen can be used when liquid hydrogen or liquid helium is used. Several points regarding cryopumping may be worth emphasizing.

1. Low temperature refrigerants are generally costly and steps should be taken to minimize losses.

   a. Use a closed cycle refrigerating system where the size of system merits this procedure.

   b. Heat insulate the various connecting lines.

   c. Design the arrangement of cooling surfaces (traps) so as to minimize losses (see Chapter 11).

2. As the pressure is reduced the cryopumping increases rapidly. This is just the opposite of what happens in a mechanical pump. Because of this behavior:

   a. Make connecting lines to the cooling surfaces as short and as large as possible to reduce pumping losses.

   b. When possible, place the cooling surface within the volume to be pumped.

3. The use of liquid hydrogen for cryopumping systems which release large quantities of neon, hydrogen, or helium is usually not practical. Similarly, liquid helium is generally not used where quantities of helium are being released.

4. Cryopumps can be combined with various other types of pumps, including mechanical diffusion pump systems, to obtain pressures in the ultra-high vacuum regions (see Chapter 14).

Liquid nitrogen is generally delivered in a Dewar flask, which is a double walled glass container with the space between the walls pumped down to a low pressure so as to reduce loss of liquid nitrogen by heat from the surroundings. The walls are usually silvered to further reduce losses. In the cases of liquid hydrogen and liquid helium the temperatures involved are considerably lower than in the case of liquid nitrogen so that more elaborate measures must be taken to reduce losses. Containers for these liquids are usually made with four concentric shells. The innermost (fourth shell) contains the liquid hydrogen or helium, and liquid nitrogen is contained between the second and third shells. The spaces between the first and second shells and the third and fourth shells are evacuated. The advent of new and better insulating materials will undoubtedly lead to simpler methods of handling and transporting cryogenic materials. When large quantities of cryogenic materials are being used it may be well worth considering installing equipment for the production of these materials. Such equipment is available from a number of manufacturers.

Certain precautions should be taken in handling cryogenic materials. They cannot be sealed off in containers because of the high pressure build-up. Liquid nitrogen, hydrogen, and helium will also cause an accumulation of liquid or frozen oxygen. A build-up of oxygen together with the presence of organic materials can result in an explosion. The solution is to periodically clean the cold surfaces. Particular attention should be paid to the handling of liquid hydrogen since this is a flammable material. For proper safety procedures, the appropriate governmental agency in each country should be consulted, such as the National Bureau of Standards in the U.S.A., or appropriate information can be obtained from the manufacturers of cryogenic materials.

5.8   Other Sorption Pumps

Any material that sorbs gases and vapors is a form of pump (or trap). Usually such materials are used only when the pressure has been reduced to a fairly low value. Getters are simply classes of sorption materials that are used to reduce the pressure to a low value (after some other type of pumping) and to maintain it at that value.

Although sputter-ion and evaporation-ion pumps are usually pumped down first to a few microns with a mechanical pump, this may result in oil vapor getting into the system. To avoid this, various types of oil-free pumps have been tried as roughing pumps. Much of the effort has been concerned with the use of sorption materials. In order that such materials reduce the pressure from atmospheric to a few microns, it is necessary that they have a high gas capacity and pump the main atmospheric constituents effectively. Principal attention has been given to *activated charcoal, zeolite,* and *activated alumina.* Some properties of these materials are discussed in Chapter 8.

Jensen, Mercer, and Callaghan (*Rev. Sci. Instr.,* **30,** 377, 1959) have described the use of activated charcoal, cooled by liquid nitrogen. They found that a volume of activated charcoal would exhaust about ten times its own volume of atmospheric air to achieve a pressure of less than 10 μ in the course of a few minutes. This pressure is low enough to start sputter-ion pumps. At this stage the roughing pump can be valved off from the main system. With large systems, in order to reduce the amounts of activated charcoal and liquid nitrogen, it is possible to do a preliminary (and clean) rough pumping with some such pump as a water aspirator or a steam jet ejector. An alternative method is to flush with a gas such as carbon dioxide, which can then be frozen out (trapped). Zeolites and activated alumina behave in much the same way as activated charcoal, with roughly the same degree of efficiency. Attempts have been made to use these materials without refrigeration but the gas sorption capacity is much reduced.

5.9   Water Aspirator, Toepler, and Sprengel Pumps

The water jet pump (or aspirator) is a form of ejector pump and finds use in vacuum work in the mm Hg ranges, as filter pumps or for pumping out contaminating vapors. The principle of operation is illustrated in Fig. 5.12a. A stream of water flows through the jet tube *J* into the suitable choke tube *C*. At the entrance to *C* the water jet from the nozzle drags along the surrounding air. As it enters the mouth of the choke tube, air forms a sheath around the jet. As the jet diverges, the air sheath grows thinner, becomes entangled in the jet, and is carried out with the water. Figure 5.12a shows a short tube *A*. This gives a whirling motion to the water as it falls through

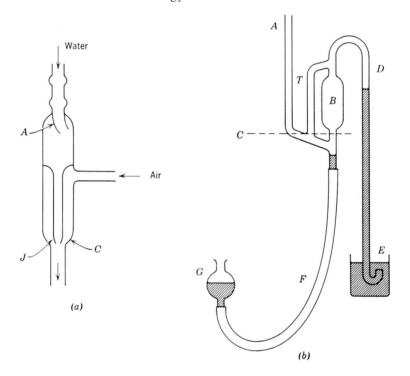

Fig. 5.12   (*a*) Form of aspirator.   (*b*) Form of Toepler pump.

the pump.   The result is higher speed and efficiency.   The limiting
pressure of an aspirator is the saturation pressure of water vapor at
the operating temperature.   At 15°C, this amounts to about 18 mm Hg.
The pressure can be reduced by using drying agents between the aspirator and the vacuum enclosure.

A Toepler pump is a form of mercury piston pump.   The principle
of this type of pump is illustrated in Fig. 5.12*b*.   When the mercury
reservoir G is lowered, bulb B and tube T are filled with gas from the
enclosure being evacuated (through tube A).   When G is raised,
mercury rises in tube F and cuts off the gas in B and T at C.   This
gas is then forced through the mercury in tube D into the atmosphere.
The end of tube D is bent upward at E to facilitate collection of gas
(or vapor).   By alternately raising G, a pumping action results.
Clearly, tubes F and D must be long enough to support mercury
columns corresponding to atmospheric pressure (76 cm at sea level).
Instead of using mercury to provide a valving action at C, it is possible to use a glass float valve.   The operation of the arrangement

shown in Fig. 5.12*b* is tedious and the speed is very low. Many attempts have been made to obtain automatic operation. One form of automatic Toepler pump which has gained fairly wide acceptance is that reported by Urry and Urry (*Rev. Sci. Instr.*, **27,** 819, 1956). This is sometimes referred to as an Urry-Toepler pump. Some improvements in design have been reported by Vratny and Graves (*Rev. Sci. Instr.*, **30,** 597, 1959).

Sprengel pumps originated from studies of water jet pumps for producing compressed air. However, various designs can be used to exhaust vessels. The basic principle of the pump involves a controlled flow of mercury into a bulb (connected to vessel) and thence into a tube of narrow bore. As the droplets of mercury fall into the tube they entrap air or gas, which passes into the atmosphere through the end of the tube. The tube must be of barometric height and is partly filled with mercury before the pumping action is begun. The reservoir must be replenished with mercury periodically, using that collected in a container at the end of the narrow tube. Various designs of automatic Sprengel pumps have been developed.

SIX

Measurement of Pressure

6.1   The Nature of Vacuum Gauges

Devices used for measuring pressures in vacuum systems are called vacuum gauges or, sometimes, pressure gauges. Various types of commercial gauges are available, which cover the pressure range from atmospheric pressure to less than 10^{-12} mm Hg. In the higher pressure region gauges are used that depend on the actual force exerted by the gas. At low pressures some specific property of gases and vapors, such as thermal conductivity, is used as the basis for several type of gauges. However, the lowest pressures are measured by ionizing the gases. In many cases the interest is in some characteristic of the gas other than the pressure, such as composition. However, gauges are generally calibrated in pressure units, such as μ or mm Hg. The various types of common vacuum gauges can be summarized as follows:

1. *Hydrostatic gauges.* These depend on the actual force exerted by the gas. Examples are: mercury and oil manometers, McLeod gauges, Bourdon gauges, and diaphragm gauges.

2. *Thermal conductivity gauges.* These depend on the change in the conduction of heat through gas with pressure. The most common examples are the Pirani and thermocouple gauges.

3. *Viscosity gauges.* These depend on the fact that the viscosity of a gas changes with pressure. An example is the Langmuir viscosity gauge.

4. *Radiometer gauges.* These depend on the fact that fast moving (higher temperature) molecules exert a larger force than slow (lower temperature) molecules. An example is the Knudsen gauge.

5. *Ionization gauges.* These depend on measuring electrical currents resulting from ionization of the gas. Common types are:

   a. Thermionic ionization gauges.

   b. Cold cathode gauges (Penning or Philips).

   c. Gauges using radioactive sources, such as the Alphatron.

6. *Discharge tubes.* These make use of certain physical characteristics, such as appearance or dimensions, of an electrical discharge in a gas.

Many new types of vacuum gauges are continually being developed, usually with the aim of extending the range to lower pressure values. However, the emphasis here is on types of gauges that are commercially available. As is pointed out later, the proper calibration of vacuum gauges is not an easy matter, particularly at low pressures. The readings obtained on commercial gauges can be considerably in error. The errors will vary from manufacturer to manufacturer, but even gauges of the same type from the same manufacturer may differ considerably in their readings. To ensure accuracy of readings, gauges should be calibrated according to procedures discussed here. It must be kept in mind that manufacturers generally calibrate their gauges for air (or dry air) and many types of gauges give different responses for various gases and vapors.

6.2   Bourdon and Diaphragm Gauges

These are mechanical gauges that are used primarily for giving an indication that a vacuum system is actually being pumped down. However, some commercial designs will read pressures down to a fraction of a mm Hg and can be used in certain vacuum processes. Most of these gauges read from atmospheric pressure down to their lower limit. They can be constructed of materials making it possible to use them in the presence of corrosive gases and vapors. Since they work on the basis of the force exerted by a gas, they will measure the total pressure of a mixture of gases and vapors.

Bourdon gauges make use of a tube which is sealed off at one end with the other end leading to the connection for the vacuum system. The tube is usually of elliptical cross section and bent into an arc. One end is rigidly fixed, so a change of pressure inside the tube makes it change its curvature. This change is transmitted through a series of levers and gears to a needle which gives a reading of the pressure on a circular scale behind the needle. The calibration of the scale is generally based on inches of mercury, from 0 to 30, where 0 represents atmospheric pressure and 30 represents a very good vacuum. Consequently, the vacuum becomes better as the needle approaches the scale mark 30 (see Figs. 6.1*a* and *b*). The lowest pressure that can be read will depend on the size of the scale divisions. Most commercial gauges are about 3 or 4 in. in diameter and it is not possible to read the scale to better than about ¼ in. Hg. This represents a pressure of about 6.3 mm Hg. Actually, the accuracies of most Bourdon gauges may not be sufficient to read a pressure this low. However, these gauges are usually used only to give an indication of the condition of a vacuum system. If such a gauge does not give a reading near 30 in. Hg in a few minutes on a laboratory size vacuum system, this probably indicates that a valve has been left open, a gasket is not seating properly, or some other leak has developed in the system. Without such a gauge, the usual practice is to listen to the forepump and assume there are no gross leaks if the pump sounds "hard" after a few minutes.

A diaphragm gauge usually consists of a flexible metal membrane whose deflection varies with the pressure in the vacuum system. The deflection is transmitted to some indicating device, usually through a series of levers and gears, to a needle which rotates over a calibrated scale. In some cases the scale is calibrated in in. Hg as in the case of the Bourdon gauge. In other cases the scale is calibrated in mm Hg with 760 mm Hg corresponding to atmospheric pressure and 0 corresponding to the lowest pressure. A simple gauge of this type will measure down to ¼ in. Hg or somewhat less. A general arrangement of such a gauge is shown in Fig. 6.1*c*. The reference vacuum is a pumped-down and sealed chamber. This allows measurement of lower pressures by getting larger changes in deflection. This type of gauge is connected to the vacuum system in the same manner as a Bourdon gauge and is often used to serve the same function, i.e., to indicate the presence of gross leaks. In other cases it can be used to measure pressures in systems operated in the mm Hg range.

Many special types of gauges making use of the deflection of a flexible membrane (or bellows) have been developed to measure pressures considerably lower than the mm Hg range. Often electrical meth-

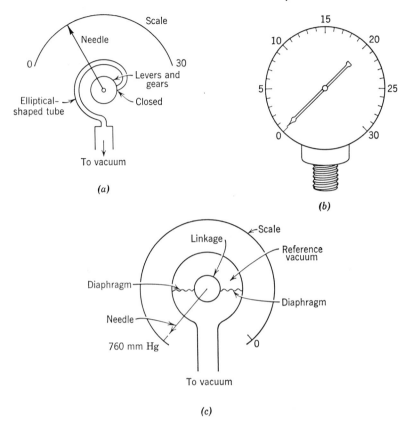

Fig. 6.1 (*a*) Principle of simple Bourdon gauge. (*b*) Appearance of Bourdon gauge. (*c*) Form of diaphragm gauge.

ods are used to magnify a mechanical deflection, e.g., in inductive or capacitance gauges. In other cases optical methods are used to produce a magnification. The most sensitive gauges are generally restricted to measuring a rather narrow range of pressure.

In connection with gauges using electrical output, the following advantages and disadvantages can be listed.

ADVANTAGES

1. The electrical output can be used for remote indication.

2. Construction materials can be chosen that allow corrosive gases to be used in the vacuum system.

3. The volume exposed to vacuum conditions is usually small.
4. Many forms of these gauges can be baked.

DISADVANTAGES

1. The sensitivity of the diaphragm may change.
2. The diaphragm may show aging effects, particularly if subjected to high pressure differences or temperatures.
3. The gauge reading may drift with temperature changes.
4. The zero point may change.

6.3   U-Tube Manometers

The most common form of manometer is a U-tube filled with a low vapor pressure liquid, usually mercury, although certain vacuum oils are sometimes used. A form of U-tube manometer with one side exposed to atmospheric pressure and the other side connected to the vacuum system is shown in Fig. 6.2a. The pressure reading will depend on the value of the atmospheric pressure, i.e., the barometric pressure. The barometric pressure will balance the pressure due to the height of liquid, h, plus the pressure due to the gas, P; or gas pressure, P = barometric pressure − pressure due to height of liquid, h. Clearly when the gas pressure is very low, the barometric pressure is essentially balanced by the height of liquid, h. This height will be greater for light liquids. Since for a low gas pressure, the height h will be 76 cm for mercury at sea level, this form of manometer becomes impractical for lighter liquids. The surface of the liquid assumes a shape shown in the insert of Fig. 6.2a. The curvature of the surface (meniscus) is due to surface tension of the liquid, i.e., "sticking" of liquid to the inside walls of the glass tubing. The glass tubing used for the two sides should have the same inside diameter, which should be uniform in order to avoid errors due to surface tension. Also, the inside of the tubing should be clean. Methods of cleaning mercury and glass are discussed in Chapter 9 and in Section 6.4. The inside diameter of the glass capillary tubing which is generally used is a few millimeters. Too small a diameter introduces greater errors due to the effect of any dirt present or nonuniformity of bore.

Another form of U-tube manometer is shown in Fig. 6.2b. Here, the reference pressure is not atmospheric but some pressure produced by an auxiliary pumping system. After pumping down to a desired

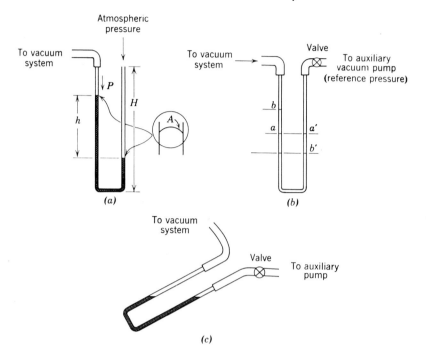

Fig. 6.2 (*a*) Barometric form of manometer. (*b*) Manometer using auxiliary pump. (*c*) Inclined manometer.

reference pressure, the valve shown can be closed and measurements made on the vacuum system of concern. When the liquid levels are at *a* and *a'* then the pressure in the vacuum system equals the reference pressure. At positions *b* and *b'* the pressure of the vacuum system is less than the reference pressure. In general the gas pressure is given by the following rules:

1. Pressure in system greater than reference pressure, i.e., level *b* below level *b'*:

Gas pressure = reference pressure + pressure due to
difference in levels of liquids

2. Pressure in system less than reference pressure, i.e., level *b* above level *b'* (as in Fig. 6.2*b*):

Gas pressure = reference pressure − pressure due to
difference in levels of liquids

This type of U-tube manometer can be used with mercury or various types of low vapor pressure liquids. When using mercury, the pressure due to the difference in liquid levels is readily found by measuring this difference in millimeters or centimeters, which gives the pressure directly in mm Hg or cm Hg, both of which are commonly accepted units of pressure. On the other hand, when using other liquids the density of the liquid must be considered. In this case the difference in liquid level can be converted to mm Hg by the following rule:

Pressure in mm Hg

$$= \text{difference in levels in millimeters} \times \frac{\text{density of liquid}}{13.6}$$

As an example, suppose a U-tube of the type shown in Fig. 6.2b is used with butyl phthalate (a diffusion pump oil), which has a density of 1.047 g/cc and the reference pressure is 0.2 mm Hg. Suppose, further, that the difference in liquid levels is 10 mm and the gas pressure is greater than the reference pressure. Then, according to the preceding rules, the gas pressure is $0.2 + 10 \times 1.047/13.6 = 0.2 + 0.77 = 0.97$ or about 1 mm Hg. Of course, if mercury had been used in the manometer, then the gas pressure would have been

$$0.2 + 10 = 10.2 \text{ mm Hg}$$

Where pressures of only a few mm Hg are to be measured, the reference pressure is often only a small fraction of a mm Hg and can be ignored in obtaining the gas pressure, i.e., the gas pressure equals the difference in liquid levels. Instead of pumping down to a particular reference pressure and holding one end of the U-tube at this value or cutting off the pumps with a valve, such as a stopcock, it is possible to seal off the end of the U-tube at the desired pressure. However, outgassing may lead to errors by producing changes in the pressure. Proper degassing can minimize this problem.

Although mercury is the most commonly used liquid for U-tube manometers, particularly because of its high density, it suffers from certain disadvantages. These are:

1. Its surface tension is high compared to various low vapor pressure liquids. This makes it difficult to obtain a reproducible meniscus and the column may stick in the glass tube.

2. In many cases it is undesirable to have mercury vapor get into the vacuum system. Mercury has a vapor pressure of about 1 μ at room temperature. To prevent mercury vapor entering the vacuum

system, a liquid nitrogen trap should be inserted between the manometer and the vacuum system.

3. If the manometer column is used to conduct electricity, the metallic vapor may be objectionable.

4. The high density makes a mercury manometer insensitive.

To overcome item 3 above, in some cases the surfaces of the mercury are covered with a few drops of low vapor pressure oil. However, a more general approach to overcoming the difficulties noted above has been to use oil manometers, which have sensitivities about 15 times those of mercury manometers, the actual gain in sensitivity depending on the density of the oil being used. Some oils which have been used are: butyl phthalate, butyl sebacate, and Apiezon B. The principal difficulty with oils is that they absorb gases. Heating the oil (and glass tubing) in vacuum will drive off a lot of gas. However, on exposure to gases there will be further absorption. As has been noted, these liquids can be used in a U-tube manometer of the type shown in Fig. 6.2b. The sensitivity can be increased by a factor of 5 or 10 by inclining the tube, as is shown in Fig. 6.2c. For a given difference in liquid levels (h), the oil will move correspondingly farther in the tube. The earlier comments regarding the glass capillaries and the cleaning of glass apply here. With extraordinary precautions, oil manometers of the type shown in Figs. 6.2b and 6.2c can be used to measure pressures down to around 50 μ.

Commercial U-tube manometers are generally calibrated in mm Hg. Reading can be done by simply sighting on the top of the meniscus (point A in insert of Fig. 6.2a). In other cases a movable magnifying glass can be attached to the manometer or a cathetometer can be used to make the measurement. With the most accurately constructed gauges, the mercury level can be read to a maximum accuracy of about 0.5 mm Hg, which corresponds to around 50 μ for oil manometers. For maximum accuracy, clean, degassed oil or mercury should be used. After the glass has been cleaned, it should be degassed by being heated to around the annealing temperature, say 300°C. Degassed, clean oil or mercury can then be introduced into the manometer. Further degassing of glass and liquid in vacuum is then advisable. Best results with mercury are obtained by vacuum distilling mercury into the degassed manometer tubing.

Many efforts have been made to increase the sensitivities of mercury and oil manometers. Small mirrors have been floated on the surface of the liquid to reflect light to a scale so as to magnify the movement of the liquid surface. Many special designs of oil manometers have been built in an effort to extend the range of accurate measurement down to 10^{-4} mm Hg or better. However, the need for extreme cleanliness, uniform glass bore, and elimination of gases from glass and liquid makes it extremely difficult to obtain consistent performance with these sensitive manometers.

6.4 McLeod Gauges

The McLeod gauge is a form of manometer which is generally accepted as the standard for absolute measurement of pressure, i.e., a device that can be calibrated from its dimensions to give a direct reading of pressure. The Knudsen absolute radiometer gauge has been used for this purpose but the advantages of the McLeod gauge have made it the most acceptable standard for calibrating other gauges. Of course, a particular type of gauge calibrated against the McLeod gauge can be used as a secondary gauge to calibrate other gauges.

The principle of the McLeod gauge is rather simple. It involves trapping a known volume of gas at the pressure which is to be measured and compressing this volume to a new pressure. The calibration is carried out by applying Boyle's law, which states that if the temperatures remain unchanged,

Original pressure × original volume = final pressure × final volume

Knowing the original volume, the final pressure, and the final volume, it is possible to obtain the pressure in the vacuum system (original pressure). Mercury is most commonly used for trapping and compressing the known volume of gas although gauges are constructed using a low vapor pressure oil. The basic elements of a McLeod gauge are shown in Fig. 6.3a. In view of the fact that the levels of mercury must be observed, the construction is of glass, usually a hard glass like Pyrex for strength. When some equilibrium pressure has been reached in the gauge and it is desired to measure this pressure, the mercury in a reservoir is forced up tube F by some method. When the mercury covers the cut-off, D, it traps the gas in the bulb A and the closed capillary B at the pressure which is to be measured. As the mercury is raised it fills bulb A and begins to enter tube E, capillary C (open), and capillary B (closed). Bulbs such as G and H are often incorporated so that the mercury doesn't suddenly hit capillary C and break into droplets. The gas in capillary B is then compressed, the amount of compression being given by the ratio of the volume of bulb A and capillary B to the volume in capillary B after compression. Since the mercury in capillary C is still exposed to the original pressure in the gauge, it will rise above the mercury in capillary B. If this pressure is very low, its effect in depressing the mercury column in capillary C can be neglected. Therefore, the pressure of the compressed gas is simply given by the difference in levels of the mercury columns. Tube E is used simply to pump out bulb A.

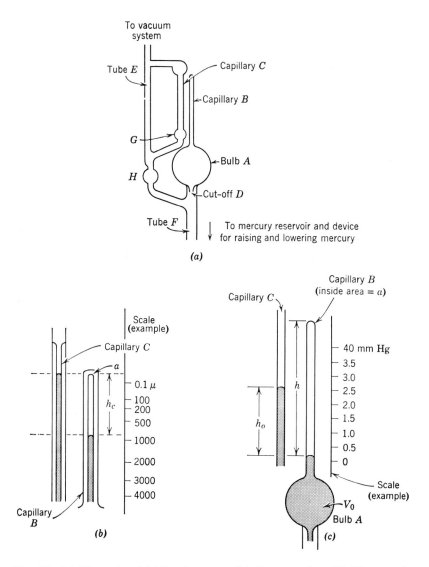

Fig. 6.3 (*a*) Elements of McLeod gauge. (*b*) Square scale. (*c*) Linear scale.

Two methods are used for making measurements with a McLeod gauge. Each of these involves raising the mercury to a fixed point, the reference point, and then measuring the difference in mercury levels in capillaries *B* and *C*.

METHOD 1. THE QUADRATIC OR "SQUARE" SCALE METHOD

This method is illustrated in Fig. 6.3*b*. When a pressure measurement is to be made, the mercury is raised in capillary *C* to a level even with the top of capillary *B*. The difference in mercury levels in capillaries *B* and *C* is then h_c. Suppose the cross-sectional area of capillary *B* is *a*. Then the volume of the compressed gas is ah_c. The pressure of this gas, neglecting the pressure on the mercury in column *C*, is simply the difference in mercury levels, h_c. If this difference is measured in millimeters, then the pressure is expressed in mm Hg. Suppose the original volume of gas, i.e., the trapped volume after the mercury reaches the cut-off *D*, is V_0. This volume is at the pressure being measured, *P*. Then by Boyle's law

$$V_0 \times P = ah_c \times h_c$$

or

$$P = \frac{ah_c^2}{V_0} \tag{6.1}$$

a and V_0 are constants for the gauge. It is evident that the pressure increases as the square of the difference in mercury levels (h_c), hence the name quadratic or square scale. This results in a scale that is spread out at low pressures. The general nature of the scale is indicated in Fig. 6.3*b*.

EXAMPLE

A McLeod gauge has a closed capillary with a diameter of 1 mm and a volume above the cut-off (V_0) of 200 cc. Suppose, when a pressure reading is taken with this gauge, that the difference in mercury levels (h_c) is 4 mm. To find the value of the pressure, millimeter units are commonly used, which give the pressure in mm Hg. The area *a* is $\pi r^2 = \pi \times 0.5^2 = 3.14 \times 0.25 = 0.785$ mm², h_c is 4 mm, and V_0 is $200 \times 1000 = 200,000$ mm³. Therefore, the pressure is

$$\frac{ah_c^2}{V_0} = \frac{0.785 \times 4 \times 4}{200,000} = \frac{0.785 \times 8}{100,000} = 6.28 \times 10^{-5} \text{ mm Hg}$$

Since *a* and V_0 are constant for a given gauge, only $k_1 = a/V_0$ need be known. This is called the gauge constant for a quadratic scale and the pressure in this case is given by $P = k_1 h_c{}^2$. For the gauge above, the constant is $0.785/200,000 = 3.9 \times 10^{-6}$ mm^{-1}.

It should be noted that the height h_c occurs as a square in eq. 6.1. However, for accurate work there are really two heights to be considered. One is the difference in mercury levels and the other is the distance from the mercury level in capillary B to the "top" of this capillary. If the top of the capillary is flat then the mercury level in capillary C can be aligned fairly accurately with the top of capillary B; the two heights are then equal and eq. 6.1 results. However, in practice the top of capillary B will be curved and the "effective" height of the capillary above the mercury level must be determined experimentally. Methods for finding this effective height are described in several of the references in Appendix E. It is only when the difference in mercury levels becomes small, say less than 2 mm, that a correction for this effective height should be made. However, many commercial gauges have a movable scale which is adjusted so the zero point is lined up visually with the top of the closed capillary. Pressure readings corresponding to differences of mercury levels of less than about 2 mm are likely to be in error, the amount of error depending on the exact shape of the end of the closed capillary and on how accurately the measurement is made.

METHOD 2. LINEAR SCALE METHOD

This method of measuring pressure results in a linear scale, i.e., a scale of the same type as an ordinary foot ruler or yardstick. The method is used for measuring higher pressures than the square scale method, usually in the mm Hg range. The method involves raising the mercury to a fixed point (reference point) on capillary B and then measuring the difference in mercury levels in capillaries B and C (Fig. 6.3c). The reference point is usually taken near the point of juncture of capillary B and bulb A. Let h represent the distance from the reference point to the end of the closed capillary B and h_0 the difference in mercury levels (see Fig. 6.3c). Again let a be the cross-sectional area of capillary B and V_0 the volume of gas being compressed at the pressure, P, being measured. Again, by Boyle's law,

$$PV_0 = ahh_0$$

where ah is the volume of compressed gas. Hence, the pressure being measured is given by

$$P = \frac{ahh_0}{V_0}$$

a, V_0, and h are constants of the gauge so the pressure is directly proportional to h_0. Figure 6.3c shows the nature of the scale resulting from this method.

EXAMPLE

A gauge has a diameter of 2 mm and a volume, V_0, of 200 cc. The height of capillary B from reference point to end (h) is 20 cm (generally an upper limit in length). Suppose the difference in mercury levels is 6 cm (h_0). The cross-sectional area of the capillary is $\pi \times 1^2 = 3.14$ mm^2 and the volume V_0 is 200,000 mm^3. The pressure is then given by

$$P = \frac{3.14 \times 200 \times 60}{200,000} = \frac{3.14 \times 12}{200} = 0.19 \text{ mm Hg}$$

Note that all dimensions are given in terms of millimeters. Often a gauge constant for a linear scale is quoted. In this case, the gauge constant is

$$k_2 = \frac{ah}{V_0} = \frac{3.14 \times 200}{200,000} = 3.14 \times 10^{-3}$$

Therefore, the pressure being measured is given by $P = k_2 h_0$. The maximum pressure that can be measured with the above gauge is about $3.14 \times 10^{-3} \times 200 = 0.63$ mm Hg. To increase the maximum pressure that can be measured, either a or h can be increased. The maximum value of h is determined by the length of capillary B and its length is limited by practical considerations. Increasing the value of a means decreasing the lowest pressure that can be measured by the gauge using the square quadratic scale method. Many gauges have both quadratic and linear scales, so a compromise has to be made. It should be pointed out that increasing V_0 will give lower pressures on both quadratic and linear scales. The previous comments regarding effective height also apply here when maximum accuracy is desired. On many commercial McLeod gauges, the maximum pressure readings are the same on both linear and quadratic scales. Also, in some cases two or three open capillaries like C are used with different reference points to cover several pressure ranges. By proper choice of capillary diameter and reference points it is possible to cover a total pressure between about 20 mm Hg and 10^{-4} mm Hg.

It is important that capillaries B and C be of the same diameter and of uniform bore. This will prevent errors due to capillary depression resulting from the surface tension of the mercury. As has been noted above, reducing the diameter of the capillaries makes it possible to measure lower pressures. However, there is a limit to how small the diameter can be made that is due to difficulties with mercury sticking. Any small amount of dirt will increase the difficulties. A practical diameter is 1 mm, although capillaries down to 0.5 mm have been used with extraordinary attention being given to cleanliness of glass and mercury. Another way of making it possible to measure lower pressures is to increase the volume of the bulb A. The limit on the size of this bulb stems from the difficulty in handling a large weight of mercury in glass apparatus. A common size of bulb is about 300 cc. However, some commercial McLeod gauges have bulbs with a volume greater than 2000 cc, making it possible to measure pressures below 10^{-6} mm Hg. Extreme care must be exercised in measuring pressures below 10^{-5} mm Hg.

Apart from the choice of matched and uniform capillaries, one of the problems in the design of McLeod gauges is the method of moving the mercury up into the capillaries. One method of doing this without the use of an auxiliary pump is shown in Fig. 6.4a. Here, control of the mercury level is done by raising and lowering the reservoir F, which is filled with mercury. Rubber tubing is usually used to connect the reservoir and gauge proper. In Fig. 6.4a, the gauge is at atmospheric pressure, so the levels of mercury in gauge and reservoir are the same. As the gauge is pumped down, the mercury will rise into bulb A unless the reservoir is lowered. Clearly, at low pressures the reservoir will have to be lowered at least 76 cm below the cut-off (at sea level) so mercury does not enter the bulb A. For this reason this type of gauge is often called a "barometric" gauge. To make a reading, the reservoir is raised until mercury fills bulb A and enters the capillaries. Sometimes a trap is incorporated in the tubing between reservoir and gauge in order to trap bubbles. Various forms of cut-offs can be used. The main thing is to trap a definite volume of gas when a pressure measurement is to be made. Often a constriction is placed in the main tube to the gauge to avoid having the mercury rise too rapidly and possibly break the glass. The bulb G is included to act as a reservoir for the mercury if it is raised too far. This bulb prevents mercury from getting into the vacuum system. Sulfur from the rubber tubing may contaminate the mercury. This can be avoided to a large degree by boiling the tubing in caustic soda or caustic potash before installing it.

The difficulties with the arrangement of Fig. 6.4a are: (1) it is awkward manipulating the mercury reservoir, (2) the overall size of the gauge is uncomfortably large, (3) even after cleaning the rubber, there is still the possibility of contaminating the mercury, (4) the rubber may be porous and allow air to get into the gauge. To avoid these

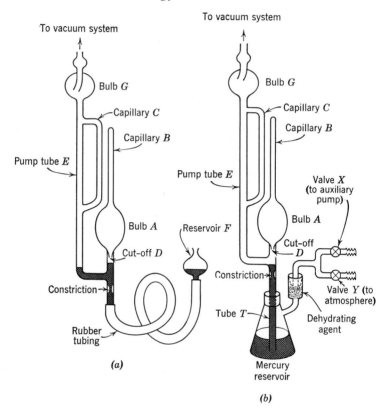

(a)

(b)

Fig. 6.4 (a) Barometric type of McLeod gauge. (Sometimes a plunger is used.) (b) McLeod gauge using an auxiliary pump.

difficulties, the most common arrangement used is one making use of an auxiliary vacuum pump in combination with atmospheric pressure to manipulate the mercury. One form of such an arrangement is shown in Fig. 6.4b. Here the mercury reservoir is connected to the gauge through a ground joint. This reservoir can be removed for filling. In some designs the reservoir is joined to the gauge by a glass seal. In this case it has to be filled through the pumping tube or an opening in the reservoir that can be made vacuum tight. Figure 6.4b shows two lines connected to the mercury reservoir through a bulb containing some type of drying agent (desiccant) such as calcium chloride. This drying agent helps to keep water vapor out of the gauge. Two valves are shown, X being in the line which is connected to some type of vacuum pump and Y being in the line which is open to atmospheric pressure. Needle valves are quite suitable for this purpose. Many commercial gauges make use of two-way stopcocks. If the gauge is at atmospheric pressure to begin with and it is pumped on by the vacuum system, then the mercury will begin to rise in tube T. Valves X and Y should first be closed, but as soon as pumping is started valve X should be opened. This will pull the mercury down from bulb A. When the gauge has been pumped down and a reading is to be made, valve X is closed and valve Y is slowly opened to allow the mercury to rise into the bulb. As the mercury nears the open end of the closed capillary, valve Y should be gradually closed so the mercury rises very slowly. When the reference point is reached valve Y is closed. Raising and lowering the mercury as necessary can be done by carefully manipulating valves X and Y. Figure 6.4b shows a constriction in tube T, a bulb G at the top of the capillaries, and a cut-off below bulb A, all of which serve the same purposes as they do in the arrangement of Fig. 6.4a. The connection to the vacuum system is often made with rubber tubing (properly cleaned). However, when vapors are to be kept to a minimum, glass tubing can be sealed directly to the gauge and connected to the vacuum system. Some manufacturers will supply suitable metal-to-glass seals so that metal tubing can be used. The auxiliary vacuum pump used to operate the gauge need not produce particularly low pressures (a few mm Hg is adequate) and its pumping speed can be quite small (usually 1 l/min or less) since the gauge volume is small.

The high compression resulting in a McLeod gauge will generally liquefy vapors unless they are present in very small quantities. When vapors are liquefied their pressures will not be measured by a McLeod gauge. Consequently, the vapor pressures of water and mercury are generally not measured. Many commercial gauges are available with different pressure ranges, determined by bulb size and capillary cross section. They are usually shipped without mercury because of the possibility of damage. Often the inside surfaces are roughened by chemical or mechanical treatment so as to minimize sticking of the mercury. Manufacturers often specify methods for setting up and operating their products. However, some general rules can be listed.

SETTING UP AND OPERATING A NEW MCLEOD GAUGE

1. Fill the gauge with clean, dry mercury (cp grade, not exposed to the atmosphere). Where the reservoir is connected to the gauge by a ground joint, it should be removed and filled with mercury. The

ground joint should be lightly coated with a good grade of vacuum grease.

2. Sometimes it is desirable to flame the gauge before adding the mercury.

3. Set up the gauge in a vertical position and secure firmly.

4. Attach to the vacuum system. Include a liquid nitrogen trap in the connecting line to prevent mercury vapor from getting into the vacuum system and vapors from getting into the gauge. The connecting line should be as short as possible.

5. To minimize contamination, degas the glass and mercury under vacuum (mercury raised into bulb) by flaming carefully, avoiding too much localized heating. In any degassing operation, care must be taken to avoid damaging the scale. Where feasible, it should be removed.

6. Set the scale (where applicable) so the zero scale unit is lined up with the top of the closed capillary. This generally takes care of the linear and square scales.

7. In taking readings, if the mercury sticks or breaks in the capillary, the capillary can be lightly tapped or flamed carefully (tap or flame above the mercury beads while pumping on the capillary).

CLEANING A MCLEOD GAUGE

When a gauge becomes contaminated it may have to be cleaned. General methods for cleaning glass and mercury are discussed in Chapter 9. Particular methods are outlined here at the risk of some repetition.

1. *Cleaning glass.*

a. Cover some potassium dichromate crystals with concentrated sulfuric acid.

b. Allow the mixture to stand for about 5 min, which will give a solution faintly green in color.

c. Pour it into the glass vessel to be cleaned. Carefully add a little water. Heat will be evolved.

d. Wash the glass carefully with distilled water. Do not use solvents at this latter stage because of the possibility of getting traces of fats on the glass.

e. Dry in air or by torching.

2. *Cleaning mercury.* Mercury can be contaminated by dust and grease, which float as a scum, and by water and metals, which form

amalgams. The more reactive metals (zinc, copper, alkali metals) can be removed by the following method:

a. Shake the mercury vigorously in a strong, stoppered vessel containing 10% nitric acid. A scum is formed.

b. Remove the scum by pouring the mercury into a separating funnel. This method does not remove metals such as tin, silver, gold, etc. These can be removed by vacuum distilling the mercury (see Chapter 9).

CALIBRATING A MCLEOD GAUGE

The calibration of a McLeod gauge is determined by the volume of the bulb and closed capillary, measured from the cut-off, and by the cross-sectional area of the capillary. One method of determining these quantities involves the following:

1. The volume can be found by filling the closed capillary and bulb (to cut-off) with mercury and weighing the mercury. The weight of the mercury divided by the density at the temperature at which the measurement is made gives the volume. Instead of mercury, water could be used.

2. The cross-sectional area of the capillary can be found by inserting a slug of mercury in the capillary and observing the length of the slug at different positions. Changes in this length will give an indication of the lack of uniformity of the capillary. The diameter can be found from the length of the slug and the density of the mercury.

Clearly the above method is most readily used before the gauge has been assembled. It must be modified to apply to assembled gauges. However, the basic procedure is still applicable in the latter situation.

OIL MCLEOD GAUGE

Instead of using mercury, it is possible to use a low vapor pressure oil. The use of oil theoretically permits measuring lower pressures than with mercury because of the low density of the oil. Most oils which can be used should give a sensitivity more than ten times that given by mercury. However, oils suffer from several difficulties, the most important of which are:

1. Oils tend to absorb gases.
2. They tend to stick at the entrances to the capillaries.

Difficulty 1 is best avoided by degassing the oil and glass in vacuum. Sometimes it is advisable to degas the oil in vacuum before putting it in the gauge, followed by further degassing. Subsequently, the oil should not be exposed to air except at low pressures, say less than 10^{-2} mm Hg. Difficulty 2 can be minimized by installing a small heater at the capillary junctures. Oil McLeod gauges cannot be considered to be industrial-type devices because of the above difficulties and their fragility.

TILTING MCLEOD GAUGES

Tilting McLeod gauges operate on the same basic principle as the type of gauge described above. However, they are generally attached directly to a vacuum system so that readings can be made as desired. Also, the space above the mercury reservoir is not exposed to atmospheric pressure. The general appearance of such a gauge is shown in Fig. 6.5. The gauge is attached to the vacuum system in a horizontal position through a connection at the rotary vacuum seal, by means of rubber, plastic, or copper tubing. The mercury has now drained out of capillary D and this capillary (as well as the rest of the gauge) is pumped down to the pressure of the system. When a reading is desired, the gauge is tilted into a vertical position (direction B). Mercury then flows into capillaries C and D, and when point E is passed a volume of gas from E to the top of capillary D is trapped. The amount of mercury used is generally such that it will rise in capillary

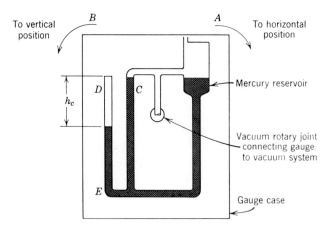

Fig. 6.5   Tilting McLeod gauge.

C to a position level with the top of capillary D. This gives a quadratic scale. The reference point can be changed by adjusting the amount of mercury so as to get a linear scale.

Some advantages of this form of gauge are:

1. It is self-contained, i.e., no electrical connections are needed.
2. It can be absolute.
3. It is compact and reasonably rugged (due to the use of a protective case).

The disadvantages are much the same as for the laboratory-type McLeod gauge. Commercial gauges of this type are manufactured by the F. J. Stokes Corporation, Philadelphia, Pennsylvania, and W. Edwards High Vacuum, Ltd., Crawley, Sussex, England. Various models are claimed to cover the range from about 0.01 μ (10^{-5} mm Hg) to 50 mm Hg. Sometimes a valve is incorporated in the gauge near the rotary vacuum seals. Also, various types of cold traps are available to go along with these gauges. Several attempts have been made to use low vapor pressure oils in such gauges, but the same difficulties experienced with oil manometers are encountered.

ADVANTAGES AND DISADVANTAGES OF MCLEOD GAUGES

Attention here is directed to mercury McLeod gauges of the laboratory type shown in Fig. 6.4. Much work has been carried out in an attempt to use some other type of gauge, such as the Knudsen radiometer gauge, as an absolute vacuum gauge, i.e., one that can be calibrated from known physical quantities. However, the wide experience with the McLeod gauge has led to its general acceptance as the primary absolute gauge; but, this type of gauge still suffers from certain disadvantages, which can be listed as follows:

1. The mercury vapor should be kept out of the vacuum system by use of a refrigerated trap, usually liquid nitrogen.
2. Mercury puts limitations on the construction materials that can be used. Glass, which is fragile, is commonly chosen.
3. The length of tubing between the gauge and the vacuum system, together with the large volume of the compression bulb, gives differences between the pressure in the vacuum system and that measured by the gauge.
4. The gauge can only be operated intermittently, so it is not possible to obtain a continuous record of pressure. Considerable time lags are involved in taking readings.

5. The gauge requires a trained observer and does not lend itself to remote or automatic operation.

6. The gauge does not generally measure the pressures of vapors. For some purposes, however, this can be an advantage.

6.5 Pirani and Thermocouple Gauges

These types of gauges are often called thermal conductivity gauges. They operate on the principle that the loss of heat from a heated object in vacuum will depend on the pressure. This is only true for a certain pressure range within which the mean free paths of the molecules are comparable to the dimensions of the container (gauge tube) in which the heated object is placed. An object heated in vacuum can lose heat in three ways:

1. By conduction through the leads.
2. By radiation to the surroundings.
3. By conduction through the gas.

It is the loss through means 3 that is measured by this type of gauge. If this loss becomes independent of the pressure or becomes very small compared to the losses through means 1 and 2, then the gauge fails. Figure 6.6*a* indicates the methods by which a heated filament loses heat. Molecules striking the filament will carry heat away to the walls of the container. The more molecules striking the filament, the more heat will be removed and, therefore, the lower will be the temperature of the filament (if heated at a constant rate by an external source, usually an electrical supply). As the number of molecules striking the filament is reduced (lower pressure), less heat is removed and the temperature of the filament will rise.

The above performance only holds for a certain range of pressure, depending on the arrangement of filament, leads, and container. At higher pressures, where the molecules make many collisions in moving from filament to the walls of the container, changes in pressure will have no effect on the rate of removal of heat from the filament. When this happens the upper pressure limit of the gauge has been reached. At low pressures, where few molecules strike the filament, losses through 1 and 2 become more important than the loss through 3 and the gauge then reaches its lower pressure limit. The rate of loss of heat from the filament will depend on the type of gas surrounding it. Some gases

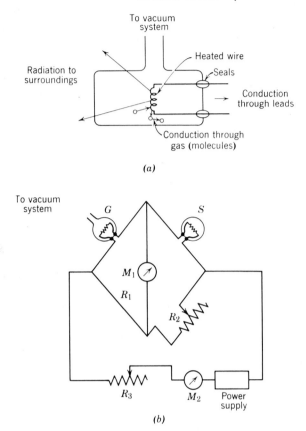

Fig. 6.6 (a) Losses of heat from heated wire. (b) Arrangement of Pirani gauge.

(and vapors) can conduct heat more readily than others. This property is known as the thermal conductivity and is expressed by a constant called the thermal coefficient of conductivity which is measured in calories per centimeter per second per degree centigrade (see Appendix A). The higher the value of the thermal coefficient of conductivity, the greater will be the rate of loss of heat and, therefore, the greater the change in temperature for a given change in pressure. From Appendix A it is found that hydrogen has a higher thermal coefficient than air and, therefore, will give a larger response for a given pressure difference. In practice, changes in temperature due to changes in pressure are measured indirectly, usually by a change in electrical resistance or an electrical voltage.

PIRANI GAUGE

This type of gauge measures pressure essentially by measuring an electrical resistance. The electrical resistance of most metals will increase as the temperature rises. Since the temperature changes with pressure (in the proper range) measuring the resistance gives a measure of the pressure. It is desirable to use wires that give a large change in resistance for a given change in temperature, i.e., wires that have a large temperature coefficient of resistance. Most commercial Pirani gauges use tungsten wire. Nickel has also been widely used. Some commercial gauges make use of semiconductor materials called thermistors, which have very large temperature coefficients of resistance. A change in the temperature of the gauge surroundings (ambient temperature) will affect the temperature of the wire and therefore its resistance. This results in a change in the pressure indication. To avoid this difficulty, a dummy gauge tube identical with the gauge being used to make pressure measurements is added. These gauges are mounted in close contact so that they assume the same temperature. Also, the dummy tube is sealed off at a pressure considerably below that to be measured. The usual arrangement for measuring pressures with a Pirani gauge is shown in Fig. 6.6b. The two gauge tubes G and S, where S is the dummy tube, are connected in a bridge circuit (Wheatstone bridge) with two resistances R_1 and R_2. Electrical current from a power supply, which can be a battery but is usually a rectifier unit, is passed through the bridge elements and is adjusted to the proper value by a variable resistance R_3, the value being read on meter M_2 (a milliammeter). Meter M_1 (a microammeter) is calibrated in pressure units, usually microns. Often this meter may have two scales or more, with an appropriate switching arrangement. Many commercial gauges omit meter M_2 and use suitable electrical circuitry to provide a constant current to the bridge circuit. The zero point is set on M_1 by adjusting the resistance R_2 while gauge G is pumped down to a very low pressure. This meter will then give a reading as the pressure is raised into the range of the gauge. Most commercial gauges are designed to operate off 115-v or 220-v a-c mains.

A Pirani gauge will read pressures of vapors as well as of permanent gases. Consequently, these gauges must be calibrated for the various gases and vapors that are to be measured. Generally speaking, these gauges are not used to obtain highly accurate pressure values. The pressure readings for nitrogen, oxygen, and water vapor are much the same. However, the presence of hydrogen, helium, and various solvent vapors will have an effect on the pressure. This makes the gauge useful for leak-hunting purposes (see Chapter 15). The pressure range covered by most commercial Pirani gauges is about 1 μ to 1 mm Hg. They are used to determine when diffusion pumps should be turned on, to measure pressure for processes operating in their pressure range, and to operate various devices such as safety interlocks (for diffusion pump cooling water, etc.). Various filament arrangements are used, such as stretched wires or coils. Sometimes the dummy tube is a separate tube and sometimes it is incorporated in a single housing with the gauge tube. Commercial units often show two scales—a lower one with a range of around 0 to 50 μ, which is essentially linear, and an upper one of around 0 to 2000 μ, which is compressed at the high pressure end. Figure 6.7 shows typical curves for the response to different gases and vapors.

The principal advantages of Pirani gauges are:

1. They have fairly rapid response to changes in pressure.
2. The electrical indication lends itself to self-recording and automatic devices.
3. The electrical circuitry is relatively simple.
4. They measure the pressures of both permanent gases and vapors.

Item 4 makes these gauges useful in leak hunting on account of their different response to various gases and vapors. However, in some cases this is considered to be a disadvantage since the reading will change if the composition of gases and vapors is changed. In general, the disadvantages of the gauge stem primarily from the fact that it depends on the transfer of heat through gases. Any change in the condition of the surface of the heated wire will change the rate of loss of heat from the wire, both through the gas and by radiation. This results in two effects:

1. A change in the calibration of the gauge.
2. A change in the zero.

Various methods are used to try to minimize these difficulties. One method is to "flash" the wire, i.e., heat it to a high temperature for a short while. This will drive off various contaminants on the surface. The danger is that certain contaminants, such as oil vapor from pumps, may "crack" and leave various deposits, particularly carbon. To eliminate these deposits it is necessary to clean the gauge tube, usually with chemicals, and then dry it. Another method of minimizing the

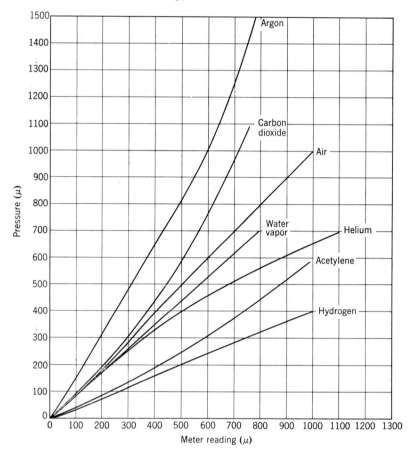

Fig. 6.7   Calibration of Pirani gauge for various gases and vapors.

above difficulties is to treat the wire so that subsequent contamination will have little effect on the characteristics of the surface. A further disadvantage of these gauges stems from the fact that certain delicate contacts must be used. Any change in these contacts will change the rate of heat loss. Nevertheless, proper construction prevents difficulties with these contacts. A general method of cleaning this type of gauge is to pour a solvent such as acetone into it and shake the gauge. However, care must be taken in this procedure because of the rather delicate wires that are often used. It is safest to consult the manufacturer on the proper procedure.

THERMOCOUPLE GAUGE

This type of gauge is really a form of Pirani gauge—the difference being that instead of measuring the change in electrical resistance of a wire, the procedure is to attach a thermocouple to a heated wire and measure the voltage produced by the thermocouple, which depends on the temperature. The basic principle governing this type of gauge is shown in Fig. 6.8. As the temperature of the heated wire changes due to a change in pressure, the temperature of the thermocouple is also affected, which results in a different voltage being produced by the thermocouple. The current resulting is measured by the meter M, which has a scale calibrated in pressure units, usually microns. Commercial instruments operating off the a-c line or batteries are available. Only a few volts are needed to operate the gauge and the necessary heating current is a few milliamperes. When the a-c line is used to provide the heating, there is no need to use a rectifier. Many types of thermocouples have been used, such as copper-Constantan, Cupron-Chromel P, etc. Commercial gauge tubes that do not use a separate wire for heating the thermocouple junction are available. This ar-

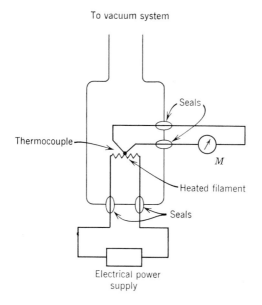

Fig. 6.8 Principle of thermocouple gauge.

rangement generally results in a smaller and more rugged gauge tube. Sometimes several thermocouple junctions are used to obtain a higher electrical output, thus making it possible to measure lower pressures. Other thermocouple junctions are sometimes added to compensate for changes in the surrounding (ambient) temperature. When several junctions are used, the gauge tube is often called a thermopile. The basic elements of such a thermopile have been described by J. M. Benson (*Vac. Symp. Trans.*, 1956).

Many of the comments regarding Pirani gauges, such as advantages and disadvantages, are applicable to thermocouple gauges. However, certain additional comments may be pertinent with regard to the latter type of gauge. This type of gauge can be made quite small and rugged. Consequently, cleaning can be done with less chance of damage than in the case of most Pirani gauges. Thermocouple gauges give different readings for different gases and vapors (same pressure), although in general the change in reading is not as great as for Pirani gauges. This feature makes the gauge useful in hunting for larger leaks. The relative response curves for various gases and vapors are much the same as for the Pirani (Fig. 6.7). Most commercial gauge tubes are made of metal and are often provided with a pipe threaded tubulation, usually standard ⅛-in. They can be soft-soldered to the vacuum system or attached with a vacuum cement or wax. If a gauge tube has a smooth tubulation it can be attached to the system with an appropriate stuffing box or Wilson seal. Commercial glass thermocouple gauge tubes are not too common because of their fragility. The highly nonlinear scale of the thermocouple gauge is often considered to be a disadvantage. However, their compactness, ruggedness, and simple electrical circuitry make them quite popular in the same applications where Pirani gauges are sometimes used. Both thermocouple and Pirani gauges can be used to drive electrical recorders, so a permanent record of the pressure (with time) can be obtained.

6.6   Thermionic Ionization Gauges

This type of gauge operates on the basis of ionizing a gas by means of electrons emitted by a heated filament. The electrons are speeded up by an electric field and the positive ions produced in the gas are collected. The number of positive ions formed will depend on the number of gas molecules present and therefore on the pressure. Consequently, the positive ion current gives a measure of the pressure. The basic principle of operation is illustrated in Fig. 6.9a. Three electrodes are shown—filament, grid, and collector. These electrodes are mounted in a tube that is connected to the vacuum system. The filament is heated by an electric current to a high temperature and emits electrons. The grid is at a positive voltage with respect to the filament and attracts the electrons. When the electrons speed up sufficiently they will have enough energy to break apart molecules into positive ions and electrons. This only happens when the pressure is low enough so that the electrons do not make too many collisions in passing between the filament and the grid. The electrons pass through the grid, which is an open wire structure, and produce ions in the region between the grid and the collector (plate). The positive ions formed are collected by the collector and the resulting current is measured by an appropriate meter. The collector is maintained at a negative voltage with respect to the filament (and grid). In some cases the grid is used as a collector. The grid is then negative and the collector (plate) positive with respect to the filament.

Two basic electrical circuits that are commonly used with thermionic ionization gauges are shown in Figs. 6.9b and c. Figure 6.9b shows an "external collector." In this case the plate (collector) is at a negative voltage with respect to both the grid and the filament. Consequently, the plate attracts positive ions and the current due to these ions gives a measure of the pressure. A, B, and C provide the necessary electrical voltages and currents (rectifier units or batteries). Unit A provides the positive (+) voltage for the grid to speed up electrons from the filament. The filament is heated by current from unit B while unit C provides the minus (−) voltage for the plate to collect the + ions. Meter M_1 is a microammeter and M_2 is a milliammeter. M_1 is the meter that gives the pressure indication, being calibrated in pressure units. Figure 6.9c shows an arrangement in which the grid is used as the collector ("internal collector"). In this case the grid has a − voltage (to collect the positive ions) and the plate has a + voltage (to speed up electrons from the filament). The power supplies A, B, and C serve the same general purpose as they did in the case of Fig. 6.9b, although now unit A gives a − voltage for the grid and unit C gives a + voltage for the plate. Often alternating current is used to heat the filament. The voltages shown in Figs. 6.9b and c are only typical values. Voltages between +100 and +300 v on the accelerating electrode (grid for external collector and plate for internal collector) and between −2 and −25 v on the ion collector are commonly used. The electron current from the filament

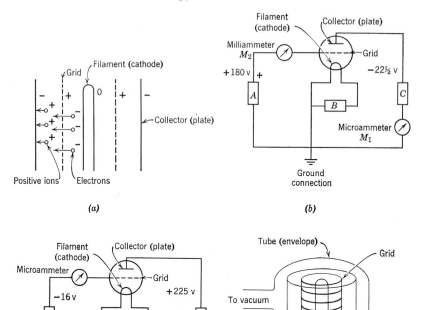

Fig. 6.9 (*a*) Elements of thermionic ionization gauge. (*b*) External collector.
(*c*) Internal collector. (*d*) Physical arrangement.

is usually a few milliamperes. This current has to be constant (well-regulated) in order to avoid changing pressure readings.

The sensitivity of a thermionic ionization gauge is given by the ion collector current for a given electron emission and given pressure change. It is usually stated in microamperes per milliampere per micron Hg. A fairly typical value is 20 microamperes per milliampere per micron (20 μA/mamp/μ). Many manufacturers quote values such as 100 μamp/μ/5-mamp electron emission. Sometimes the 5-mamp electron emission is left out but is listed elsewhere as the recommended electron emission. The physical arrangement of electrodes in a typical gauge tube is shown in Fig. 6.9*d*. The tube itself is usually around

3 in. in length and 1½ to 2 in. in outside diameter (OD) although various sizes are available commercially. Also, various tubulation diameters and lengths are available, although an OD of ½ in. and a length of 3 in. is fairly typical. Ionization gauges are usually made of a hard glass such as Pyrex. When attaching to a glass system it may be necessary to use a graded seal, depending on the kind of glass in the system. Attachment to a metal system is usually done by means of a vacuum connector such as a stuffing box or by means of a glass-metal seal.

The major problems occurring when using ionization gauges arise from: (1) outgassing, (2) filament failure, and (3) electrical leakage. Outgassing is not peculiar to this gauge but is quite often aggravated in a hot filament system by the operation of the filament at a high temperature. The effect of outgassing is to indicate a fictitiously high pressure for the system. Various methods are used to outgas ionization gauge tubes. Common methods are:

1. Electron bombardment of grid and plate.
2. Heating of grid, plate, and filament directly with an electrical current.
3. Induction heating.
4. Furnace heating.
5. Torching.

Method 1 is the most common. Usually the grid and plate are connected together (outside the tube), a voltage is applied between them and the filament, and the filament is heated. Electrons from the filament bombard the plate and grid and raise them to a high temperature. This method is usually accompanied by torching of the glass envelope. This is also true of method 2. Most manufacturers include a "degas" button, which simplifies the outgassing procedure. The greatest failure of ionization gauges is probably connected with the design of the hot filament. Some manufacturers use two or more filaments in the gauge tube, each of which can be used successively so as to extend the life of the tube. Materials commonly used for the filament are: pure tungsten, thoriated tungsten, oxide coated metals, and lanthanum boride, which operates at low temperature and does not react with hydrogen.

Tungsten filaments erode rapidly in the presence of water vapor. The use of an appropriate cold trap will minimize this problem. Tungsten filaments must be operated at very high temperatures to get enough electron emission. Exposure to atmospheric pressure will burn out these filaments when they are hot. Even at pressures of a few microns the life of the filament will be drastically reduced. The filament should not be turned on until the pressure is around 1 μ or less, as indicated by another gauge, e.g., the Pirani or thermocouple gauge. Thoriated and oxide coated filaments can be operated at lower temperatures than pure tungsten filaments. However, such filaments are subject to "poisoning" by various vapors, such as pump oils. Even with tungsten filaments, cracking of pump oil vapor can cause difficulties by changing the electron emission characteristics. The use of a cold trap between the gauge tube and the system (or in the system near the point of connection of the tube) will help avoid these problems. The filament burnout problem can be minimized by using thoria coated iridium cathodes. Iridium is not easily oxidized and can be safely exposed to air at operating temperatures for short periods of time. There is apparently no problem with activation of the thoria. An example of such a gauge is the RG-75, which is manufactured by the Veeco Vacuum Corporation, New Hyde Park, Long Island, New York.

Electrical leakage between the tube electrodes (inside and outside) can be a problem, particularly when the tube is operated near the lowest pressure which it will measure. Leakage problems occur when there is a sputtering and chemical transfer of filament materials to the walls. The most common method of reducing leakage is to shield the lead-in wires at the press in the glass envelope by means of a mechanical guard ring. Leakage is minimized also by spacing the lead-in wires through the envelope at widely different points rather than making all press seals in the same region. From the standpoint of leakage, some commercial gauge designs are considerably better than others. In using an ionization gauge, the outside surface, particularly the insulating surfaces on the connectors, should be periodically cleaned with a solvent such as acetone.

The upper pressure limit of an ionization gauge, with an electrode arrangement as in Fig. 6.9d, is a few microns, being determined by the point where the gauge breaks down into a glow discharge and the ion current becomes nearly independent of the pressure. The filament life is also shortened at high pressures, particularly if chemically active gases are present. The lower pressure limit under the best possible circumstances is between 10^{-7} and 10^{-8} mm Hg. Pressures below 10^{-8} mm Hg cannot be measured with this type of gauge because of various effects which will not be discussed here. A different type of thermionic ionization gauge which makes it possible to measure very low pressures (down to around 10^{-10} *or* 10^{-11} mm Hg) is the *Bayard-*

Alpert gauge. This is discussed below. A thermionic ionization gauge will give different pressure readings for different gases and vapors. Some idea of the change in pressure reading with the gas being measured can be gained by examining Table 6.1. These values were obtained by Consolidated Vacuum Corporation, using their VG-1A gauge tube and a particular vacuum setup. It will be noticed from Table 6.1 that the sensitivity for helium is much lower than for air. At low pressures much of the gas left in a gauge tube is helium and the pressure reading obtained will consequently be too low because of the reduced sensitivity. For accurate measurements a correction must be made. The connection between the region of the vacuum system where the pressure is desired and the gauge tube should be kept as short as possible since there will be a pressure rise in this connection near the gauge. Of course, the gauge tubulation itself will cause the pressure in the envelope to be higher than the pressure at the far end of the tubulation. The effect of the tubulation on pressure is discussed in Section 6.12. This discussion applies to any vacuum gauge tube that is connected to a system by a tubulation (round). It should be pointed out that outgassing will add to the pressure reading and the pumping action of the gauge will tend to lower the pressure. The pumping can be both ion pumping and gettering, the gettering action depending on the materials of construction. Thorough outgassing will minimize the first effect. The pumping action can be minimized by operating the gauge intermittently. The calibration of a thermionic ionization gauge (and other types of gauges) is discussed in Section 6.12. It is difficult

Table 6.1 Change of Sensitivity with Gas

Gas	Sensitivity (μamp/μ pressure/ 15-mamp electron current)
Helium	14
Hydrogen	16
Oxygen	85
Nitrogen	110
Air (dry)	100
Carbon monoxide	112
Carbon dioxide	120
Argon	162

The figures for hydrogen, oxygen, and carbon dioxide are influenced by the gettering action of the gauge tube.

to clean the inside of this type of gauge because of the fragile electrode structure and the manufacturer's recommendations should be followed.

The general arrangement of the electrodes in a Bayard-Alpert gauge, together with a simplified electrical circuit, is shown in Fig. 6.10. It will be noted that the ion collector is a small diameter wire. The modulator wire is included to get to lower pressures and is at ground potential. Such a modulator was not included in earlier designs (*Rev. Sci. Instr.*, **21**, 571, 1950). Some pertinent dimensions for this particular gauge are shown in Fig. 6.10. The filament is heated by alternating current. Typical operating conditions are: grid voltage = 105 v, electron emission current (filament) = 8 mamp, collector voltage = 25 v. With this type of gauge it is possible to measure pressures as low as about 10^{-12} mm Hg. However, to measure such low pressures it is nec-

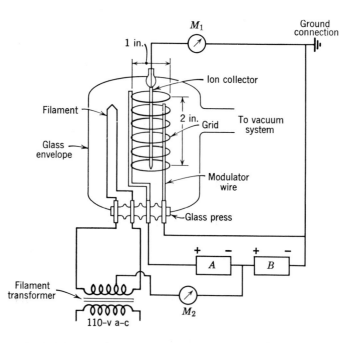

Fig. 6.10   Modified Bayard-Alpert gauge (P. A. Redhead, *Rev. Sci. Instr.,* **31,** 343, 1960).

Diameter of ion collector wire = 7 mils
Diameter of modulator wire (tungsten) = 10 mils
A = grid voltage supply
B = plate voltage supply
M_1 = ion current meter (pressure scale)
M_2 = electron emission meter

essary to thoroughly outgas the tube. The grid, collector, and modulator can be connected together outside the envelope and heated by electron bombardment. The envelope can be outgassed by torching or in a furnace, the latter method being preferred. The currents involved in measuring very low pressures are extremely small, being around 10^{-12} amp at a pressure of about 10^{-11} mm Hg. Sensitive, stable d-c amplifiers are available for measuring such currents. Some commercial Bayard-Alpert gauges have tubulations as large as 1 in. in diameter. Thermionic ionization gauges (Bayard-Alpert and regular) have linear pressure scales. Gauges based on principles different from those of this type of gauge have been developed to measure extremely low pressures (down to around 10^{-14} mm Hg) (see Section 6.11).

Several points regarding installation and operation of ionization gauges may be worth mentioning.

1. If accurate measurements at low pressures are required, the gauge should be calibrated. The calibrations should be carried out for whatever gases are to be used in the system.

2. Install the gauge tube in the vacuum system as near as possible to the point where pressure measurements are desired.

a. For pressures greater than about 10^{-6} mm Hg a vacuum connector with rubber gaskets can be used; to get to somewhat lower pressures, Teflon gaskets can be used.

b. For very low pressures (less than 10^{-8} mm Hg), the gauge tubes can be sealed directly to glass systems, using graded seals if necessary. Sealing to metal systems can be done using metal-glass seals, the metal being brazed to the vacuum system. Gaskets made of such materials as rubber or Teflon cannot be used in this application because of the necessity for baking.

3. Use a liquid nitrogen trap in the line connecting the gauge tube to the vacuum system or in the system near the tube.

4. Use as short and large diameter connecting lines as possible.

5. Outgas the tube. If the envelope is torched and a connector using some type of gasket is used, care must be taken to avoid damaging the gasket. The outgassing procedure should be repeated whenever the tube has been let down to air for a considerable period of time, depending on the accuracy desired.

6. Do not turn on the gauge until the pressure is down to 1 μ or less and the liquid nitrogen trap is filled. Do not allow the liquid nitrogen to get too low.

7. Turn off the gauge if there is a sudden rise in pressure. Automatic controls can be devised for this, making use of a Pirani or thermocouple gauge.

Thermionic ionization gauges are the most commonly used type of gauge for measuring low pressures, offering such advantages as a linear pressure scale, being fairly readily outgassed, and having the capability of measuring quite low pressures. However, they do suffer from the following disadvantages:

a. A considerable amount of auxiliary electronic equipment is necessary for precise operation of the gauge.

b. Filament emission may change radically in the presence of certain gases.

c. Sensitivity varies for different gases and vapors. This is an advantage in leak hunting.

d. The filament is susceptible to burning out if exposed to air while hot (except the thoria-iridium type).

e. The hot filament decomposes certain gases, particularly hydrocarbons.

f. The gauge acts as a pump. This action must be thoroughly understood, particularly in small systems.

6.7 Cold Cathode Ionization Gauge

This type of gauge is sometimes called a Penning or Philips gauge. It operates on the same basic principle as the thermionic type, i.e., positive ions are produced by electrons and the current due to these ions gives a measure of the pressure. Actually the current is the sum of positive ions arriving at the collector and electrons leaving it. The difference between this kind of gauge and the thermionic type is that the electrons are not produced by a hot filament but originate at the cathodes. Also, a magnetic field is used. The manner in which the gauge operates can be seen from Fig. 6.11*a*. The anode, *A*, is between two cathodes, *C*, and a magnetic field is produced by the horseshoe magnet, *M*. The magnetic field is perpendicular to the planes of the cathodes and the anode. A high voltage, usually between 2000 and 4000 v, is applied between *A* and *C*, with *A* positive and *C* negative. The meter, M_1 (microammeter), reads the current between cathode and anode and is calibrated in pressure units. Referring to Fig. 6.11*b*, electrons leaving the cathodes will head for the anode but the magnetic field has the property of forcing them to move in circles. Consequently, the electrons will move in spiral paths between the cathodes except for those that started off directly for the anode ring. Many electrons will not reach the cathodes because of inelastic collisions with gas molecules. They will make many passages between the cathodes

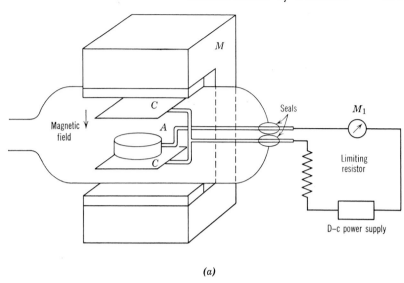

(a)

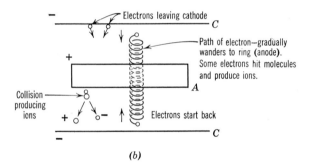

(b)

Fig. 6.11 (a) Physical setup of cold cathode gauge. (b) Motions of electrons.

before finally working their way over to the anode ring, where they are finally collected. While making these passages, the electrons will ionize gas molecules. The positive ions will be collected by the cathodes while the electrons formed in the ionization process will behave like the electrons that leave the cathodes. The long paths of the electrons result in many more ions being formed than if they were able to go directly from the cathodes to the anode. This gives larger currents and simplifies the whole problem of measuring electric currents. As a matter of fact, with this type of gauge an amplifier is not required.

The most common commercial cold discharge gauges do not use separate cathode plates, as seen in Fig. 6.11a. Rather, the trend has been to all-metal construction with the inside walls of the tube acting as the cathode. The anode is usually in the shape of a ring as shown in Fig. 6.11a and may be round, square, or rectangular. In some cases a wire loop is used, sufficiently heavy wire being employed to prevent vibration and sagging. The tube body is at ground potential, so the only high voltage present is on the anode connector. A high strength alloy magnet such as Alnico is used, being designed to be as compact as possible. Usually the magnet and gauge tube are made as a single unit. The materials of construction do not appear to be critical. Stainless steel, aluminum, and nickel-plated copper are used in commercial gauges for the tube body (cathode). Theoretically, the cathode material should not sputter readily so that it will not produce a conducting layer on the insulator through which the anode connection is brought in. Usually some form of shield, often Teflon, is used to protect the insulator from sputtering. A-c voltage can be used but the sensitivity is then less than with d-c voltage. Consequently, commercial gauges use a d-c power supply. Instead of using a magnetic field which is perpendicular to the plane of the anode, sometimes the field is made parallel to the axis of the tube. This can be done readily by using a coil which fits over the tube and a cylindrical anode with its axis parallel to the magnetic field direction. Examples of gauges of this type are those manufactured by Elliott Brothers (London) Ltd., Borehamwood, Hertfordshire, England.

The general form of calibration curve for a cold cathode discharge gauge is shown in Fig. 6.12. It will be noted that the curve is linear up into the higher pressure end, in this case about 10 μ (10^{-2} mm Hg).

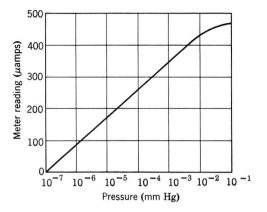

Fig. 6.12 Calibration curve for a cold cathode discharge gauge (air).

The usual range of this type of gauge is about 10^{-2} to 10^{-5} mm Hg, although some commercial gauges claim ranges from as high as 10^{-1} mm Hg to as low as 10^{-7} mm Hg. As a general rule, it is sometimes difficult to start a discharge in these tubes at low pressures. Also, sometimes the discharge becomes unstable, causing unpredictable jumps in the calibration curve. This type of gauge does not have the accuracy of a thermionic gauge (when read properly). The sensitivity of a cold cathode gauge depends on the nature of the gases and vapors in the system, just as in the case of a thermionic ionization gauge. As a matter of fact, the dependence on gas or vapor is much the same as for the thermionic gauge (see Table 6.1). Table 6.1 will give some idea of the effect of a particular gas on the reading of a cold cathode gauge. However, this type of gauge is generally used to give some idea of the overall pressure in a system without attempting to get really accurate values.

A cold cathode gauge will act as a pump through both ion and getter pumping. Because of the high voltages used, the pumping speed of this type of gauge is perhaps ten to a hundred times greater than that of a thermionic gauge. Consequently, the size of connection between gauge tube and vacuum system is very important. A connecting line with high resistance to gas flow will result in too low a pressure reading. Usually outgassing effects are not too important since the gas discharge tends to drive gases out of the tube structure, particularly in the higher pressure range of the gauge, where the currents are higher. However, contaminants such as oil vapors can cause trouble since they can be "cracked" and will cause erroneous readings, as well as possible electrical leakage paths. The best way to avoid errors due to the pumping action is to use as large a connecting line as possible. Many commercial gauge tubes are provided with tubulations of 1 in. diameter. These tubulations can be brazed directly to the vacuum system or can be connected with a vacuum cement or wax. Some manufacturers provide vacuum adaptors for this purpose. Some gauge tubes are provided with flanges for direct attachment to the vacuum system (bare or "nude" gauge tubes). Also, metal gaskets have been used to avoid vapors from such materials as rubber.

The advantages and disadvantages of cold cathode discharge gauges can be summarized as follows:

ADVANTAGES

1. Simple electronic circuitry.
2. Rugged construction.

3. Operation in a pressure range covering parts of the ranges covered by thermionic ionization gauges and Pirani or thermocouple gauges.

4. Exposure to atmosphere causes no damage.

5. Highly resistant to mechanical shock.

6. Long life.

7. Relatively easily cleaned. This is particularly true of gauge tubes that can be disassembled. Solvents and abrasive agents such as fine emery cloth are used. It is important to keep iron filings out of the tube.

8. The gauge reads total pressure.

DISADVANTAGES

1. The accuracy is not as great as that achieved with thermionic gauges.

2. A magnet is required. With the availability of high strength, compact alloy magnets, this presents no great problem except that precautions must be taken to keep magnetic particles out of the tube.

3. A high voltage is used, which can be hazardous. Exercise of reasonable care will avoid injury. In most commercial gauges, the high voltage connector is well insulated and the gauge tube is at ground potential. Parts of the high voltage power supply (and gauge tube) should not be touched for a few minutes after turning off the power supply because of the electrical charge on the condenser. Shorting this condenser with a screwdriver (insulated handle) will help.

4. The gauge tube acts as a higher speed pump than a thermionic gauge.

Certain precautions in installing and operating these gauges should be taken, which will vary somewhat from gauge to gauge. Most commercial gauges operate off the 115-v a-c lines but some can be obtained that operate at 220 v a-c.

LOCATION OF COLD CATHODE DISCHARGE TUBE IN
VACUUM SYSTEM

1. The gauge tube will give readings of all vapors present in the vacuum system. If a correct dry air reading is required, use a liquid nitrogen trap between the tube and the system (or in the system adjacent to the tube connection).

2. Position the tube so small steel chips, steel wool, etc., cannot fall into it or be easily attracted by the permanent magnet (or coil). A

horizontal position is usually most practical. Whenever possible, the tube should be installed so that its tubulation (when used) extends into the vacuum system.

3. Organic material will be broken down in the tube. Precautions to take are:

a. Position the tube so that it does not face any refrigerated traps. Loss of refrigerant allows the trap to become warmer and trapped material can evaporate. This material can reduce the insulation in the tube.

b. Do not leave the gauge tube "on" for extended periods of time if the system pressure is above a few microns. If this rule is not observed the tube will contaminate very rapidly. On the other hand, extremely intermittent operation will minimize the effect of the pumping action.

c. Never operate the tube during roughing-down periods or when the traps are warming up.

CONNECTING TUBE TO VACUUM SYSTEM

The principal methods for connecting a tube (unless flanged) are:

1. Solder or braze.
2. Use a vacuum wax such as Apiezon W or Dekhotinsky Cement.
3. Use a vacuum adaptor, as appropriate. Rubber tubing is not recommended.

MAINTENANCE AND SERVICING

Follow the instructions of the manufacturer.

6.8 Gauges Using Radioactive Materials

Both the thermionic and cold cathode gauges operate on the principle of measuring currents due to ions produced by electrons. The difference between these gauges is that in the former type the electrons are obtained from a heated filament while in the latter type they arise from a cold cathode. With enough energy any charged particle can produce ions in a gas. The number of ions produced will be proportional to the gas pressure for a given particle energy and, therefore, can be used to measure pressure. Radioactive materials send out high energy charged particles, alpha particles, which are helium ions (positive) or beta particles (electrons). An early form of gauge using a radioactive material to produce ions was developed by National Research Corporation, Newton, Massachusetts. This gauge is called the Alphatron and makes use of a sealed radium source which emits alpha particles. Consequently, no electric circuitry is required to produce and speed up the charged particles used to ionize the gas. However, sensitive amplifiers are needed to amplify the ion current produced so as to get a reading on a meter. The rapid advances being made in electronics do not make this problem particularly difficult although electronic difficulties can still present problems in maintenance and service. The pressure region from atmospheric (or higher) to about 10^{-4} mm Hg (or less) can be covered by this gauge in several ranges.

Because of the lower pressure limit, the Alphatron gauge does not in general have to be outgassed. Also, there is no problem with pumping action. It responds differently for various gases and vapors, behaving

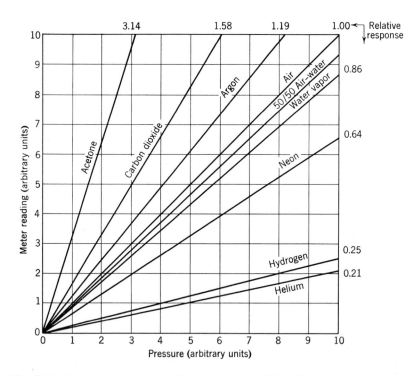

Fig. 6.13 Response curves for Alphatron gauge (NRC Equip. Corp., Newton, Mass., Model 520).

similarly to the thermionic and cold cathode ionization gauges. Calibration curves for several gases and vapors are shown in Fig. 6.13. Note how linear the scales are, i.e., how the meter reading increases uniformly with pressure. Some of the advantages and disadvantages of an Alphatron gauge for the pressure range covered are as follows:

ADVANTAGES

1. Exposure to atmospheric pressure will not harm it.
2. There is no filament to burn out or sag.
3. The response is linear within certain working limits.
4. The gauge is easily cleaned.
5. The use of a long life radioactive material ensures an unchanging calibration.
6. The gauge can be used as a leak detector and is particularly useful for large leaks where the pressure is high.
7. The gauge reads total pressure.

DISADVANTAGES

The principal disadvantages stem from the need for measuring very small electrical currents. These are:

1. The very sensitive amplifier system needed may develop instabilities.
2. The use of a radioactive material may introduce some hazard to personnel. Relatively simple precautions can be taken, however.
3. Relatively high cost.
4. Presently available models cannot be baked.

Various attempts have been made to develop ionization gauges that use some radioactive material other than radium, on the basis that radium can be more hazardous than some other radioactive materials. The principle of operation is basically the same as in the case of the Alphatron. A material that has been studied in some detail is tritium. This radioactive material emits beta particles (electrons). Compared to radium, tritium presents less health hazard, and a higher sensitivity can be obtained because a high intensity source can be employed. At normal temperatures tritium is a gas; it is therefore usually combined with titanium to form titanium tritide (a solid). For certain purposes, gauges using radioactive materials may well find more general use.

6.9   Viscosity- and Radiometer-Type Gauges

The *viscosity type of gauge* is only used for special purposes. Its operation depends on the fact that when the mean free path of a gas becomes large compared to the dimensions of the containing vessel, the viscosity becomes proportional to the pressure of the gas. A change in the viscosity of a gas can then be used as a measure of the pressure. In a general sense viscosity refers to the drag of a gas on the motion of an object. The simplest form of gauge based on this principle consists of a quartz fiber or strip that is set in motion. The time for the total movement (amplitude) of the fiber or strip to drop to one-half its value at the initial time of measurement is taken as a measure of the pressure. Figure 6.14*a* shows one form of this type of gauge, sometimes called a "decrement" form of gauge. Here, a thin glass disk *D* is suspended by means of a thin wire or quartz fiber *F*. The movement of the wire or fiber can be observed by means of the mirror *M*. The wire (or fiber) is set in vibration by bringing a magnet near a slug of soft iron *S* (magnet outside gauge tube). The oscillations produced by this initial pulse are then observed by a galvanometer system. Gauges of this type are often made with just a quartz fiber with no glass disk or iron slug. In such cases the fiber is set in vibration by tapping the gauge tube. The decrease in the swing (amplitude) of the fiber is again observed through the use of a mirror. The quartz fiber is typically a few centimeters long (from 5 to 10 cm) with a diameter of a few tenths or hundredths of a millimeter (0.05 to 0.2 mm). The time for the amplitude of the fiber to decrease to one-half its initial value will depend on its diameter and length, as well as the pressure. For a fiber 9 cm long with a diameter of 0.2 mm, these times will vary from about 25 sec at a pressure of 80 μ to over 700 sec at a pressure of 2×10^{-5} mm Hg. Clearly, taking readings with this type of gauge is time-consuming, particularly at lower pressure values. Various forms of the decrement gauge will cover the pressure range of 0.1 to 10^{-5} mm Hg.

Another form of viscosity gauge is shown in Fig. 6.14*b*. This is often called a molecular gauge. Disk *A* is rotated by means of a rotating magnetic field with a speed up to 10,000 rpm. The motor doing this is shown by the coils and armature, with the armature inside the gauge tube and the coils outside. *F* can be a thin wire or quartz fiber. The deflection of the wire due to the drag of the gas between *A* and *B* is measured by means of the mirror *M*. The higher the deflection,

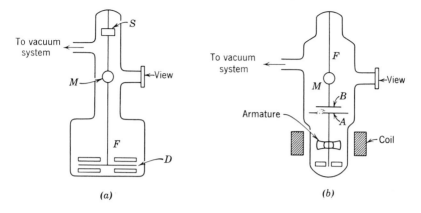

Fig. 6.14 (*a*) Decrement form of viscosity gauge. (*b*) Molecular form of viscosity gauge.

the greater will be the turning of the wire (or fiber). This type of gauge can be designed to be baked out at around 300°C and will measure pressures in the range of 10^{-3} to 10^{-7} mm Hg. A further modification of the quartz fiber gauge has resulted in the development of the quartz membrane gauge. This gauge consists of a quartz fiber holding a relatively large membrane in suspension. The membrane is set into vibration and the amplitude of the oscillation is observed. A considerable increase in sensitivity is obtained through the additional area.

Several advantages can be cited for the various types of viscosity gauges.

ADVANTAGES

1. They can be made with no exposed metal parts so they can be used with corrosive gases and vapors.
2. There are no electrical discharges or electron beams to cause gaseous disintegration.
3. They can be made with small volumes.

However, these gauges also have some fairly evident disadvantages.

DISADVANTAGES

1. Trained observers are necessary.
2. These gauges are fragile. The consequences are:
a. They must be mounted on a vibration-free support.

b. They must be connected to the vacuum system by a flexible connection.

c. They must be let down to air slowly to avoid damage to internal parts.

3. They must be calibrated for the gas with which they will be used.

4. They are not adaptable to remote reading.

5. Taking readings is time-consuming.

The *radiometer type of gauge* works on the same basic principle as the radiometers which can be observed in optometrists' shops. The principle involves using a heated surface against which molecules collide and leave with increased velocities. Parts of a vane are exposed to these molecules while other parts are exposed to molecules at a lower temperature, normally room temperature. This results in a rotation of the vane, the amount depending on the pressure of the gas. The rotation can be observed with a mirror, light, and scale. Gauges depending on this basic principle are often called *Knudsen gauges*. The elements of such a gauge are shown in Fig. 6.15. Two fixed heaters, *H* and *H′* are located on opposite sides of the gauge and near corre-

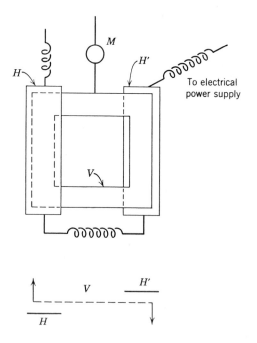

Fig. 6.15 Elements of radiometer (Knudsen) gauge.

sponding edges of a square vane V. Molecules hitting these parts of the vane will be at a higher temperature than molecules hitting the opposite edges and therefore will exert a greater force and cause the vane to rotate. Mirror M, with a suitable optical system, is used to measure the rotation of the vane. The operation of the gauge depends on the mean free paths of the molecules being large compared to the distance between heater and vane. Commercial instruments assume various forms, often including permanent magnets to damp the motion of the vane. In any case, to achieve high sensitivity the vane must be very lightweight. Also, the suspension for the vane must be of very small diameter. Often a quartz fiber is used. The advantages of viscosity gauges also apply here, except perhaps for the matter of small volume. However, by far the most important advantages of the radiometer gauge are:

1. It measures total pressure and its response does not depend on the nature of the gases present (or vapors, if not condensed).

2. It can be made an "absolute" gauge, i.e., the pressure can be calculated from physical quantities (dimensions, mass, angle). Therefore, it can be used to calibrate various "relative" gauges, such as ionization gauges, instead of using a McLeod gauge.

3. Models can be made to cover a wide range of pressure, say 10^{-2} to 10^{-9} mm Hg. This makes it possible to calibrate ionization gauges well below the pressures achievable with McLeod gauges (usually around 10^{-5} mm Hg).

The disadvantages of this type of gauge are much the same as for the viscosity gauge except for the matter of its independence of the nature of the gases. These disadvantages make it generally unsuitable for industrial-type vacuum systems. However, it is finding increased use in various specialized applications.

6.10   Use of Discharge Tube to Measure Pressure

This method of measuring pressure is not widely used. Generally speaking it is used to get a fairly rough indication of vacuum conditions. Besides giving some indication of the pressure, it will give some idea of the gases and vapors present in the system and, therefore, can be used in leak hunting.

The usual form of discharge tube consists of a cylindrical glass tube containing two electrodes which can be attached to the system (sealed

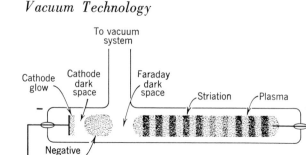

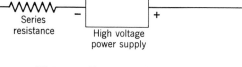

Fig. 6.16   Features of a discharge.

directly or with a graded seal to glass systems and with a glass-metal seal to metal systems). The tube is usually ½ in. or more in diameter (inside) and several inches long. The electrodes are often simple disks (of smaller diameter than the inside diameter of the tube) though in some cases one of them is simply a metal rod. A metal that doesn't sputter readily is usually used, such as molybdenum, nickel, or chromium. Sometimes aluminum is used. The general nature of a simple discharge tube is shown in Fig. 6.16. The high voltage supply shown provides direct current. As the pressure is reduced, eventually a point is reached where an electrical discharge occurs in the gas in the tube. This discharge will first occur at a few mm Hg, the exact value depending on the dimensions of the tube and on the voltage applied. The voltage used is usually several kilovolts. The discharge starts as a narrow streamer between the electrodes but as the pressure is reduced it spreads out to fill most of the tube. The overall color will depend on the gas in the tube, being reddish with air. Distinct changes in the appearance of the discharge occur as the pressure is reduced. The regions that show up as the pressure is reduced are indicated in Fig. 6.16. Next to the negative electrode (cathode) there is a bright glow called the cathode glow. This is followed by a dark region called the Crooke's (or cathode) dark space. Next comes a bright region called the negative glow, followed by a dark region called the Faraday dark space. Finally there is a general bright region extending to the positive electrode (anode) called the positive column or plasma. Alternate bright and dark parts, called striations, appear in this part of the discharge. The spacings of these striations will depend on the pressure, the voltage being used, and the geometry of the tube. At higher pressures, say 0.5 mm Hg or so, the positive column fills most of the tube and the Crooke's dark space is quite narrow. As the pressure is reduced, the Crooke's dark space becomes more extensive, finally filling the tube entirely. After this the tube begins to fluoresce and there is a "black discharge." The color of the fluorescence will depend on the kind of glass used in the tube. With lead glass, the fluorescence is blue and with Pyrex or soda glass, it is yellow-green.

With dry air, a tube of around ½ in. ID and a few inches long (say 4 in.), and a voltage of several kilovolts, the changes occurring in the discharge are indicated in Table 6.2. For a particular tube and voltage, the width of the Crooke's dark space can be calibrated against pressure. Also, the appearance of a black discharge can be used to indicate a particular pressure (given tube and voltage). With a variable voltage supply, the voltage can be calibrated against pressure using the black discharge as the indication, since a higher voltage is needed for a black discharge at lower pressures. Some common discharge colors (determined by the positive column) are: air, red or pink; helium, violet-red; carbon dioxide, white; water vapor, faint (transparent) blue; mercury vapor, greenish. Colors for additional gases and vapors are shown in Appendix D. In practice, an induction

Table 6.2   Change in Appearance of Discharge

Appearance	Pressure (mm Hg)
General glow discharge	7–10
Beginning of striations, closely spaced	1–1.5
Striations 1 cm apart	0.5
Crooke's dark space 2.5 mm long	0.55
Crooke's dark space 5 mm long	0.27
Crooke's dark space 10 mm long	0.12
Crooke's dark space 15 mm long	0.07
Crooke's dark space 20 mm long	0.05
Crooke's dark space 30 mm long	0.03
Fluorescence	0.01–0.001
No visible discharge ("black")	less than 0.001

Tube with ID about ½ in. and length 4 in.
Dry air.
Spark coil giving spark ⅜–½-in. in air.

(spark) coil is usually used as the high voltage supply because of simplicity and low cost. Instead of incorporating a discharge tube in a system, it is possible to bring one connection from a spark coil, such as a Tesla coil, to a part of a glass system and get a discharge which can be used to get an indication of pressure. This procedure is mostly used for leak hunting since various solvents such as acetone or ether will change the appearance of the discharge.

6.11   Gauges to Measure Very Low Pressures

The lowest pressure measurable with a Bayard-Alpert gauge is limited by a residual current to the ion collector which is independent of pressure. The residual current is the result of photoemission from the ion collector caused by (a) soft x-rays from the electron bombarded grid, and (b) ultraviolet radiation from the hot filament. This residual current normally limits pressure measurements to around 10^{-10} mm Hg. To measure lower values it is necessary to eliminate this current or to know it accurately.

Developments in connection with space exploration, various electronic devices (transistors, tantalum capacitors), solid state physics, etc., have led to an increasing interest in the measurement of very low pressures. Devices that have been developed in recent years include:

1. Modified Bayard-Alpert gauges.
2. Cold cathode inverted magnetron gauge.
3. Hot cathode magnetron ionization gauge.
4. Mass spectrometers.

Modulation of a Bayard-Alpert gauge has been mentioned in Section 6.6. Modifications suggested by Nottingham (*Vac. Symp. Trans.,* 1954, p. 76) and Alpert (*J. Appl. Phys.,* **24**, 860, 1953) have led to improved sensitivity. The cylindrical grid is closed at top and bottom and a second grid, acting as a screen grid, is installed around all the electrodes. The closed grid prevents ions from escaping to the negatively charged glass wall. This results in longer paths for the ions before they are collected. The screen grid shields the gauge from the wall charges. Operating this grid at negative voltage causes the electrons to oscillate several times through the positive grid before being collected. This increases the path length, which results in more ionization and, therefore, greater sensitivity. With such modifications, it is possible to measure pressures down to about 10^{-12} mm Hg. A commercial version of such a gauge is manufactured by the NRC

Equipment Corporation, Newton, Massachusetts. It is sometimes referred to as the Nottingham gauge.

The cold cathode inverted magnetron gauge is basically a modified form of the common cold cathode (Philips) gauge. The basic modification is the use of crossed electric and magnetic fields to increase the path lengths of the electrons so as to get more ionization at low pressures (Hobson and Redhead, *Can. J. Phys.*, **36**, 271, 1958). The nature of such a gauge is shown in Fig. 6.17*a*. The geometry provides efficient electron trapping in the discharge region, and the auxiliary cathode provides the initial field emission for starting and allows the positive ion current to be measured independently of the field emission current. This cathode also acts as an electrostatic shield for the ion collector. The two short tubular shields, which project 2 mm into the ion collector from the auxiliary cathode, protect the end plates of the ion collector from the high electric fields and provide the field emission which initiates the discharge. Hobson and Redhead's gauge operated with a magnetic field of 2060 oersteds and a potential of 6 kv on the anode. Pressures between 10^{-3} and 10^{-12} mm Hg were measured with it, and the lower limit may be well below 10^{-12} mm Hg. The pressure-current curve is linear when plotted on log-log graph paper. Hobson and Redhead noticed that there was a time lag between application of anode voltage and the initiation of the discharge at lower pressures (around 10 min near 10^{-12} mm Hg). A commercial version of this gauge (sometimes called the Redhead gauge) is manufactured by NRC Equipment Corporation, Newton, Massachusetts.

The hot cathode magnetron ionization gauge is a form of thermionic ionization gauge, modified so that the electrons travel in longer paths before they are collected by the positive grid or anode (Lafferty, *J. Appl. Phys.*, **32**, 424, 1961; U.S. Patent 2,884,550, filed Oct. 17, 1957). The arrangement used by Lafferty is shown in Fig. 6.17*b*. A cylindrical magnetron is operated with a magnetic field greater than cut-off. Two end plates, at a negative potential relative to the cathode, prevent the escape of electrons. One or both of these plates may be used to collect the positive ion current generated in the magnetron. Electrons emitted by the tungsten filament spiral around the axial magnetic field in the region between the negative end plates. If the magnetic field is sufficiently high, most of the electrons fail to reach the anode. Some of the electrons make many orbits around the cathode before being collected. This results in a higher probability of ionization. Lafferty found that stable operation was possible when the gauge was operated at very low electron emission levels. Theoretically, pressures almost as low as 10^{-14} mm Hg should be measurable with this type of

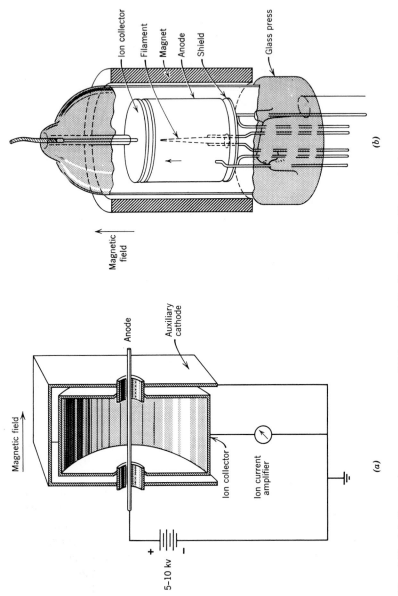

Fig. 6.17 (*a*) Inverted magnetron gauge (Hobson and Redhead). (*b*) Hot cathode magnetron ionization gauge (Lafferty). (Adapted from S. Dushman, *Scientific Foundations of Vacuum Technique*, 2nd ed., J. M. Lafferty, editor, John Wiley and Sons, New York, 1962.)

gauge. In practice, its ability to read low pressures is limited by the sensitivity of the external circuit used to measure the ion current.

The Houston ionization gauge (*Bull. Am. Phys. Soc.*, **11**, 1, 301, 1956) is an ultra-high vacuum ionization gauge used to measure pressures less than 10^{-8} mm Hg. It is similar to the Philips (cold cathode) gauge except that a hot cathode is used. The gauge consists of two end plates at a negative potential with respect to a hot filament. An anode cylinder, between the end plates, is at a positive potential of several hundred volts with respect to the filament. The gauge is operated in a magnetic field (parallel to the axis of the anode). Typical operating parameters are: emission current $= 10^{-8}$ amp, magnetic field $= 1700$ oersteds, potential on anode $= 1$ kv, sensitivity $= 10$ amp/mm Hg. Pressures down to about 10^{-12} mm Hg have been measured with this gauge. Theoretically, the lowest pressure measurable is about 10^{-15} mm Hg.

Mass spectrometers are used to obtain information about the residual gases in an ultra-high vacuum system. They will measure the partial pressures of the various gases present. The Omegatron is a mass spectrometer that operates on the principle of a cyclotron (see Chapter 15). It is used fairly extensively in analyzing residual gases up to a mass of around 40. Mass spectrometers of the magnetic deflection type have also been used to measure residual gases. Reynolds (*Rev. Sci. Instr.*, **27**, 928, 1956) has described a high sensitivity mass spectrometer using a nine-stage electron multiplier with magnesium-silver dynodes to give a gain in the range of 10^3 to 10^6 electrons per ion. Partial gas pressures of the order of 10^{-12} mm Hg were measured. A total pressure of 5×10^{-10} mm Hg was achieved during operation of the spectrometer after rigorous bake-outs. Davis and Vanderslice (*Vac. Symp. Trans.*, 1960, p. 417) have described a small, portable magnetic deflection type of mass spectrometer with a secondary emission electron multiplier. It can be sealed directly to the tube or system being investigated. Partial pressures as low as 10^{-13} mm Hg can be measured with this type of spectrometer. Cooling the multiplier in liquid nitrogen makes it possible to measure pressures of the order of 10^{-15} mm Hg. Time-of-flight mass spectrometers and the Paul quadrupole spectrometer have also come into use recently.

6.12   Calibration and Ranges of Vacuum Gauges

No standard calibration procedures have been adopted by any countries at the time of this writing. However, committees in various countries, including the American Vacuum Society in the United States, are actively working on this matter. In addition, the International Union for Vacuum Sciences, Techniques, and Applications (IUVSTA) has a standards committee.

The only absolute gauges generally available are the McLeod and Knudsen (or radiometer). These gauges can be calibrated from their dimensions. Once a gauge has been calibrated against an absolute gauge (absolute calibration), it can then be used to calibrate other gauges (relative calibration). Certain general points regarding calibration should be noted:

1. Connect the gauge being calibrated and the reference gauge to the same point in the vacuum line.

2. Make the connections to the gauges as identical as possible (for equal pressure drops).

3. Use identical liquid nitrogen traps in the connecting lines to the gauges. These traps are not needed when using a condensable gas.

One general method of calibrating against a McLeod gauge involves the use of a *constant leak* (small). The gauge being calibrated will indicate the pressure and when it is sufficiently high, the McLeod gauge can be used to get a check point. It can be assumed that the rate of pressure rise is constant. The *pipette* method, which involves the use of known volumes, is sometimes used. In making accurate calibration measurements, the system should be well-evacuated and degassed (by baking).

No attempt will be made here to cover the specific steps involved in calibrating the various types of gauges. However, details of a procedure for calibrating a thermionic ionization gauge are included. Many of the steps involved are applicable to other types of gauges. The procedures outlined here are intended to provide as high a degree of accuracy as possible. Figure 6.18 shows the type of setup that can be used. It should be noted that the conditions under which the gauge is operated are just as important as the actual calibrating procedure. This means that the operating voltages and emission current must be measured accurately and the values obtained apply only to the pressure calibration data obtained. Various test gases can be used and they should all be of high purity. A standard gas that is often used is pre-purified nitrogen (less than 0.003% oxygen). The ambient temperature should be 20°C $\pm$ 2°. The specific steps involved in preparing the test system (Fig. 6.18) are:

1. Place McLeod gauge under vacuum.

2. Valves A and B are fully open and valve C is lightly closed.

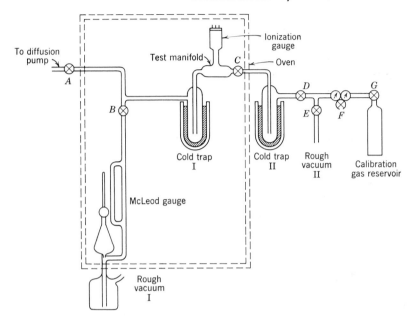

Fig. 6.18   Recommended gauge calibration setup.

Start rough pump II to evacuate cold trap II and tank regulator *F*. Tank valve *G* should be shut tightly.

3. All parts of the test system are now under vacuum. The calibration gas admittance system is now under rough vacuum. The McLeod gauge, test gauge, and cold trap I are under high vacuum (a mercury diffusion pump is recommended).

4. Degas the test gauge for a few minutes to drive off water vapor, etc.

5. Bake the test cabinet (dotted lines) for about 16 hr at 350°C min.

6. Cool oven to 150°C before removing access door.

7. Install the Dewar on cold trap II and partially fill when the trap is warm to the touch.

8. Rigorously degas the test gauge.

9. After degassing, fill cold trap I. Allow cabinet to cool to ambient temperature (20°C).

10. The observed (not calibrated) vacuum in the test gauge should be in the middle or low 10^{-8} mm Hg range.

11. Close valve *C* tightly after cooling.

12. Install the Dewar for cold trap II and fill.

13. Close valve E and open valves F and G to fill gas inlet manifold to about 1 psig.

14. Slowly open valve C until the highest pressure for which a calibration is desired is reached inside the test gauge. Close valve A prior to admitting test gas.

15. Close valve C. The entire test system inside the dotted lines is now at maximum calibration pressure (microns) and pressure should be very stable if the ion gauge under test is not pumping (low emission current).

In general, a McLeod gauge cannot be used to calibrate below about 1 μ with any high degree of accuracy. However, it can be used down to about 10^{-5} mm Hg if the resulting loss of accuracy can be tolerated. For highest accuracy, a cathetometer readable to within 0.05 mm or better should be used. Some precautions which should be taken during calibration are:

1. Achieve high vacuum in the test gauge before admitting the calibration gas.

2. Avoid ion pumping by the gauge. Use low emission values except where required for specific tests. Provide the test control with a switch that will remove grid voltage without changing any other test conditions. Turn on this switch only at the moment the McLeod gauge pressure is to be read (mercury reaches cut-off).

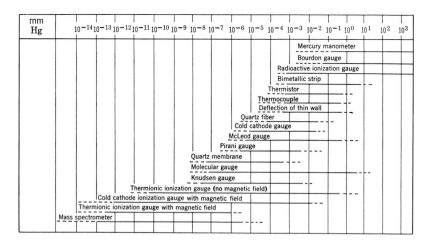

Fig. 6.19 Pressure ranges for various gauges. (Adapted from S. Dushman, *Scientific Foundations of Vacuum Technique,* 2nd ed., J. M. Lafferty, editor, John Wiley and Sons, New York, 1962.)

3. Cold trap temperatures must be varied if calibration is desired for gases other than the noncondensables (helium, nitrogen, dry air, etc.). Cold trap levels must be maintained to avoid loss of trapped material.

4. Accurate read-out of ion currents is highly desirable. Digital voltmeter readings of the control amplifier output or self-balancing recording potentiometers are useful.

The test setup of Fig. 6.18 can be modified to make relative calibrations by replacing the McLeod gauge with a calibrated gauge having the right pressure range.

Some information regarding the pressures measured by various types of gauges is contained in previous sections of this chapter. However, it is often convenient to show the behavior of gauges in a simple chart. Figure 6.19 shows the pressure ranges for several types of gauges. The solid lines apply to the usual type of gauge while the dotted lines apply to various specialized forms.

SEVEN

Measurement of Pumping Speed

7.1   Why Measure Pumping Speed?

Although manufacturers provide speed-pressure curves for their pumps, the exact circumstances under which these curves are obtained are not always specified. Information about the following factors needs to be known. The pump speed is theoretically proportional to the square root of the absolute temperature—did the manufacturer take this into consideration? Also, as will be seen, the physical setup the manufacturer used is very important. In addition, the vacuum gauge may give different readings for different gases. Many manufacturers use a gauge calibration for air (or dry air), thus getting pumping speeds for this mixture of gases only. In practice it may be necessary to measure speeds for other gases. Also, the performance of a pump may change with time due to such factors as the condition of the oil or heater units, the amount of cooling water, etc. In any case whenever a vacuum system is not performing properly, it is important to know whether or not the pumps are operating correctly.

As was indicated in Chapter 2, the pumping speed is defined in general as follows:

$$\text{Pumping speed} = \frac{\text{throughput}}{\text{pressure}}$$

Any real pump will reach its ultimate or base pressure after a sufficient length of time. The value of this pressure will depend on the nature of the vacuum system, i.e., on leaks through the walls of the system and sources of gas or vapor inside the system. With commercial pumps an artificial leak is usually used in order to measure the pumping speed and the only throughput considered is that which passes through this leak. According to the general definition of pumping speed, the pumping speed is zero when the ultimate pressure is reached. Naturally, the test setup that is used should be leak tight.

7.2 The Metered-Leak Method

This method is sometimes called the constant pressure method. Basically it involves admitting air at atmospheric pressure into a test chamber above the pump being tested and measuring the pressure in the chamber. The rate at which air is admitted must be measured by some type of metering device (flowmeter). Knowing the rate of flow and the resulting pressure, the pumping speed is given by

$$\text{Pumping speed} = \frac{\text{atmospheric pressure} \times \text{flow rate}}{\text{pressure in chamber}} \qquad (7.1)$$

If the atmospheric pressure and the pressure in the chamber are measured in the same units, then the pumping speed will have the same units as the flow rate. It must be kept in mind that the atmospheric pressure will depend on the locality and on the weather conditions on any particular day. For accurate results the barometric pressure should be used and the temperature maintained within $\pm 1°C$. Often the flowmeter is calibrated so as to indicate the number of cubic centimeters of air at atmospheric pressure passing into the test chamber. By measuring the time required for a certain number of cubic centimeters to flow into the chamber, the flow rate is obtained. The pumping speed can then be expressed as

$$\text{Pumping speed} = \frac{\text{atmospheric pressure} \times 10^{-3}}{\text{pressure in chamber} \times \text{time for 1 cc to flow (sec)}}$$

$$(7.2)$$

Both pressures must be measured in the same units, and the pumping speed will be in liters per second.

EXAMPLE

In checking the pumping speed of a 4 in. oil diffusion pump, it is observed that when 3.2 cc of air at an atmospheric pressure of 750 mm Hg are admitted to the system in 100 sec, the pressure becomes 10^{-4} mm Hg.

The flow rate is 3.2/100 = 0.032 cc/sec. In liters per second this is $0.032 \times 10^{-3} = 3.2 \times 10^{-5}$. By using eq. 7.1, the pumping speed is found to be

$$\frac{750 \times 3.2 \times 10^{-5}}{10^{-4}} = 750 \times 3.2 \times 10^{-5} \times 10^4 = 750 \times 3.2 \times 10^{-1}$$

$$= 75 \times 3.2 = 240 \text{ l/sec at } 10^{-4} \text{ mm Hg}$$

The time required for 1 cc of air at atmospheric pressure to flow into the system is 100/3.2 = 31.25 sec. By using eq. 7.2, the following pumping speed is obtained:

$$\frac{750 \times 10^{-3}}{10^{-4} \times 31.25} = \frac{750 \times 10^{-3} \times 10^4}{31.25} = \frac{750 \times 10}{31.25} = \frac{7500}{31.25}$$

$$= 240 \text{ l/sec at } 10^{-4} \text{ mm Hg}$$

This is the same value that was obtained by use of eq. 7.1. Clearly, as the air leak is reduced in value the pressure will drop. Finally, when the ultimate pressure is reached, the air leak will be zero and the pumping speed will be zero. This is shown in the various curves of Chapters 3 and 4. As we have noted, the only throughput considered in this method is the leak. The pumping speed obtained is the *measured speed* and the method is applicable to any type of pump.

Suppose the above measurements were made at a temperature of 30°C. Take 25°C as a standard temperature for pumping speeds. Then 25°C = 273 + 25 = 298°K and 30°C = 273 + 30 = 303°K. The pumping speed at 25°C (298°K) can be found from the following relationship:

$$\frac{\text{pumping speed at 25°C}}{\text{pumping speed at 30°C}} = \sqrt{\frac{298}{303}}$$

or the pumping speed at 25°C $= 240\sqrt{298/303} = 240\sqrt{0.9835} = 240 \times 0.992 = 238$ l/sec at 10^{-4} mm Hg. Clearly a few degrees change in temperature from a standard of 25°C doesn't change the value of pumping speed substantially.

If it were possible to measure the total throughput (exclusive of pump fluid vapor) through the inlet port of a vapor pump, then the speed obtained by dividing this total throughput by the pressure at the point of measurement could be called the *intrinsic speed.* In some cases the speed is quoted as a small change in throughput (due to a controlled leak) divided by the corresponding small change in pressure. This is called the *operational speed.* Suppose a test setup is pumped down to its ultimate pressure and then air is admitted at a controlled rate. If the final pressure is considerably greater than the ultimate pressure, say ten times as large, then the measured and operational speeds will be essentially the same. It is only when the initial and final pressures are very nearly the same that the measured and operational pumping speeds differ substantially. Suppose the initial pressure is 10^{-5} mm Hg and the final pressure is 2×10^{-5} mm Hg. Then the pressure difference is $2 \times 10^{-5} - 10^{-5} = 10^{-5}$ mm Hg. The measured speed is found by using the final pressure, 2×10^{-5} mm Hg, in the denominator of either eq. 7.1 or eq. 7.2. On the other hand, the operational speed can be found by using the pressure difference, 10^{-5} mm Hg, in the denominator of one of these equations. Consequently, in this particular case the operational speed comes out to be twice the value of the measured speed.

7.3 Positioning of Vacuum Gauge and Air Leak

The relative positions of vacuum gauge and air leak in making metered-leak measurements of pumping speed are extremely important. Figure 7.1*a* shows a case where too low a value of the measured pumping speed is obtained. Here the air stream is beamed toward the inlet of the gauge, which results in too high a pressure reading. Since pumping speed equals throughput divided by pressure, this leads to too low a value of measured pumping speed. A second case is shown in Fig. 7.1*b*. Here the pressure reading is too low, which gives too high a value for the pumping speed. To obtain a reasonably accurate value of the pumping speed, the vacuum gauge tubulation should be placed at right angles to the leak and low on the wall of the test chamber near the mouth of the pump. The tubulation should extend across the chamber so as to be in the main stream entering the pump. The leak should be placed high on the side of the chamber and directed across to the opposite wall or dispersed by a screen or baffle near the top of the chamber. Two recommended arrangements are shown in Fig. 7.1*c* and Fig. **7.1***d*.

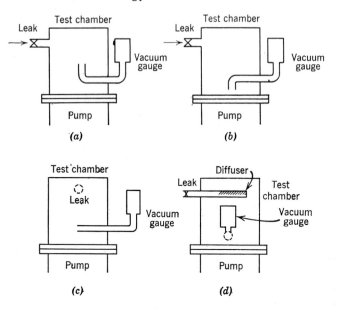

Fig. 7.1 (*a*) Too low a pumping speed. (*b*) Too high a pumping speed. (*c*) Recommended arrangement 1. (*d*) Recommended arrangement 2.

7.4   Admission of Air

To make metered-leak measurements, a calibrated flowmeter and a vacuum gauge are needed. Many different types of flowmeters have been used but only some of the more common types will be discussed here. The usual form of flowmeter is a manometer or burette. Figure 7.2*a* shows the case of a calibrated capillary containing a mercury pellet. With the needle valve closed, the chamber will pump down to some steady state value as determined by an appropriate gauge. The needle valve is then adjusted until the desired pressure reading is obtained. Air at atmospheric pressure will now be entering the chamber from the capillary. The rate at which the mercury pellet moves through the capillary can be used to determine the rate of flow of air into the chamber. Suppose the needle valve is adjusted until the pressure reading is 2×10^{-4} mm Hg. The capillary being used has an average ID of 2 mm and is calibrated in millimeters. If the pellet moves 30 mm in 60 sec then the volume of air entering the chamber in this time is $30 \times \pi \times 1^2 = 30 \times 3.14$ mm^3 = 94.2 mm^3 = 0.0942 cc. The flow rate is then $0.0942/60 = 0.00157$ cc/sec or 0.00157×10^{-3}

1/sec. The pumping speed according to eq. 7.1 is then (760 × 0.00157 × 10⁻³)/10⁻⁴ = 760 × 0.00157 × 10 = about 12 1/sec at 10⁻⁴ mm Hg, when the atmospheric pressure is 760 mm Hg. Once a measurement has been made, the mercury must be returned to a convenient starting point. Arranging matters so the needle valve–capillary assembly can be removed from the chamber and then holding the capillary vertical with the needle valve open and tapping the capillary is one way of doing this. Sticking of the pellet in the capillary can be avoided by making sure both the mercury and the inside surface of the capillary are clean. In this method mercury vapor will get into the system and the type of gauge used will determine whether or not the pressure of this vapor is measured.

Figure 7.2*b* shows one arrangement of oil manometer for admitting air. A low vapor pressure oil is used, such as a silicone oil, an Apiezon oil, or butyl phthalate. Butyl phthalate is often used because it has less tendency to wet glass. These oils contain dissolved gases so it is best to heat them before making measurements. In the arrangement

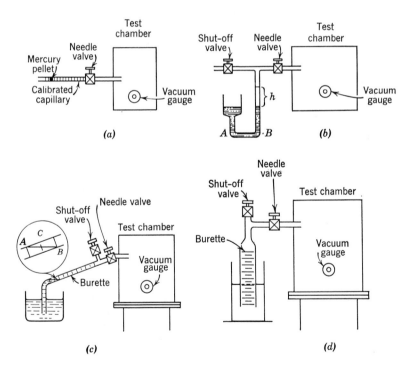

Fig. 7.2 (*a*) Calibrated capillary with mercury pellet. (*b*) Oil manometer. (*c*) Use of standard burette. (*d*) Burette arrangement for large throughputs.

of Fig. 7.2*b*, the shut-off valve should be open at first, and the needle valve should be closed while the chamber is being pumped down. After reaching a pressure well below the pressure at which the speed measurement is to be made, the needle valve can be opened until the latter pressure is reached. The shut-off valve is then closed and the oil level will rise in side *B* of the U-tube. The time for the oil to rise some height, *h*, is measured. Suppose the ID of tube *B* is 3 mm and the oil level rises 10 cm in 100 sec. Then the volume of air entering the chamber is $\pi \times \frac{3}{2}^2 \times 100 = \pi \times \frac{9}{4} \times 100 = 90\pi \times 2.5 = 707$ mm³ or about 0.71 cc in 100 sec. The flow rate is $0.71/100 = 0.0071$ cc/sec $= 0.0071 \times 10^{-3}$ l/sec. It is assumed that this air is at atmospheric pressure. This is not strictly true because of the difference in oil levels on the two sides of the U-tube. However, the error introduced by assuming atmospheric pressure can usually be neglected as long as the difference in levels is not more than about 15 or 20 cm. Suppose the atmospheric pressure is 750 mm Hg and the pressure at which the measurement is made is 10^{-5} mm Hg. Then, by eq. 7.1, the pumping speed $= (750 \times 0.0071 \times 10^{-3})/10^{-5} = 750 \times 0.0071 \times 10^2 = 750 \times 0.71 = 533$ l/sec at 10^{-5} mm Hg.

Figures 7.2*c* and 7.2*d* show two methods of using burettes in making metered-leak measurements. Figure 7.2*c* shows the case of a standard burette of relatively small bore (in the millimeter range) for measuring relatively small throughputs. The method of using this arrangement is the same as for the oil manometer. Again, a low vapor pressure oil is used. The inclination of the burette to the horizontal depends on the bore, being less as the bore size is decreased. A typical angle of inclination is 10°. The oil level can be read at any part of the calibration mark as long as the same part is consistently used, e.g., the top (*A*), the bottom (*B*), or the middle (*C*). The effect of the difference in levels between the oil in the reservoir and the oil in the burette on the pressure of the air being admitted can be neglected as long as the straight part of the burette above the oil in the reservoir is kept at a minimum, say less than 15 cm. The arrangement of Fig. 7.2*d* is used with pumps which handle large throughputs (such as ejector pumps). Here a large burette (1 to 3 in. in diameter) is used so that large air flow rates (throughputs) can be used. In using this arrangement, the measurement is started with the oil levels the same in the reservoir and the burette. After adjusting the needle valve so the desired pressure is obtained, the rate of rise of oil in the burette is observed. When the throughput is too high to be measured with the arrangement of Fig. 7.2*d*, it is possible to use some form of commercial gas flowrator or dry gas meter.

7.5 Vacuum Gauges

The importance of the location of the vacuum gauge tubulation with respect to the air leak has been discussed in Section 7.3. The type of gauge and its calibration are also extremely important. The gauge used will be determined by the pressure range to be covered. In the micron range a Pirani (or thermocouple) gauge is most commonly used. For lower pressures an ionization gauge is used, usually the hot filament (thermionic) type. In many cases a McLeod gauge is used. Various models can be selected to cover different pressure regions. In selecting a vacuum gauge several points must be kept in mind:

1. A McLeod gauge does not measure vapors and, therefore, it gives pumping speeds only for permanent gases.

2. Difficulties associated with thermionic ionization gauges are:

a. Oxygen and organic vapors may affect the filament and therefore the reading.

b. Errors may arise due to incorrect calibration for air and pump fluid vapor.

c. Errors may arise due to outgassing and electrical clean-up, which depend on the length and diameter of the tubulation.

3. Any air leak involving the use of mercury will admit mercury vapor to the gauge. The only gauge which will not measure the pressure of this vapor is the McLeod gauge (unless a cold trap is used).

4. Gauges other than the McLeod gauge can be trapped so they will only read the total pressure of permanent gases. However, this will give too low a pressure and therefore too high a pumping speed according to the methods of Section 7.2.

5. A McLeod gauge is an intermittent type of gauge and a certain amount of time is required to take a reading. It must be assured that the pressure does not change while the reading is being taken.

6. Thermionic and Philips ionization gauges act as pumps and therefore read incorrect pressures (too low).

Some specific steps to be followed in making metered-leak measurements of pumping speed are:

1. In general make the gauge connection as short and large as possible, consistent with the proper positioning of the end of the tubulation.

2. Where the pumping speed for all gases and vapors is required, use an untrapped gauge.

3. To measure the pumping speed for the permanent gases in the atmosphere, use a McLeod or Knudsen gauge. If mercury vapor is not desired in the test setup, use a trap with the McLeod gauge. Other types of gauges can be used as long as they are trapped.

4. To measure the pumping speed for a particular permanent gas, such as hydrogen or helium, use a trapped McLeod gauge or a Knudsen gauge.

7.6 Test Dome

In previous sections, the use of a test chamber has been discussed briefly. This type of arrangement can only be used on an isolated pump. When the pump is incorporated in a system, pumping speeds can be measured by using an air leak and vacuum gauge positioned with respect to each other as shown in Fig. 7.1c. The size of the part of the system in which the gauge and air leak are placed is important. In general, the dimensions of the region in which the pumping speed measurement is made should be large compared to the dimensions of the connection to the pump (or to the pump itself). When a manifold is included above the pump, a good value of the measured pumping speed of the pump can be made in the manifold. On the other hand, if the measurement is made in the vacuum process chamber then the value obtained will be the speed in the chamber rather than the speed of the pump.

When a pump can be tested by itself, the arrangement of components is very important. The relative positions of leak and vacuum gauge have already been mentioned (Fig. 7.1c). However, nothing has been said so far about the size of the test chamber or dome. For reasonably accurate values of pumping speed this dome should have certain dimensions compared to the dimension of the inlet port of the pump. Some general rules regarding this matter can be stated as follows. The diameter of the test dome should be greater than the inside diameter of the pump being tested. Also, the dome height should be at least one and a half times its diameter. For smaller pump diameters, say up to 4 in., the height should be up to three or four times the diameter. If these dimension specifications are followed, the impedance of the dome can usually be neglected. A typical setup for measuring pump speeds is shown in Fig. 7.3a. The dimensions shown are roughly in the proper proportion. An ionization and a Pirani gauge are shown just to indicate an arrangement which will cover a wide pressure range. In some

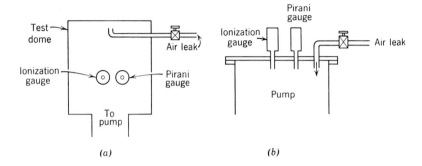

Fig. 7.3 (*a*) Typical setup for measuring pumping speed. (*b*) Blank-off method.

cases a blank-off method of measuring pumping speed is quoted. In this method the air leak and the vacuum gauge are inserted in a plate used to blank off the pump. This method is illustrated in Fig. 7.3*b*. The pumping speeds obtained are too high since the pressures measured are too low, as in the case of Fig. 7.1*b*. Consequently, the blank-off type of pumping speed measurement should be avoided.

7.7   Rate-of-Rise Measurement

This method of measuring pumping speed is often called the constant volume method. A calibrated leak is not needed but the volume of the system being used must be known and a vacuum gauge must be used. A typical setup for making rate-of-rise measurements is shown in Fig. 7.4*a*. Here an appropriate valve, usually a needle valve, and a suitable vacuum gauge are mounted in the manifold above a pump. A valve between the pump and the manifold can be used to isolate the pump from the manifold and the rest of the system. When this valve is closed the pressure in the system will start to rise since the average vacuum system will release gases and vapors and there may be small leaks. This arrangement effectively measures the pumping speed at the pump. However, the speed can be measured at any point in the system, the point of measurement being determined essentially by the position of the end of the vacuum gauge tubulation. To find the speed at the pump it is then necessary to take into consideration the pumping losses in the connecting lines. The positioning of the air inlet valve and the vacuum gauge should follow the recommendations of Section 7.3.

The procedure to be followed in making a rate-of-rise measurement is as follows:

1. With the needle valve closed, pump down the system to its equilibrium pressure value.
2. Record the equilibrium pressure.
3. Close the pump valve and use a stopwatch to time the rate of pressure rise. Letting the pressure rise to at least ten times its initial value is usually adequate.
4. Open the pump valve and pump down to the original equilibrium pressure value.
5. Open the needle valve until the pressure rises to a suitable value, say a few times the equilibrium value.
6. Close the pump valve a second time and use a stopwatch to time the rate of pressure rise. Again at least ten times the initial value is usually satisfactory.
7. Open the pump valve and close the needle valve.

The rate of rise in pressure is simply the change in pressure divided by the time involved. Suppose the initial pressure were 10^{-5} mm Hg and it took 50 sec for the pressure to rise to 5×10^{-4} mm Hg. Then the rate of rise of pressure would be

$$\frac{5 \times 10^{-4} - 10^{-5}}{50} = \frac{50 \times 10^{-5} - 10^{-5}}{50} = \frac{49 \times 10^{-5}}{50}$$

$$= 0.98 \times 10^{-5} = 9.8 \times 10^{-6} \text{ mm Hg/sec}$$

In order to find the pumping speed, it is necessary to know the volume of the system, up to the pump. This can be estimated from the external dimensions. It must be kept in mind that whatever error occurs in this estimate will also occur in the value of the pumping speed. The pumping speed can be obtained from the following relationship:

$$\text{Pumping speed} = \text{volume} \times \frac{\text{difference in rates of pressure rise}}{\text{difference in pressures}} \qquad (7.3)$$

EXAMPLE

A vacuum system (up to the pump) is estimated to have a volume of 1200 l. The initial pressure is 10^{-5} mm Hg and when the pump valve is closed the pressure rises to 2×10^{-4} mm Hg in 90 sec. The rate of pressure rise is then $(2 \times 10^{-4} - 10^{-5})/90 = (20 \times 10^{-5} - 10^{-5})/90 = (19 \times 10^{-5})/90 = 0.21 \times 10^{-5}$ mm Hg/sec. After the pump valve is opened and the system pumped down to the initial pressure value,

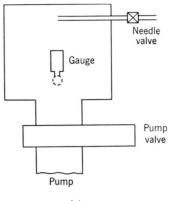

(a)

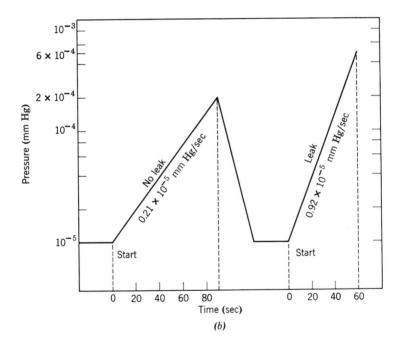

(b)

Fig. 7.4 (a) Rate-of-rise arrangement. (b) Rate-of-rise curves.

air is admitted until the pressure becomes 5×10^{-5} mm Hg. When the pump valve is closed, the pressure rises to 6×10^{-4} mm Hg in 60 sec. The rate of pressure rise now is

$$\frac{6 \times 10^{-4} - 5 \times 10^{-5}}{60} = \frac{60 \times 10^{-5} - 5 \times 10^{-5}}{60}$$

$$= \frac{55 \times 10^{-5}}{60} = 0.92 \times 10^{-5} \text{ mm Hg/sec}$$

The difference in pressures is $5 \times 10^{-5} - 10^{-5} = 4 \times 10^{-5}$ mm Hg. The pumping speed is then given by

$$1200 \times \frac{0.92 \times 10^{-5} - 0{:}21 \times 10^{-5}}{4 \times 10^{-5}} = 1200 \times \frac{0.71 \times 10^{-5}}{4 \times 10^{-5}}$$

$$= 1200 \times 0.18 = 216 \text{ l/sec}$$

The pumping speed obtained by this method is the operational speed. The pumping speeds in various pressure ranges can be found simply by increasing the flow of gas (usually air) into the system and making successive measurements. The manner in which the pressure behaves during a rate-of-rise measurement is illustrated in Fig. 7.4b. Here the pressure is plotted against time. A McLeod gauge cannot be used because of the time involved in making measurements. Values of pumping speed obtained by this method will agree quite well with values obtained by the metered-leak method as given by eq. 7.1 or 7.2 as long as the pressures are considerably larger than the ultimate pressure, say by a factor of 10. At lower pressures, reasonable agreement can be obtained by using the pressure difference instead of the final pressure in the denominators of these equations.

It is possible to obtain a value of the pumping speed by making only one measurement of the rate of pressure rise. One way of doing this is to pump down the system to the lowest possible pressure, ensuring that the system is clean and that there are no leaks in the system. This pressure is then the ultimate or base pressure. Now use an air leak to raise the pressure to a convenient operating value. Next make a rate-of-rise measurement at this operating pressure. This is done by closing the pump valve and measuring the time required for the pressure to rise to some higher value. The pumping speed is then given by

$$\text{Pumping speed} = \text{volume} \times \frac{\text{rate of rise of pressure}}{\text{operating pressure} - \text{base pressure}}$$

$$(7.4)$$

The volume is the volume of the system up to the pump. The pumping speed found by this method is the operational speed. If the operating pressure were used in the denominator of eq. 7.4 then the measured speed would be found. Again, if the operating pressure is considerably higher than the base pressure, the two speeds are essentially the same.

EXAMPLE

A system is pumped down to an ultimate pressure of 2×10^{-6} mm Hg. The air leak is opened so the operating pressure is 2×10^{-5} mm Hg. When the pump valve is closed the pressure rises to 5×10^{-4} mm Hg in 60 sec. Suppose the volume of the system is 600 l. The rate of rise of pressure at 2×10^{-5} mm Hg is

$$\frac{5 \times 10^{-4} - 2 \times 10^{-5}}{60} = \frac{(50 - 2) \times 10^{-5}}{60} = \frac{48 \times 10^{-5}}{60}$$

$$= 0.8 \times 10^{-5} \text{ mm Hg/sec}$$

The change in pressure is $2 \times 10^{-5} - 2 \times 10^{-6} = (20 - 2) \times 10^{-6} = 18 \times 10^{-6}$ mm Hg. Therefore, the pumping speed, by eq. 7.4, is

$$600 \times \frac{0.8 \times 10^{-5}}{18 \times 10^{-6}} = \frac{480}{18} \times 10^{-5} \times 10^{6} = \frac{4800}{18} = \text{about 267 l/sec}$$

This is the operational pumping speed. The measured pumping speed is given by the throughput (volume times rate of rise of pressure) divided by the operating pressure, or

$$\frac{600 \times 0.8 \times 10^{-5}}{2 \times 10^{-5}} = 600 \times 0.4 = 240 \text{ l/sec}$$

In this case the two speeds do not differ greatly.

7.8   Use of a Known Conductance

Another method of measuring pumping speed involves the use of a tube of known dimensions. Two arrangements for doing this are shown in Figs. 7.5a and b. Figure 7.5a involves the use of a long tube attached to the inlet port of the pump (same diameter). An air leak is used at the end of the tube farthest from the pump. The inlet tube from this leak is bent so that the air stream is directed toward the wall of the tube. Doing this prevents the air from striking the inlet of the vacuum gauge shown at A. In order to use this arrangement to measure the pumping speed it is necessary to find the conductance of the tube between the two gauge positions (A and B). The curves in Fig. 3.7 can be used for this purpose. These curves include a correction for the impedance due to end effect. However, as long as the distance between A and B is at least about five times the diameter of the tube, these curves can be used directly. It is a good general rule to use a tube at least this long. The curves in Fig. 3.7 apply to the molecular flow region. The arrangement of Fig. 7.5a is generally used with vapor pumps or getter- and sputter-ion pumps at pressures giving molecular flow. When it is used in the viscous flow region it is necessary to apply the corrections which were discussed in Chapter 3.

In using the setup of Fig. 7.5a certain precautions must be taken in order to obtain a reasonably accurate value of the pumping speed. The method involves estimating the conductance of the tube between A and B and measuring the pressures at these points. Consequently the vacuum gauges which are used are quite important. The main thing is that the gauges read in the same way. They do not have to read accurately since the method involves the ratio of the pressure readings. Thermionic ionization gauges are often used but other gauges

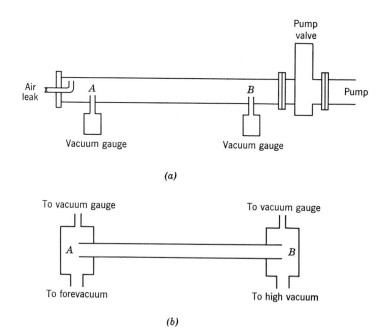

Fig. 7.5   (a) Tube with air leak. (b) Tube pumped at both ends.

such as cold cathode or McLeod could also be used. If the tube is thoroughly clean (preferably baked for low pressures) and dry air is admitted through the air leak, it is not necessary to trap the gauges. Of course, a McLeod gauge does not measure vapors. Since this type of measurement involves an equilibrium condition the time involved in using this type of gauge is not a disadvantage. Of course, the pressure range it covers is limited. The steps involved in making a pumping speed measurement are:

1. Pump down the tube until the pressure is steady. This means that the bulk of the gases and vapors being released have been removed.

2. Compare the readings of the gauges at A and B. Commercial vacuum gauges can differ considerably in their readings even when obtained from the same manufacturer and even when they are from the same production batch. Compare readings at several pressures by closing the pump valve and admitting dry air through the air leak. Set up correction factors at various pressures for one gauge in terms of the other.

3. If thermionic ionization gauges are used, operate them at the lowest possible filament current. This will reduce their pumping action.

4. Record the pressure readings at A and B as the amount of dry air admitted through the air leak is increased. This makes it possible to obtain a pumping speed curve.

5. Assume either of the two gauges to be the standard. Use the calibration factor to correct the readings of the other gauge.

6. Use the following relationship to obtain the pumping speed:

$$\text{Pumping speed} = \text{conductance of tube} \times \frac{\text{pressure at } A}{\text{pressure at } B} - 1 \quad (7.5)$$

In symbols this can be written

$$S = U\frac{P_A}{P_B} - 1 \quad (7.6)$$

where S is the pumping speed, U is the tube conductance, and P_A and P_B are the pressures at points A and B respectively.

EXAMPLE

A 4 in. oil diffusion pump is connected to a tube of diameter 4 in. and length 4 ft with a needle valve and diffuser tube connected at the end farthest from the pump. Two thermionic ionization gauges are connected to the tube at points 3 ft apart. Let A be the gauge position farthest from the pump and B the gauge position nearest the pump (as in Fig. 7.5a). When the pump valve is closed and the needle valve gradually opened the following gauge readings are obtained:

Gauge at A	Gauge at B	Calibration Factor for Gauge at A
2×10^{-6} mm Hg	1.8×10^{-6} mm Hg	0.90
8×10^{-6}	7×10^{-6}	0.88
2×10^{-5}	1.9×10^{-6}	0.95
7×10^{-5}	6.2×10^{-5}	0.90
6×10^{-4}	5.5×10^{-4}	0.92

The average calibration factor is about 0.91. Now the pump valve is opened and the tube is again pumped down to the ultimate pressure. With the valve open, air is admitted through the needle valve until the gauge at B reads 2×10^{-5} mm Hg. The gauge at A then reads 5×10^{-5} mm Hg. Using the calibration factor of 0.91, the corrected reading for the gauge at A is $0.91 \times 5 \times 10^{-5} = 4.55 \times 10^{-5}$ mm Hg. From Fig. 3.7, the conductance of the tube between A and B is about 190 cfm. The pumping speed at point B is then, from eq. 7.6

$$190 \left(\frac{4.55 \times 10^{-5}}{2 \times 10^{-5}} - 1 \right) = 190 \left(\frac{4.55}{2} - 1 \right) = 190(2.28 - 1)$$

$$= 190 \times 1.28 = \text{about 243 cfm}$$

or

$$\frac{243 \times 28.3}{60} = 115 \text{ l/sec}$$

Actually this is a conservative figure since the conductance of a 4 in. diameter tube as given by Fig. 3.7 includes an end effect. If the straight part of this curve were extended down (case of no end effect) the conductance of a 3 ft length would come out to be about 300 cfm. This gives a pumping speed of about 384 cfm or 181 l/sec. An average value of $243 + 384/2 = 627/2 = 314$ cfm or $115 + 181/2 = 296/2 = 148$ l/sec is probably a reasonably conservative value of the pumping speed at point B at a pressure of 2×10^{-5} mm Hg. Naturally the same method can be used to obtain pumping speeds at other pressures.

In the case of the arrangement shown in Fig. 7.5b, the tube is not closed. One end is pumped on by the vacuum system being tested for pumping speed while the other end is pumped on by a forepump. This results in a pressure difference across the tube which is determined by the conductance of the tube. In this case there is an end effect that has to be taken into consideration. Therefore, the curves of Fig. 3.7 can be used directly. The same comments regarding the use of vacuum gauges apply here as in the case of Fig. 7.5a. The procedure discussed in the above example applies here.

EIGHT

Properties of Some Vacuum Materials

8.1   The Elements

An element is a material that cannot be broken down into a simpler form by chemical means. Examples are: oxygen gas, argon gas, nickel, tungsten, bromine, and copper. When elements are combined together chemically they form chemical compounds (more simply referred to as just compounds). One of the most familiar compounds is water. This can be made by combining (burning) hydrogen gas and oxygen gas, two elements. When ordinary charcoal (carbon) is burned in air it combines with the oxygen in the air to form a gas (or gases). When little oxygen is available the gas carbon monoxide is formed. With a plentiful supply of oxygen the gas carbon dioxide is produced.

Table 8.1 lists some properties of elements used in vacuum practice. This list is confined to the more commonly used elements. Other elements may be used in specialized applications and some information concerning the vapor pressures of these elements will be found in the general references in Appendix E.

Information regarding the vapor pressures of other materials, including liquids, alloys, elastomers, cements and waxes, greases, etc., will be found in other sections of this chapter and in other chapters. In Table 8.1 it will be noted that all elements tabulated are metals

Table 8.1   Vapor Pressures of Some Elements at Various Temperatures

Vapor Pressure (mm Hg)

Element	Ref.	10^{-5} at °C	10^{-4} at °C	10^{-3} at °C	10^{-2} at °C	10^{-1} at °C	1 at °C	mp (°C)
Aluminum	1	882	972	1082	1207	1347	1547	659
Antimony	1	382	427	477	542	617	757	630
Barium	1	417	467	537	617	727	867	710
Beryllium	1	902	987	1092	1212	1367	1567	1283
Bismuth	1	450	508	578	661	762	892	271
Cadmium	1	149	182	221	267	321	392	321
Calcium	1	402	452	517	592	687	817	850
Carbon	1	1977	2107	2247	2427	2627	2867	—
Cesium	1	46	75	110	152	206	277	30
Chromium	1	1062	1162	1267	1392	1557	1737	1903
Cobalt	1	1162	1262	1377	1517	1697	1907	1495
Copper	1	942	1032	1142	1272	1427	1622	1084
Gold	1	987	1082	1197	1332	1507	1707	1063
Indium	1	670	747	837	947	1077	1242	156
Iridium	1	1797	1947	2107	2307	2527	2827	2454
Iron	1	1107	1207	1322	1467	1637	1847	1539
Lead	1	487	551	627	719	832	977	328
Lithium	1	348	399	460	534	623	737	181
Magnesium	1	287	330	382	442	517	612	650
Manganese	1	697	767	852	947	1067	1227	1244
Mercury	1	−28	−8	16	45	81	125	−39
Molybdenum	1	1987	2167	2377	2627	2927	3297	2577
Nickel	1	1142	1247	1357	1497	1667	1877	1452
Osmium	2	2101	2264	2451	2667	2920	3221	2697
Palladium	1	1157	1262	1387	1547	1727	1967	1550
Platinum	1	1602	1742	1907	2077	2317	2587	1770
Potassium	1	91	123	162	208	266	341	64
Rubidium	1	64	95	133	176	228	300	39
Silicon	1	1177	1282	1357	1547	1717	1927	1415
Silver	1	757	832	922	1032	1167	1337	961
Sodium	1	158	195	238	290	355	437	98
Strontium	1	342	394	456	531	623	742	770
Tantalum	1	2397	2587	2807	3067	3372	3737	2997
Thorium	2	1686	1831	1999	2196	2431	2715	1827
Tin	1	882	977	1092	1227	1397	1612	232
Tungsten	1	2547	2757	3007	3297	3647	—	3377
Uranium	1	1442	1582	1737	1927	2157	2447	1130
Zinc	1	208	246	290	342	405	485	420
Zirconium	1	1837	2002	2187	2397	2647	2977	1852

1. R. E. Honig, "Vapor Pressure Data for the More Common Elements," *RCA Rev.*, **18**, 195 (1957).

2. S. Dushman, *Scientific Foundations of Vacuum Technique*, 2nd ed., J. M. Lafferty, editor, John Wiley and Sons, New York, 1962.

except for carbon, which has both metallic and nonmetallic properties. It is included here because of its wide usage in vacuum practice (see Section 8.14).

A great deal of information concerning the properties of many materials is contained in these books: Saul Dushman, *Scientific Foundations of Vacuum Technique,* 2nd ed., J. M. Lafferty, editor (John Wiley and Sons, New York, 1962) and K. Diels and R. Jaeckel, *Leybold Vakuum-Taschenbuch* (Springer-Verlag, Berlin, 1958). No attempt is made here to show the details to be found in these books. Emphasis is on the more common vacuum materials and their pertinent vacuum properties.

8.2　Mercury and Water

It may at first be thought peculiar to treat mercury and water together. However, both these materials are liquids at room temperature and are commonly encountered in vacuum practice. Mercury is used in: diffusion pumps, vacuum gauges (manometers, McLeod gauges), cut-offs, switches, discharge lamps, etc. Water is present in the normal atmosphere and will get into the vacuum system unless the system is let down to dry air (or some other dry gas). The

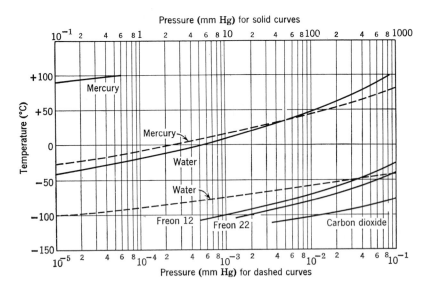

Fig. 8.1　Vapor pressure versus temperature for several materials.

absorption and adsorption of water in vacuum materials is one of the main causes of slow pump-down. The pump-down time can be reduced by using proper cleaning techniques (Chapter 9) and by heating parts (baking). Cold traps with the proper refrigerant will also reduce the vapor pressure to tolerable values when water has contaminated the vacuum materials.

The vapor pressures of both mercury and water are listed for several temperatures in Tables 5.2 and 5.4. It will be noted that to reduce the vapor pressure below 10^{-7} mm Hg, it is possible to use dry ice $(-78°C)$ for mercury, whereas this is entirely inadequate with water. In the latter case, liquid nitrogen is generally used to reduce the vapor pressure below 10^{-7} mm Hg. Figure 8.1 shows curves of vapor pressure of mercury and water versus temperature, over the range $-150°C$ to $100°C$. Curves for Freon 12, Freon 22, and carbon dioxide are also shown.

Some properties of mercury and water which are often important in vacuum work are summarized below:

MERCURY

Density = 13.6 g/cc (room temperature). Melting point (mp) = $-39°C$. Boiling point (bp) = $357°C$ (at normal atmospheric pressure). Readily combines with most metals at room temperature to form amalgams. Does not, however, readily affect materials such as iron, cobalt, nickel, platinum, rhodium, molybdenum, tungsten, stainless steel, and various alloys. Proper electroplating of copper, brass, and bronze with nickel will protect them against mercury vapor. Cleaning (see Chapter 9). The vapor is poisonous—try to avoid getting vapor into the air and use in a well-ventilated area.

WATER

Density = 1 g/cc (room temperature). Melting point (ice) = $0°C$. Boiling point = $100°C$ (at normal atmospheric pressure).

8.3 Liquids Used in Vacuum Practice

Apart from mercury and water, most liquids encountered in vacuum practice are solvents, which are primarily used for cleaning purposes.

Some of them are used in leak-hunting procedures (see Chapter 15). Of course, pump oils (mechanical and diffusion) are also liquids but they will not be included here since their properties are summarized in Appendix B.

The vapor pressures of a number of liquids at 20°C (room temperature) are shown in Table 8.2. Figure 8.2 shows the boiling points of several solvents at different degrees of vacuum (in inches of mercury). The value 0 on the abscissa represents normal atmospheric pressure. Some properties and uses of several solvents can be summarized as follows.

ACETONE

Volatile solvent used extensively for cleaning and degreasing. Miscible with water, alcohol, and oils. Flammable.

ALCOHOLS

Ethyl alcohol (ethanol, grain alcohol). A volatile solvent used for cleaning and drying. Miscible with water in all proportions. Flammable.

Table 8.2   Vapor Pressures of Some Liquids

Liquid	Vapor Pressure (mm Hg at 20°C)
Acetone	184.8
Benzene	74.65
Carbon disulfide	297.5
Carbon tetrachloride	91.0
Chloroform	159.6
Ethyl alcohol	43.9
Methyl alcohol	96.0
Ethyl ether	442.2
Ethyl bromide	386.0
Turpentine	4.4
Water	17.535
Mercury	0.0012

Handbook of Chemistry and Physics, 43rd ed., Chemical Rubber Publishing Co., Cleveland, Ohio, 1961.

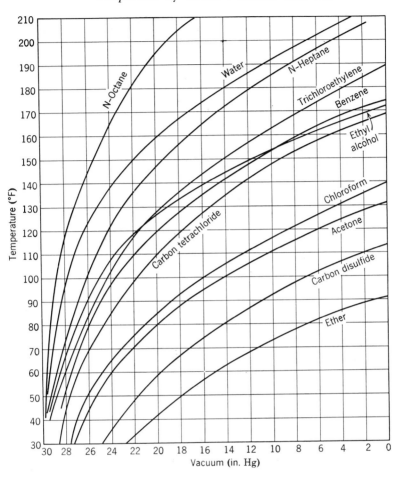

Fig. 8.2 ` Boiling points of some solvents under vacuum. (F. J. Stokes Corp., Philadelphia, Pa.)

Methyl alcohol (methanol, wood alcohol). In pure form, a useful volatile solvent for cleaning and drying parts (preferred to ethyl alcohol because of lower water content and greater availability). Miscible with water in all proportions. Poisonous and flammable. Avoid inhaling.

Isopropyl alcohol. In pure form, carbon dioxide is bubbled through it in brazing and soldering techniques. At brazing temperatures, the isopropyl alcohol decomposes to form carbon monoxide and hydrogen, which provide a reducing atmosphere.

BENZENE

A volatile, flammable solvent used for degreasing. Not recommended due to its toxicity.

CARBON TETRACHLORIDE

A volatile, heavy, inflammable solvent for cleaning and degreasing. The vapors are toxic so use good ventilation. Its inflammable nature makes it suitable for leak hunting.

TRICHLOROETHYLENE

An inflammable solvent useful as a degreasing agent. Do not inhale the vapors. Parts to be cleaned should be handled in a tray or basket, and the hands should be protected with synthetic rubber gloves because the degreasing action is sufficiently strong to cause painful drying of the skin. This material is commonly used as the heat transfer liquid for dry ice traps.

The above liquids are probably the most commonly used solvents. Sometimes materials such as diethyl ether (ether), chloroform, and carbon disulfide are used. These materials are highly volatile and flammable. Because of their high volatility, chloroform and ether are often used in leak hunting (see Chapter 15).

8.4   Gases and Vapors

In a sense, gases and vapors are the most important substances encountered in vacuum practice. The whole function of a vacuum system is to remove these materials from an enclosure. The most common gases encountered in vacuum work are those in the atmosphere. The gases argon, helium, krypton, neon, radon, and xenon are called *noble* or *rare* gases. Helium is commonly used in vacuum practice—for helium leak detectors and, in the liquid form, for trapping (or pumping) gases and vapors. Liquid air, liquid nitrogen, and solid carbon dioxide (dry ice) are most commonly used in traps. Liquid hydrogen is also used for pumping. Melting point and boiling point data for several gases are given in Table 5.3. Common vapors are those of mercury and water, pump oils, and various solvents.

Gases and vapors are sorbed by vacuum construction materials and are only gradually released. This sets one limit on the lowest pressure that can be reached in a particular vacuum system. The usual method of overcoming this problem is to degas the materials, usually by baking, i.e., raising the system to a high temperature while pumping. The temperature to use will depend on when the material begins to change its properties. Sometimes an electrical discharge is used to drive gases off of surfaces (adsorbed gases). Also, instead of driving out gases and vapors, sometimes a low temperature is used to hold them in the vacuum material (at a low vapor pressure). The gases and vapors sorbed by any particular material depend on the nature and history of that material. Water vapor is the prime offender with glasses. This is followed by carbon dioxide. For certain types of glasses, sulfur dioxide may constitute a large percentage of the gas. To remove essentially all gases and vapors from a glass it is necessary to go well above the melting point and this causes some decomposition of the glass. In practice, it is usually necessary to keep the bake-out temperature below the softening point, which is determined by the particular glass being used. Most glasses used in vacuum practice can be baked at about 450°C. At this temperature, a high percentage of the gases and vapors in a thin surface layer (mostly adsorbed) are evolved in a matter of an hour or so. A suitable trap can be used to remove the water vapor. The most effective procedure is to carry out the heating during exhaust in two or more stages of gradually decreasing temperatures. It should be kept in mind that heating a glass to around 500°C at atmospheric pressure removes 80 to 90% of the gas that would otherwise be evolved in the bake-out.

Water vapor is also the prime offender with metals. However, except for the rare gases, many gases react with metals to form compounds or effect solution of the gas in solid or liquid metal. Under special conditions, the rare gases will go into solution (when speeded up as ions in a discharge, etc.). Hydrogen reacts with or forms solutions with most metals. It is not absorbed by gold, zinc, cadmium, indium, or tellurium. Oxygen is chemically active and forms oxides with most metals. Nitrogen dissolves only in those metals which form nitrides at higher temperatures, e.g., molten aluminum, zirconium, tantalum, manganese, and iron. Carbon monoxide forms carbonyls with some metals, e.g., nickel and iron. Besides being present in the interior of metals in solution or in the form of chemical compounds, gases are adsorbed on the surface. If such metal parts are not previously degassed, they will gradually evolve gas in a vacuum, especially if they are heated to even a slight extent, because of increased rate of diffusion from the inside to the surface. Metal parts can be pre-heated in pure, dry hydrogen whenever the nature of the metal permits. A preliminary heating in vacuum by means of high frequency is valuable, since, after such treatment, metal may be stored for at least a few days in a dry atmosphere at 100 to 200°C without reabsorption of more than an almost negligible amount of gas. Removal of essentially all sorbed gas means going to high temperatures, often above the melting point. Gas-free metals can be prepared in this manner (vacuum fused metals) and they find use in various aspects of vacuum work. However, it is impractical to heat whole metal systems to very high temperatures, due to changes in mechanical properties and the limitations imposed by constructional materials. Consequently, metal systems are degassed at fairly modest temperatures (say 400 to 500°C) for several hours while being pumped. This eliminates a substantial part of the adsorbed gases and vapors (including water vapor) and some of the absorbed material.

Since small parts, such as filaments, anodes, heater elements, etc., can be degassed at high temperatures (but still below the melting point) it may be pertinent to list the degassing characteristics of a few metals.

MOLYBDENUM

Degas at about 1800°C. Gases evolved, in order of decreasing amounts: nitrogen, carbon monoxide, carbon dioxide, hydrogen. Avoid handling and wrap in condenser paper. Percentage of gas removed increased by degassing in pure hydrogen, preferably diffused through palladium.

TUNGSTEN AND NICKEL

Degas tungsten at 2300 to 2430°C. Gases evolved: about 67% nitrogen, 30% carbon monoxide, 3% hydrogen. Degas nickel at around 1100°C or a little less. Gases evolved (decreasing amounts): carbon dioxide, hydrogen, nitrogen. Sometimes more nitrogen than hydrogen is evolved, depending on the sample.

GRAPHITE

Fire in a hydrogen furnace at 1500°C. Heat in vacuum furnace at 1800°C. Gases evolved: hydrogen, carbon monoxide, nitrogen— much smaller percentage of nitrogen than the others. It is possible to degas graphite at 2150°C so that subsequent heating at a higher temperature gives no further gas. Absorbs oxygen on subsequent exposure to air. Must heat to 2150°C to remove.

Note: With any metal, do a preliminary firing in *very pure hydrogen* at as high a temperature as possible, except in cases where hydrogen produces deleterious effects on the mechanical or electrical properties of the metal.

A property of a gas that is very important in vacuum work is the *diffusion rate*. This term refers to the relative ease with which a gas passes through a material. The rate depends on the material, the pressure differential across it (this is usually atmospheric pressure for vacuum systems), and the thickness and area of the material. It increases as the pressure differential, area, and temperature are increased and the thickness is decreased. The characteristics of the material that are pertinent are the *permeability* and *diffusion* coefficients. The diffusion rate increases as the values of these quantities increase. These coefficients are often tabulated to show the relative diffusion rates of various materials. Table 8.3 shows the temperatures needed to obtain certain diffusion rates for quartz, Pyrex, and porcelain while Table 8.4 shows comparable data for some metals. It should be noted that the flow rates in Table 8.3 are per *hour* whereas in Table 8.4 they are per *minute*. The values shown in these tables are quite

Table 8.3   Diffusion of Gases through Quartz (Silica), Pyrex Brand Glass, and Porcelain

| System | Temperature Needed for Gas Flow of the following values in μl (0°C) per cm^2 per *hr* per mm Thickness at 1 atm Pressure | |
	0.1	1
Helium-silica	160°C	430°C
Neon-silica	686	—
Hydrogen-silica	410	835
Nitrogen-silica	815	1110
Helium-Pyrex brand glass	314	—
Air-porcelain	1440	—

Dushman, *Scientific Foundations of Vacuum Technique*, John Wiley and Sons, 2nd ed., L. M. Lafferty, editor, New York, 1962.

Table 8.4   Diffusion of Gases through Some Metals

Temperature Needed for Gas Flow of the following values in μl (0°C) per cm^2 per *min* per mm Thickness at 1 atm Pressure

System	0.1	1	10	100
Hydrogen-nickel	310°C	450°C	675°C	1100°C
Hydrogen-platinum	490	660	920	1470
Hydrogen-palladium	104	180	290	480
Hydrogen-copper	640	900	—	—
Hydrogen-iron	230	375	685	1230
Hydrogen-aluminum	Gas flow: 1.03×10^{-3} at 500°C, 1.02×10^{-2} at 600°C			
Hydrogen-molybdenum	598	813	1167	1862
Nitrogen-molybdenum	1362	1687	2174	2444
Nitrogen-iron	822	1114	—	—
Carbon monoxide-iron	692	993	—	—
Oxygen-silver	590	780	—	—

Dushman, *Scientific Foundations of Vacuum Technique*, John Wiley and Sons, 2nd ed., L. M. Lafferty, editor, New York, 1962.

rough, being averages of several reported values, which have been tabulated by Dushman (*Scientific Foundations of Vacuum Technique*, 2nd ed., J. M. Lafferty, editor, John Wiley and Sons, New York, 1962). However, the tables do show the relative ease with which the materials listed pass the gases listed. A material requiring a high temperature to pass gas at a given rate is more impervious to that gas than one passing the gas at the same rate at a lower temperature. It is evident that helium diffuses through silica much more readily than the other gases shown. This gas also diffuses fairly readily through Pyrex brand glass. As a matter of fact, helium diffuses through most borosilicate glasses rather readily at room temperature, which accounts for the use of such glasses (also quartz) in helium leaks. Although not shown in Table 8.3, argon diffuses through silica at a rate much lower than neon (roughly by a factor of 10).

8.5   Elastomers and Some Other Materials

Elastomers are natural or synthetic materials that can be or have been vulcanized to a state in which they have an inherent ability to accept and recover from extreme deformation. Table 8.5 shows elastomers that are commonly used in vacuum work. Elastomer types and common names are listed. Trade or type names are often used, such as *Hycar* for a Buna N rubber and *Fluorel* and *Kel-F* for fluorinated elastomers. These are trade names of the Minnesota Mining and Manufacturing Company, St. Paul, Minnesota. *Viton* and *Teflon* are trade names for fluorinated elastomers made by E. I. duPont de Nemours and Co., Wilmington, Delaware. Some other elastomers are also used in vacuum work, such as *organic polysulfide rubbers* (Thiokol or GR-P), *urethane elastomers, chlorosulfonated polyethylene,* and *styrene butadiene rubber* (Buna S or GR-S). The most commonly used designations for the rubbers used in vacuum practice are: natural, butyl, Buna N, silicone, and neoprene. Sometimes the word rubber is used as a designation for natural rubber and the word silastic for silicone rubber. *Silastic* is the trade name of Dow Corning Corp., Midland, Michigan. Elastomers are made up in various formulations, which can have a considerable effect on properties. Consequently, the properties listed in Table 8.5 should be considered of a very average nature.

Natural and synthetic rubbers are most commonly used in vacuum work. These materials generally contain volatile oils, plasticizers, and coloring pigments with absorbed moisture and gases. Most of the chemicals used have low vapor pressures at room temperature but some, such as sulfur, light oils, etc., are not in this class. Sulfur is used as a primary vulcanizing agent for most natural and synthetic rubbers. Elemental sulfur is usually used, but certain organic sulfur-containing accelerators that do not have the comparatively high vapor pressure of the elemental form may be used. When elemental sulfur is used, a certain amount of residue remains after the vulcanizing process, which can be removed by boiling the rubber in a hot caustic solution.

A number of properties of rubbers, natural and synthetic, must be considered before using them as gasket materials. Some important considerations are:

1. Rubber is essentially an incompressible material. Deformation in one direction must be compensated for in other directions.

2. All rubbers experience a certain amount of permanent (compression) set. This results from flow when stressed over a period of time. The permanent set increases rapidly with temperature.

3. A change of hardness occurs with age.

4. All rubber compounds flow at temperatures well under their so-called "melting points."

Table 8.5 General Properties of Elastomers

Elastomer Type	Natural Rubber	Isobutylene Isobrene Rubber	Nitrile Butadiene Rubber	Chloroprene Rubber	Silicone Rubber	Fluorinated Elastomers	Styrene Butadiene Rubber
Common name	Crude rubber	Butyl or GR-1	Buna N	Neoprene or GR-M	—	Fluorel, Kel-F, Viton	Buna S or GR-S
Properties							
Durometer range	30–100	40–75	20–100	40–95	45–60	55–90	40–100
Low temperature range of rapid stiffening	$-20°F$ to $-50°F$	$0°F$ to $-20°F$	$+30°F$ to $-20°F$	$+10°F$ to $-20°F$	$-60°F$ to $-120°F$	$+20°F$ to $-30°F$	$0°F$ to $-50°F$
Compression set	Good	Fair	Good	Fair to good	Good to excellent	Good to excellent	Good
Gas permeability resistance	Good	Outstanding	Good	Very good	Good	Excellent	Good
Resistance to petroleum products	Poor	Poor	Excellent	Good	Poor to fair	Good to excellent	Poor

5. Neoprene cannot be used above about 90 to 100°C because of permanent set. Natural and most synthetic rubbers can be used to about 120°C. (Some silicone rubbers are good to 250°C.)

6. Synthetic rubbers do not swell as much under the influence of oils and most solvents as natural rubber.

7. Synthetic rubbers tear more easily than natural rubbers.

8. All rubbers lose their elastic properties at low temperatures, becoming brittle at liquid nitrogen temperature. This is true of silicone rubbers also, although some of them can be used at temperatures as low as −85°C (−120°F).

The outgassing rates for various elastomers depend on the formulation used, the area exposed, the operating temperature, and the treatment of the elastomer before use. As a rule, there is no way to control the formulation since this is determined by the manufacturer. The area exposed becomes more critical as the operating pressure is lowered. Proper gasket groove design can help considerably in reducing the exposed area (see Chapter 10). The outgassing rate increases as the temperature is raised. This indicates that the ultimate pressure can be reached more rapidly if the elastomeric gaskets are heated. However, there is a limit to how much an elastomer can be heated; all are damaged when heated too much. Also, the compression set increases rapidly with temperature. It is because of these factors that elastomeric gaskets are not normally used in ultra-high vacuum systems. Such systems are baked at temperatures well above the damage point of all known elastomers. It is possible to approach the ultra-high vacuum region by combining bake-out with cooling of gaskets. Elastomeric parts should be clean and dry when installed in systems. Vacuum grease or oil (when permissible) should be used sparingly.

The measurement of outgassing rate is not an easy matter since the results depend not only on the factors noted above but also on the exact nature of the experimental setup. Consequently, it is difficult to compare the results obtained by various observers. One important problem is in distinguishing between gases and vapors from the elastomer and those from other sources. These other sources include real leaks from the atmosphere and outgassing of other materials in the system. Outgassing data are generally presented in one of the following ways:

1. Pressure-time curves.

2. Rate of outgassing per unit area of elastomer, often in $\mu l/sec/cm^2$.

Figure 8.3 shows pressure-time curves for several elastomers. The pressure values shown should not be taken too seriously since they de-

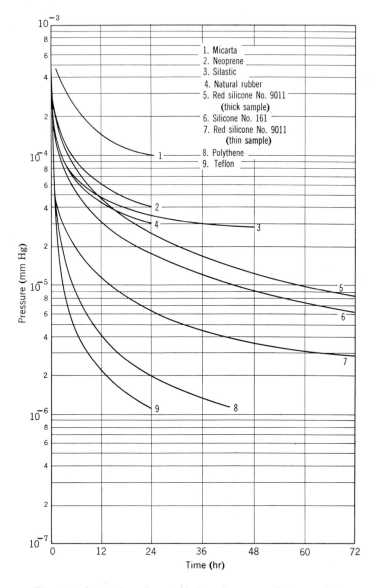

1. Micarta
2. Neoprene
3. Silastic
4. Natural rubber
5. Red silicone No. 9011 (thick sample)
6. Silicone No. 161
7. Red silicone No. 9011 (thin sample)
8. Polythene
9. Teflon

Fig. 8.3 Outgassing characteristics of some gasket materials.

pend so much on the samples used, the pumping speed, and the release of gases and vapors from other vacuum materials. However, the shapes of the curves shown are fairly typical of the outgassing behavior of elastomers. It will be noted that Teflon makes it possible to reach lower pressures than natural rubber or synthetic rubbers. Other fluorinated elastomers behave similarly to Teflon, although the formulation used can have a considerable effect. Figure 8.3 shows that many hours of pumping are required before the pressure curve begins to flatten out. Once an elastomer has been pumped on for many hours, subsequent pump-down times will be reduced considerably after short exposures to the atmosphere (preferably dry air). It is difficult to quote specific values for outgassing rates because of the many variables involved. Values ranging between 10^{-3} and 10^{-4} μl/sec/cm^2 for initial total outgassing of various elastomers at room temperature are typical. After 24 hr of pumping, these rates will usually drop by at least a factor of 10. The condensables (vapors that are removed by a liquid nitrogen trap) are much more abundant in elastomers than non-condensables.

The choice of a particular elastomer depends on the nature of the vacuum system. Factors usually considered are:

1. Temperature of operation.
2. The types of vapors present in the vacuum system.
3. The ultimate pressure required.

Natural and synthetic rubbers (usually neoprene or Buna N) are commonly used in systems that are operated at room temperature and at pressures around 10^{-5} or 10^{-6} mm Hg. Because of its temperature tolerance, silicone rubber is commonly used for low and high temperature operation. The fluorinated elastomers are highly resistant to most corrosive materials found in vacuum practice. Kel-F and Viton A are commonly used. Teflon is very good but suffers from "cold flow," i.e., flowing under pressure at room temperature. Suitable means for containing the Teflon (spring-loaded gaskets, etc.) will eliminate this difficulty. Fluorinated elastomers can produce toxic vapors at high temperatures, so good ventilation should be used. In trying to reach very low pressures, the *permeability* of the elastomer must be considered as well as its outgassing characteristics. Permeability is the property which determines how readily gases will pass through a material—the higher the permeability value, the more readily gases pass through the material. The following points can be made with regard to the permeability of elastomers:

1. Light gases (hydrogen and helium) pass through them more readily than heavy gases (oxygen, nitrogen, etc.).

2. There is a time lag involved in the gas passing through the elastomer, the value depending on the thickness of material. For a $\frac{1}{8}$ in. thickness, the time lag ranges from a few minutes for the light gases to an hour or more for the heavier gases. This behavior can complicate leak-hunting procedures.

3. Neoprene and butyl are much less permeable than Buna N, natural rubber, or silicone rubber. According to the British report A.E.R.E. GP/R 1875, neoprene is about ten times less permeable than natural rubber and butyl is somewhat less permeable than neoprene. This report gives a leakage rate of about 5×10^{-5} μl/sec for air across a 1 cm cube of natural rubber at a pressure differential of 1 atm. The fluorinated elastomers generally have low permeability values, depending on the formulation.

4. The permeability increases with temperature.

To reach low pressures at room temperature, elastomers with low vapor pressures and low permeabilities are desirable. Consequently, considerable work has been done with fluorinated elastomers, and particularly with Viton A. Baking an elastomer at a temperature that doesn't damage it will reduce pump-down time. However, it will still release some vapor after many hours of pumping. The lowest pressures have been reached by cooling elastomeric gaskets, which slows down the rate of outgassing and reduces the permeability. Farkas and Barry (*Vac. Symp. Trans.*, 1960) have reported reaching pressure values nearly as low as 10^{-10} mm Hg with butyl rubber and neoprene. The values reached with Buna N and Viton A were not much higher. Their measurements were made with the gaskets cooled to $-25°$C. At this temperature, some of the gaskets were damaged. Neoprene, Viton A, Buna N, and butyl rubber appeared to be damaged least.

The usual methods for joining elastomeric stock (cord) to make gaskets are: (1) vulcanizing and (2) cements. Vulcanizing is accomplished by heating the ends to be joined at a temperature appropriate for the material. Often the ends are first cemented together. A heating element made to fit the gasket is needed. Special techniques are required for handling some gasket materials.

Occasionally it is necessary to evacuate a device in which a cellulosic material such as cotton, paper, or wood is present. The problem with such materials is that water vapor, as well as carbon dioxide and many organic compounds, are strongly adsorbed on cellulose. Consequently, the degassing problem is quite serious. Some polymers, such as nylon, pliofilm, polyethylene, etc., are used in vacuum work (electrical insulation, thin windows, etc.). Little information is available concerning the adsorption properties of such materials. However, they are all highly permeable to gases and vapors (much more than elastomers). The permeabilities of pliofilm, polyethylene, and Saran are quite comparable. Ethyl cellulose has a considerably higher permeability. Ebonite and bakelite are sometimes used as electrical insulators in vacuum systems, but it is difficult to degas these materials. They also have high permeabilities, although not as high as those of polyethylene or pliofilm. In general, the above materials should be used in vacuum systems only when absolutely necessary.

8.6   Glasses, Ceramics, and Metals Sealing to Them

There is considerable variation in the composition of glasses manufactured in different countries. Fortunately, the manufacturer generally provides considerable detail regarding the properties of his glasses, such as specific gravity, composition, thermal expansion coefficient, annealing point, etc. The principal constituent of most glasses is sand (silica), to which various oxides are added. The chief glasses used in vacuum work are borosilicate glasses (Pyrex, Hysil, Wembley W1, etc.), although soda glasses and lead glasses find some use. Occasionally a high silica glass (Vycar, quartz, etc.) is needed. Some uses of these kinds of glasses are:

Borosilicate glasses: tubing and piping, chambers, diffusion pumps, vacuum tubes, seals to tungsten, molybdenum, Kovar, Fernico, etc.

Soda glasses: radio tubes, small vacuum chambers, seals to platinum, Dumet wire, etc.

Lead glasses: vacuum seals (lamps, tubes, etc.).

It must be kept in mind that the thermal coefficients of the metal and the glass must match pretty well. Most glasses used in vacuum work are clear. Colored glasses are made by the addition of various metal oxides. All glasses contain gases and vapors in various proportions. Water vapor and carbon dioxide are usually most prominent, although a large proportion of sulfur dioxide occurs in some types. These gases and vapors can be removed by heating, the effectiveness of removal increasing with the temperature.

Table 8.6 gives a listing, with some properties, of a few representative glasses. The thermal expansion coefficient is the important factor

Table 8.6   Some Representative Glasses

Type	Name and Code	Annealing Temp. (°C)	Thermal Expansion Coefficient (per °C)
Soda	Corning 0080	—	92×10^{-7}
	Wembley X8	520	90×10^{-7}
	Chance GW1	520	87×10^{-7}
Lead	Corning 0010	—	91×10^{-7}
	Corning 0120	—	89×10^{-7}
	Wembley L1	430	93×10^{-7}
	Chance GW2	420	86×10^{-7}
Borosilicate	Corning 7050	—	46×10^{-7}
	Corning 7052	—	46×10^{-7}
	Hysil GH1	590	33×10^{-7}
	Wembley W1	570	40×10^{-7}
	Intasil GS1	580	39×10^{-7}
	Chance GS3	480	48×10^{-7}
Silica	Corning 7900	—	8×10^{-7}
	Corning 7910	—	8×10^{-7}

in making glass-metal seals, since it must match that of the glass over a wide temperature range. A few common glasses and metals which can be sealed together are listed below:

Platinum: Thermal expansion coefficient = about $89 \times 10^{-7}/°C$. Seals to lead and soda glasses, e.g., Corning 0010, 0080, and 0120, Wembley X8, and Chance GW1 (see Table 8.6).

Dumet (or *copperclad*): (Cleveland Wire Works, General Electric Co., Cleveland, Ohio.) A copper-covered nickel steel alloy (about 42% nickel and 58% iron). Seals to lead and soda glasses. Maximum wire diameter = about 40 mils. Commonly used in "pinch seals."

Tungsten: Thermal expansion coefficient = about $40 \times 10^{-7}/°C$. Seals to borosilicate glasses, e.g., Corning 7740, Chance GS1, etc.).

Molybdenum: Thermal expansion coefficient = about $50 \times 10^{-7}/°C$. Seals to borosilicate glasses, e.g., Corning 7740, Chance GS4, etc.

Kovar: (Stupakoff Ceramic and Manufacturing Co., Latrobe, Pa.) A nickel-cobalt-iron alloy. Composition: nickel, 29%; cobalt, 17%; iron, 53.7%; manganese, 0.3%; mp = 1450°C. After annealing, thermal coefficient of expansion is 57 to $62 \times 10^{-7}/°C$ (30 to 500°C). Seals to borosilicate glasses, e.g., Corning 7052, Chance GS3, etc. The surface of the Kovar should be smooth, bright, and free from defects. After machining and polishing, degrease, rinse in methyl alcohol, dry, hydrogen fire to outgas, and anneal at 900°C for 15 to 30 min. Torch soldering operations should be done after hydrogen firing, the Kovar having first been copper plated at the site of the joint. Can be brazed to steels or nickel with pure silver or pure copper in a hydrogen furnace, after initial copper plating. No separate annealing is needed. Do not heat Kovar above 1100°C. It can be soft soldered with the same fluxes used for nickel, e.g., Nokorode salt solution. The surface must be clean. With brass or copper, apply heat to these rather than to the Kovar (previously annealed).

Fernico: (General Electric Co., Schenectady, N.Y.) Similar to but not identical with Kovar. Seals to borosilicate glasses.

Copper: OFHC (oxygen free, high conductivity) copper, often borated, seals to hard and soft glasses (see Chapter 10).

Ceramics: Ceramics are insulating materials that are commonly used in tube structures and to carry electrical power into vacuum systems. They are also used as heat insulators (furnaces, etc.). Most ceramics used in vacuum work are quite refractory, having softening points usually above 1400°C and a safe continuous operating temperature of around 1000°C or more. Some general types of ceramics are: steatites, aluminas, zircon porcelains, porcelains, natural stones (lava, etc.), magnesium silicate, titanium dioxides, and titanate. No attempt will be made to cover the numerous ceramics available commercially. Manufacturers' data and technical publications should be consulted. In general, drilling and machining are difficult because of the hard, brittle nature of these materials. Some of these materials, however, can be worked readily before firing. Examples are *lava* and *pyrophyllite*. Lava is a natural talc or soapstone, low in iron oxide and suitable for vacuum tube insulators. It can be machined with standard machining tools. After machining, a firing operation is needed to harden the material. There is some shrinkage during the firing, so careful heating and cooling are necessary. Porcelain tubes can be sealed to glass by glassblowing techniques.

Vacuum-tight seals between ceramic and metal are important in tube structures and elsewhere. After metallizing, normal brazing procedures can be used. The oldest and probably most reliable process is the *sintered metal process.* The area on the ceramic to be brazed is painted with a lacquer suspension of a refractory metal powder such as molybdenum or tungsten plus certain additives. The powders are sintered on the ceramic by firing in a controlled atmosphere furnace to a temperature slightly less than the softening point of the ceramic. There are many variations of the process, one being the *moly-manganese* process. One painting mixture used consists of: 40 g of manganese powder, 160 g of molybdenum powder (200 mesh), 4.0 g of 30 sec nitrocellulose, 120 ml of solvent composition, consisting of 20% cp acetone, 30% absolute methanol, and 50% cellusolve solvent. The solvent and metal powders are ball-milled for 5 days and nights. Nitrocellulose is added and milled for an additional 2 hr. The firing temperature and time depend on the ceramic, e.g., 1500°C for ½ hr for high alumina ceramics, in a hydrogen atmosphere with water vapor added to bring the dew point up to a minimum of 40°C. Parts must be heated and cooled slowly (minimum of 1 hr for each). After cooling the parts, they are often nickel-plated (Watts bath, etc.), followed by brazing to an appropriate metal. No cleaning should be done between metallizing and plating and parts should not be stored more than 3 days before plating. Brazing is done with eutectic silver-copper solder in a controlled atmosphere. Various painting mixtures have been used. The process is sensitive to the composition of the ceramic. The *hydride* process and the use of *"liquid"* gold and platinum are discussed briefly in Chapter 10.

High alumina ceramics have high strength, very good thermal shock properties, fairly good electrical properties, and are vacuum tight. However, it is difficult to join these ceramics to most metals or alloys because of differences in expansion coefficients. *Kovar* comes nearest to matching the expansion coefficients of these ceramics. Kovar must be nickel plated before silver soldering. Copper and nickel can also be brazed but the resulting seals are not too rigid. *Chrome iron* alloys contain chromium and iron and can be obtained with various thermal expansion coefficients to match specific glasses and ceramics. Number 430 chrome iron (distributor: Brown-Wales, Boston, Mass.) contains 14 to 18% chromium and around 80 to 84% iron, with small quantities of carbon, manganese, silicon, phosphorus, and sulfur. Its expansion coefficient is $105 \times 10^{-7}/°C$ between 0 and 600°C, and $120 \times 10^{-7}/°C$ between 0 and 1000°C (averages). It can be brazed to a ceramic such as Alsimag No. 243. Pressed powder techniques are also used for making ceramic-metal seals.

8.7 Copper and Copper Bearing Alloys

Commercial copper is electrolytically refined and contains only traces of impurities. The melting point is about 1084°C. Copper is a very good heat and electrical conductor and this property makes it useful for many vacuum applications, such as electronic tube parts, electrical leads, and cooling lines. It is practically nonporous and is widely used as a general vacuum material, for chambers, tubing or piping, base plates, etc. It is easily soldered and brazed (see Chapter 10). Ordinary electrolytic copper contains enough oxide inclusions so that in firing operations the hydrogen that diffuses into the metal reacts with the oxide to form water vapor. The water vapor (under pressure) produces ruptures at grain boundaries. This can result in brittleness, cracks, and porosity. In some cases, copper surfaces are tinned or soft-soldered, since the metal corrodes and oxidizes rapidly in air. Its high melting point makes it possible to effectively bake out copper parts for ultra-high vacuum work. For certain specialized applications, such as electronic tube bodies, gaskets, and sometimes chambers, OFHC (oxygen free, high conductivity) copper is used.

There are many alloys of copper, but only those most commonly used in vacuum practice will be considered here. The most widely used alloys are *brasses* and *bronzes*. Strictly speaking, brasses are copper-zinc alloys, while bronzes are copper-tin alloys. However, in practice, many brasses contain various other metals, and many commercial bronzes contain zinc. Alloys containing zinc, cadmium, lead, antimony, or bismuth should not be used in vacuum systems which are to be baked because of the high vapor pressures of these metals (see Table 8.1). Also, alloys containing these metals should be run through hydrogen furnaces used for these metals only (see Section 10.6). Vacuum firing is likely to alter the composition and therefore the properties of such alloys. Brasses are widely used for vacuum parts, such as diffusion pump parts (except in mercury pumps), chambers, base plates, valves, fittings, etc., in high speed dynamic systems. They are readily worked (machined, drilled, etc.), soldered, and brazed. The workability depends on the composition. Bronzes are not as widely used as brasses in vacuum practice. Many small, commercial valves are made of bronze. In general, it is difficult to get vacuum-tight cast bronze parts. Leaky parts can be heavily painted with a sealant (such as glyptal) or covered with a coating of soft solder. However, this procedure is not generally recommended.

Constantan, also known as *Advance, Cupron, Copel, Eureka*, and *Ideal*, is a nickel-copper alloy having a very low temperature coefficient of resistivity (0.00001). Its composition is: nickel, 40 to 45%; copper, 60 to 55% (Advance contains 0.5% iron). It can be readily soft soldered or brazed, and can also be spot welded. Its approximate melting point is 1210°C. It is useful for making thermocouples with iron, copper, and other metals. *Tombac* is a special alloy of copper (72%) and zinc (28%). It can be obtained as flexible, corrugated wall tubing, which is useful for making vacuum connections. *Beryllium-copper* is a heat treated alloy, containing from a fraction of a percent to around 2% of beryllium. Up to 1% of cobalt is included, and some silver is sometimes added. This type of alloy is widely used in making springs, electrical contacts, diaphragms, bellows, Bourdon gauge tubes, etc. Its resistance to fatigue is greater than that of spring steel. The life of the alloy is about twenty times that of phosphor bronze. It should be vacuum fired for degassing since the beryllium content is very sensitive to oxidation by the small amounts of oxygen and/or water vapor present in the tank hydrogen of the furnace. It can be soft soldered if this is done after final heat treatment. Otherwise, brazing is necessary. Zinc chloride and resin are satisfactory fluxes for soft soldering. It is easy to machine and in the processes of drawing, bending, or pressing, it should be softened by quenching. *Phosphor bronze* is an alloy consisting of 90 to 95% copper, with a few percent of tin and a fraction of a percent of phosphorus. It is used for the same purposes as beryllium-copper but does not have as long a life (see above). The melting point is about 1000°C.

8.8   Iron and Some of Its Alloys

A particularly pure form of *iron* known as SVEA metal is sometimes used in tube structures. However, because of low cost, availability, and excellent machining properties, *cold rolled* (low carbon) steel is widely used in vacuum practice—pump barrels, piping, chambers, base plates, tube structures. So-called "cold rolled" steel is not any particular type, although generally the low carbon steels are received from the mill in a cold drawn or cold finished condition. Perhaps *low carbon* steel is a better term since it covers hot rolled, cold drawn, cold finished, and forged parts. Most rolled and forged steel alloys are sufficiently nonporous for use in vacuum systems. Rolled steel should be examined for slag inclusions and cracks, particularly along the axis of rolling. The oxides and hydroxides formed on the surface of common iron alloys absorb and adsorb considerable quantities of water when exposed to humid atmospheres. It is difficult to remove the water. Heating will speed up removal but this method is often impractical. A better method is to stop formation of oxides and hydroxides by plating the surfaces with nickel or chromium. The plating should completely cover the surface. Cadmium plating should not be used for parts that are to be heated (tube parts, for example) because of the high vapor pressure of cadmium.

For parts that must be heated (not plated) low carbon steels that have a low sulfur content, so as to minimize gassing, are used. Some common steels used for this purpose are: 1010, 1019, 1020, and 1022. A general purpose alloy is 1020. Cast iron parts are also fairly commonly used in vacuum practice. These are often permanent mold castings. Sand castings are often porous. Of course, some sealant such as glyptal could be used although this is not generally recommended. A suitable impregnation process is more reliable. However, considerable care must be taken in the impregnation process. *Tool steels* (drill rod, ground stock) are not as easily machined as low carbon steels. However, they can be heat treated to obtain various degrees of hardness and temper, so that they are useful for making punches, dies, mandrels, etc. Such steels are often classified as *regular, moly* (molybdenum), and *cobalt*. It should be noted that all common low carbon steels are magnetic.

Stainless steels have come into fairly common use in vacuum practice, for pumps, piping, and chambers. Essentially all stainless steel ultra-high vacuum systems are fairly common. Also, such steels are useful in tube work, not only as component parts, but also for making brazing jigs, where the protective layer of chromium oxide which is formed in the hydrogen furnace prevents the work from being brazed to the jig. *Austenitic* stainless steels, such as 302, 302B, 303, and 304 are commonly used in vacuum work. These are often called 18-8 stainless steels, since they contain about 18% chromium and 8% nickel (balance iron). These steels are nonmagnetic. The melting points of the 18-8 steels are somewhat over 1400°C. Other types of stainless steels (400 series, etc.) contain mostly chromium and iron (little or no nickel). These are generally magnetic. Stainless steels are more difficult to machine than low carbon steels, the difficulty depending on the composition. Type 303 is often used because of its relatively easy machinability. Stainless steels offer several advantages over low carbon steels. Smooth surfaces remain smooth since oxides and hydroxides of the type found with low carbon steels do not occur. This means that the "effective" area is less and water is not taken up in large quantities. This leads to much easier degassing. In many cases, the nonmagnetic properties of certain stainless steels are important.

Invar is an alloy consisting of iron (64%), and nickel (36%). It has a low coefficient of thermal expansion (1×10^{-6}/°C between 0 and 100°C). Sometimes it is referred to as a nickel-iron alloy. It is magnetic and can be soft soldered, brazed, and spot welded. *Chrome iron* can be obtained with various thermal expansion coefficients to match certain glasses and ceramics (see Section 8.6).

8.9 Molybdenum, Tantalum, and Tungsten

Molybdenum is a nonmagnetic refractory metal, machinable in some forms. It is somewhat more ductile than tungsten. Wires can be drawn down to 0.025 mm and can be sealed to borosilicate glass (up to about 8 mm diameter). It has a high melting point (2620°C) and a low vapor pressure. It can be gold-nickel or platinum brazed (nickel plate the parts for best results in gold-nickel brazing), and may be readily spot welded to most metals used in electronic tubes except to itself and to tungsten, to both of which it can be spot welded only with great difficulty. If the clean metal is given a thin coating of stopcock grease, some adhesion can be obtained. Molybdenum can be either vacuum or hydrogen fired for outgassing. It readily absorbs oxygen at 1000°C and is cleaned by heating in contact with sodium nitrite at around 800°C. This metal is used for heaters, grids, anodes, heat shields, and supports. A tungsten-molybdenum alloy wire (about 51% molybdenum) is used as a standard tube filament or cathode heater wire. It has greater strength than the pure metal and is more readily shaped than tungsten.

Tantalum is a metal with a high melting point that is commonly used in tube construction. It is very inert to chemical attack, withstanding all acids except hydrofluoric and fuming sulfuric at room temperature. It is attacked by strong alkalis. This metal is ductile, malleable, and machinable, being much like mild steel in these respects. It spot welds readily to other metals and to itself. Tantalum parts for tubes must be thoroughly outgassed in vacuum since tantalum acts as a getter at elevated temperatures (see Table 5.1). This destroys the structure of the metal, making it extremely brittle. Parts should not be passed through a hydrogen furnace. Tantalum is used for tube parts and is a good electron emitter. Tantalum carbide is very hard and is used with other carbides (such as tungsten) for cutting tools, wear resistant parts, and dies for wire drawing apparatus.

Tungsten has the highest melting point (3377°C) and the greatest tensile strength of all metals. It is difficult to bend without splitting and cannot be machined. However, it can be formed and cut by grinding with carborundum or other abrasives. It is available in thin sheet and wire forms. Small wires can be formed readily into various shapes, provided there are no sharp bends, and seals to borosilicate glasses (see Chapter 10). Tungsten is almost universally used as a filament or heater material. It is also used for support and press wires because of its rigidity and glass sealing properties. Heaters and filaments should always be annealed and mounted strain-free, i.e., the ends should be welded to their supports or lead wires in the positions they take naturally, if possible. Otherwise, only a small amount of bending is allowable. Tungsten spot welds readily to nickel, tantalum, and other metals, but with difficulty to itself and to molybdenum. It can be brazed in a hydrogen furnace if first nickel plated, or if certain high temperature solders are used, e.g., gold, 82.5%, nickel, 17.5%; pure platinum; copper, 75%, nickel, 25%, and pure nickel. Tungsten oxidizes easily even at room temperature.

8.10   Nickel and Its Alloys

Nickel is a very useful metal for vacuum tube construction. It is easily degassed, has a fairly high melting point (1455°C), has good corrosion resistance, is very ductile when annealed, spot welds, hard and soft solders, machines, and forms easily. It is a very satisfactory base for oxide coated cathodes. For most purposes, a high nickel alloy is used (around 99% nickel). This type of alloy is commonly used for tube parts. Other alloys with somewhat less nickel are used for spring parts.

Monels is a name given to a series of nickel-copper alloys having various properties. The Monels containing aluminum are subject to oxidation in the hydrogen furnace, but can be nickel plated like stainless steel to protect them for brazing (see Chapter 10). Generally, these alloys contain between 60 and 70% of nickel plus cobalt. Small amounts of iron and manganese are included, plus traces of several other elements. The melting point depends on the composition but is usually between 1300 and 1350°C. Some Monels are magnetic, while others are nonmagnetic. Also, only certain of these alloys can be soldered and brazed. The machinability will vary considerably from alloy to alloy. Examples of alloys that can be soldered and brazed, and have good machinability properties are *wrought Monel* and *Monel R* (International Nickel Company, New York, New York).

Inconel is a nonmagnetic alloy of nickel (79.5%), chromium (13%), and iron (6.5%), having fair machinability and good corrosion resistance at high temperatures. It is used for heating elements and sometimes for tube elements. Brazing to Inconel is accomplished in the same manner as for stainless steel. The melting point is about 1400°C. It is difficult to solder or braze and tough to machine. *Nichrome* is an alloy of nickel and chromium, with a few percent of iron often added. This alloy is commonly used in heating elements. Most of the alloys classified as Nichrome are difficult (or impossible) to soft solder. Hard soldering is generally difficult but spot welding can be readily carried out. *Chromel* is also an alloy of nickel and chromium, with iron often included. This type of alloy is used in thermocouples and can usually be hard soldered and spot welded. The melting point is ordinarily about 1400°C. *Alumel* is a high nickel alloy, containing small amounts of iron and manganese. It can be hard soldered and spot welded, the melting point being about 1400°C. It is used in thermocouples. *Constantan* and *Invar* are often classified as nickel alloys but are considered elsewhere (see Sections 8.7 and 8.8). Various nickel-iron alloys, containing various proportions of other elements, are used to make glass-metal and ceramic-metal seals (see Section 8.6).

8.11   Vacuum Greases, Oils, Cements, and Waxes

Vacuum greases are commonly used to effect seals in such devices as stopcocks and gasket joints (static, rotating, and sliding). In some cases, vacuum oils are used, including diffusion pump oils. Oils are generally not as satisfactory as greases for most types of seals, since they are more readily squeezed out, leaving a dry seal. However, they do find uses, such as in oil-sealed stopcocks and some rotating or sliding seals (continuous lubrication). In general, vacuum greases should not have a vapor pressure of more than about 10^{-4} mm Hg at 30°C, and should maintain adequate viscosity at this temperature.

Table 8.7 gives some properties of commonly used vacuum greases and oils. This listing is not complete but is representative of commercially available products. It should be noted that the true vapor pressure of such a material is difficult to determine, since the volatile constituents and occluded gases should first be removed. Consequently, the results reported by different observers vary according to the test methods used. The data shown in Table 8.7 include manufacturers' values and values from the technical literature. The vapor pressures shown are not strictly comparable because the test methods used were not identical and the degree of degassing may not be the same for all materials. It takes many hours to degas a grease by pumping to the

Table 8.7   Some Vacuum Greases and Oils

Material	Vapor Pressure (mm Hg at $t°C$)	Melting Point (°C)	Remarks
Apiezon Oil J	10^{-3} at 200°C about 10^{-8} at 20°C	—	Used where a moderately viscous oil is needed—oil-sealed stop-cocks, etc.
Apiezon Oil K	10^{-3} at 300°C 10^{-9} to 10^{-10} at 20°C	—	Used where an exceedingly vis-cous oil is needed.
Apiezon Grease L	10^{-3} at 100°C 10^{-10} to 10^{-11} at 20°C	47	Well-fitting ground joints (not stopcocks). Safe temp. = 30°C
Apiezon Grease M	10^{-3} at 200°C 10^{-7} to 10^{-8} at 20°C	44	General purpose grease. Safe temp. = 30°C.
Apiezon Grease N	10^{-3} at 200°C 10^{-8} to 10^{-9} at 20°C	43	Glass stopcocks. Safe temp. = 30°C.
Apiezon Grease T	about 10^{-8} at 20°C	125	Safe temp. = 110°C. Used where temperatures up to this value are encountered.
Celvacene, light	10^{-6} at 20°C	90	Vacuum seals and taps where heating is encountered.
Celvacene, medium	less than 10^{-6} at 20°C	120	General vacuum grease. Useful where heating is encountered.
Celvacene, heavy	less than 10^{-6} at 20°C	130	Rubber gaskets and metal to rub-ber joints where heating is encountered.
Lubriseal	less than 10^{-5} at 20°C	40	General vacuum grease.
Silicone stopcock grease	less than 10^{-5} at 20°C	215	Good at high temperatures.
Vacuseal, light	10^{-5} at 20°C	50	General vacuum grease.
Vacuseal, heavy	10^{-5} at 20°C	60	General vacuum grease.

Note: Apiezon oils and greases: James G. Biddle Co., Philadelphia, Pa. Celvacene greases: Consolidated Vacuum Corp., Rochester, N.Y. Lubriseal grease: Arthur H. Thomas Co., Boston, Mass. Silicone grease: Dow Corning Corp., Midland, Mich. Vacuseal greases: Central Scientific Co., Chicago, Ill.

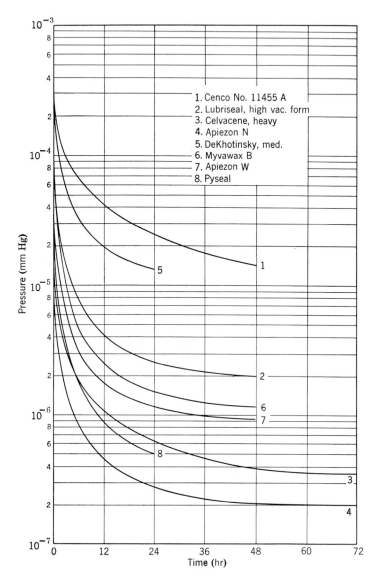

1. Cenco No. 11455 A
2. Lubriseal, high vac. form
3. Celvacene, heavy
4. Apiezon N
5. DeKhotinsky, med.
6. Myvawax B
7. Apiezon W
8. Pyseal

Fig. 8.4 Outgassing characteristics of some greases, waxes, and cements.

point where its vapor pressure approaches the ultimate value (see Fig. 8.4). It is possible to reduce the required pumping time by keeping exposure of the grease to the atmosphere to a minimum. Table 8.7 can be used as a guide for the selection of an appropriate grease, which is usually determined by vapor pressure and operating temperature. Greases are normally applied at room temperature and can be used up to a few degrees below the melting point. In general, greases should be applied sparingly, the surplus being wiped off, since they absorb gases and vapors and are dirt catchers.

Cements and *waxes* are primarily used to make temporary seals, although some of them are used for "permanent" seals when other methods are not applicable. A listing of some cements and waxes is given in Table 8.8. Apiezon Q is actually not a cement or wax. It is a plasticene-like material that can be easily worked over surfaces. It is particularly valuable for temporarily sealing leaks. Other materials of a similar nature are on the market. The materials shown in Table 8.8 are general purpose cements and waxes and have been used extensively in vacuum work. New cements and waxes are continually being made available commercially. The value of each new material can only be determined after extensive use. Eastman Resin 910 (Eastman Kodak Co., Rochester, N.Y.) is a fast drying cement that is liquid at room temperature and produces strong seals with various materials. Pyroceram Cement No. 45 (a patented product of Corning Glass Works, Corning, N.Y.) is claimed to be suitable for joining various glasses, ceramics, and metals, including tungsten and molybdenum. Epoxy resins are available from a number of manufacturers. When properly applied, these materials will make strong, permanent seals. In general, surfaces should be cleaned thoroughly and the instructions of the manufacturer should be followed in detail. Also, it is usually advisable to degas such seals before using them in vacuum. Clean surfaces are also necessary for other types of cements (and waxes). Figure 8.4 shows outgassing curves for several greases, cements, and waxes. The rapid drop in pressure at early times is typical of materials that have not been degassed. It will be noted that many hours of pumping are required to reduce the vapor pressure to near its ultimate value.

Cements are commonly used to join elastomeric materials, particularly for gaskets. No attempt will be made to list cements that can be used with presently available elastomers. The formulations used can have a considerable effect on the success or failure of a particular cement. Also, new elastomers and cements are continually being developed. It is best to consult the manufacturer, local fabricators, or

Table 8.8   Some Cements and Waxes

Name or Trade Name	Composition and/or Properties	Preparation or Source	Temperature Properties	Application and Remarks
Alundum cement	Aluminum oxide and water. A refractory cement—sets on firing. Can be reworked at lower temperatures.	Mix to consistency of thick mud.	Firing temp. = 1100°C	Setting Nichrome or other wires on furnace muffles, etc.
Apiezon Q	Modeling clay consistency. $vp = 10^{-4}$ mm Hg at 20°C	J. G. Biddle Co., Philadelphia, Pa.	Do not use above 30°C (liquefies).	Temporary sealing of leaks. Sealing bell jars, etc.
Apiezon W-40	Medium soft wax. $vp = 10^{-3}$ mm Hg at 20°C.	J. G. Biddle Co., Philadelphia, Pa.	Apply at 40–50°C. Safe temp. = 30°C.	Semipermanent joints.
Apiezon W-100	Medium hard wax. $vp = 10^{-3}$ mm Hg at 20°C.	J. G. Biddle Co., Philadelphia, Pa.	Apply at 80°C. Safe temp. = 50°C.	Semipermanent joints.
Apiezon W	Hard wax.	J. G. Biddle Co., Philadelphia, Pa.	Apply at 100°C. Safe temp. = 80°C.	Permanent joints.
Beeswax and rosin	Equal parts beeswax and rosin. Adheres to clean, cold metal. Remove with putty knife. Somewhat soluble in equal parts carbon tetrachloride and ethyl alcohol.	Smoking indicates vaporization of beeswax, making it harder. Add more beeswax to retemper.	mp = 57°C. Softens at 47°C.	Not very strong. Use for sealing bell jars, etc. (temporary). Apply smoking hot with eye-dropper or spatula.
DeKhotinsky cement	Shellac + 20–40% wood tar. Not affected by water, benzol, benzene, or turpentine. Only slightly affected by ether, chloroform, and the strong mineral acids. Slightly soluble in alcohol. Grades: hard, med., soft. $vp =$ about 10^{-3} mm Hg at 20°C.	Central Scientific Co., Chicago, Ill.	Softens at around 100°C	Sticks well to clean, hot (150°C) surfaces. For semipermanent seals.

Dennison's sealing wax	Hard wax. Soluble in alcohol. vp = about 10^{-5} mm Hg at 20°C.	Softens at 60–80°C.	J. G. Biddle Co., Philadelphia, Pa.	Semipermanent seals.
Glycol phthalate (glyptal)	Phthalic anhydride condensed on glycol or glycerol, etc. Long heating causes embrittlement. Low vp.	Do not heat above 150°C.	As glyptal (General Electric Co.), a paint either with or without a pigment (ZnO) plus a solvent such as xylol or toluol.	Adheres well to clean metals. Used to seal quartz windows to metal or glass and to seal small leaks.
Litharge and glycerin	Lead monoxide and glycerin. Inert to water, most acids and alkalies.	Holds well up to 260°C.	The pulverized litharge is heated thoroughly at 400°C. Mix with pure glycerin to consistency of a paste.	Tube basing cement. Not used in vacuum.
Picein	Black wax. vp = 10^{-6} mm Hg at 20°C. Soluble in benzol and turpentine. Unaffected by alcohol.	One variety softens at 50°C, melts at 80°C. Another variety melts at 105°C.	Shrader and Ehlers, New York, N.Y.	Very good electrical insulator. Adhesive. Used to seal mica windows and to close small leaks.
Plaster of Paris	Calcium sulfate. Irreversible.		Suspend in water to consistency of paste. Salt speeds setting, vinegar retards it.	To support glass bulbs containing mercury in wood or metal box. Tube base cement.
Myvawax S	Soft hydrocarbon wax. Low vp. Dark amber color.	mp = 72.5°C.	Consolidated Vacuum Corp., Rochester, N.Y.	Adheres well to glass and metal. Soluble in petroleum, ether, carbon tetrachloride, and benzene. Unaffected by acetone, alcohol, or Octoil.

Table 8.8 (Continued)

Name or Trade Name	Composition and/or Properties	Preparation or Source	Temperature Properties	Application and Remarks
Sauereisen cement. Insalute cement	Suspension of ceramic powders in sodium silicate solution. Somewhat refractory. Sets very hard. Soluble in boiling water and alkali solutions.	Central Scientific Co., Chicago, Ill. Thin with sodium silicate solution (water glass).	Can be used up to 590°C.	Anchoring structures in tubes. Cementing glass or ceramics to metal. Not vacuum tight.
Sealstix	A modified DeKhotinsky cement. Higher strength.	Central Scientific Co., Chicago, Ill.	Not as flammable as De-Khotinsky. Greater working temp. range.	Same as DeKhotinsky.
Shellac (pure orange only)	A natural gum. Soluble in alcohol. High tensile strength.	Hold gum in a cloth and soak in alcohol for many hours.	Polymerized by heat. Get higher softening temp. (150°C for 3 hr).	As varnish in alcohol. Temporarily sealing small leaks.
Silver chloride	Insoluble in water, alcohol, benzol, and some acids. Soluble in solution of sodium thiosulfate. Most metals are wet by the fused salt.		mp = 455°C.	Sealing optical windows. Cool slowly to prevent cracking.
Araldite	Two forms—powder and stick (light brown or "silver").	Aero Research, Ltd., England.	Curing times: 240°C—10 min 220°C—20 min 190°C—1 hr 180°C—2 hr	Heat without pressure. Good for glass, metals, mica, porcelain, and ceramics. Flat or well-fitting surfaces.

local suppliers regarding the best method of cementing a particular elastomer.

8.12   Some Drying and General Sorption Agents

The term *drying* agent is used here to designate materials whose primary function is to absorb water vapor. The common drying agents are *selective* absorbers, i.e., they absorb water vapor but not permanent gases and most other vapors. On the other hand, many *sorption* agents will absorb various permanent gases and vapors. Such agents are *nonselective*. There are also some selective sorption agents, which are used to remove some specific gas or vapor. Drying agents are used to protect some part of the vacuum system from water vapor and to let a system down to dry air (small systems). A drying agent is usually used with a McLeod gauge. Of course, drying agents are commonly used outside vacuum systems for drying various vacuum materials (or keeping them dry). A listing of various drying materials is given in Table 8.9. The choice of material is often a matter of convenience. Silica gel is included in Table 8.9. Actually this material is a nonselective absorber and can be used to trap permanent gases and vapors or as a pump (see Chapters 5 and 11).

The most commonly used sorption agents (for dynamic systems) are *activated charcoal, zeolite,* and *activated alumina.* These materials are commonly used as refrigerated or nonrefrigerated traps in a pumped-down system to reach the lowest possible pressures. Refrigeration (usually by liquid nitrogen) greatly increases the sorption capacity of these materials. They are also used as roughing (sorption) pumps for pumping from atmospheric to a few microns pressure (see Chapter 5). Of course, the sorption of gases and vapors under any circumstances constitutes a pumping action. *Activated charcoal* is often made from coconut charcoal. One method of preparation is to heat the soft part of the coconut shell in a muffle furnace for several hours just below a red heat until there is no more evolution of vapor. The temperature is then raised to a dull red heat for 30 sec. Before using, it is heated to 450°C or so for several hours in vacuo (activated). After use, the sorbed gases and vapors can be driven off by heating and the charcoal is reactivated. Theoretically, activated charcoal should have an area of 2500 m²/g, which leads to an absorption capacity of about 2000 cc (STP—standard temperature and pressure) of gas per gram. However, the actual capacity is considerably less than this. Other sources of charcoal, such as briquetted coal and sugar, can be used instead of

Table 8.9   *Drying (Dehydrating) Agents* *

Material	Volume of Air per Milliliter of Desiccant (liters at STP)	Residual Water per Liter of Air (milligrams)
Calcium sulfate (anhydrous)	0.45–0.7	2.8
Calcium chloride (granulated)	6.1–24.2	1.5
Calcium chloride (tech. anhydrous)	4.0–5.8	1.25
Zinc chloride (sticks)	0.8–2.1	0.98
Barium chlorate (anhydrous)	2.3–3.7	0.82
Sodium hydroxide (sticks)	2.3–8.9	0.80
Sulfuric acid (conc.)	—	—
Magnesium chlorate (hydrated)	4.0–7.2	0.031
Potassium hydroxide (sticks)	3.2–7.2	0.014
Silica gel	2.1–5.2	0.006
Calcium oxide	7.6–10.1	0.003
Magnesium chlorate (anhydrous)	2.8–5.9	0.002
Aluminum oxide	5.6–6.2	0.001
Barium oxide	10.6–25	0.00065

Note: Most measurements indicate that phosphorus pentoxide is somewhat better than aluminum oxide (alumina) as a dehydrating agent. Also, sulfuric acid (conc.) is comparable to magnesium chlorate (hydrated).

* Bower, *Natl. Bur. Std., J. Res.,* **12,** 241 (1934) and **33,** 199 (1944).

coconut. Sorption values comparable to those for coconut can be obtained but the source of supply and the method of preparation are extremely important. Activated charcoal will sorb the atmospheric constituents, including the rare gases, but with different degrees of efficiency. At room temperature, its sorptive capacity for oxygen, carbon dioxide, and nitrogen is much greater than for the rare gases (except argon) or for hydrogen. At liquid nitrogen temperature, its sorptive capacity for all gases goes up, but faster for some than others. At this low temperature, the value for hydrogen approaches that of nitrogen. Water vapor is also sorbed effectively down to its vapor pressure at the temperature involved.

Certain aluminosilicate minerals are important as absorbents. Chief among these are the zeolites, both natural and artificial (synthetic). The choice of material and method of preparation are important in determining sorption properties. Zeolites are dehydrated for use and are usually used in pellet form. After use, they can be degassed by heating. Synthetic zeolites are marketed under various trade names. These materials (natural and synthetic) sorb the atmospheric constituents and other gases and vapors, like activated charcoal, but with different relative efficiencies. Zeolites are particularly effective for sorbing water vapor and carbon dioxide at room temperature. Refrigeration increases the sorption efficiencies for all gases by various amounts. It appears that zeolites effectively sorb gases and vapors of smaller molecules (water, oxygen, nitrogen, and hydrogen) but vapors or gases with large molecules are sorbed with difficulty (or not at all). Zeolites can be heated to around 600°C without harm. *Activated alumina* has a similar structure to that of the zeolites and is used in much the same manner. It can be heated to around 1000°C. Reports indicate that this material is comparable to the zeolites in its ability to sorb the atmospheric gases and vapors.

Silica gel has already been mentioned as a drying agent. It is made by dehydrating silicic acid. Not only does it sorb water vapor, but it will also sorb the permanent atmospheric gases (as well as other gases and vapors). However, it is not as effective for the permanent gases as activated charcoal, which, in turn, is generally somewhat less effective than the zeolites and activated alumina. It should be noted that considerable work has been done with other types of silicates, besides the zeolites, and also with quartz, particularly for sorbing water vapor. *Palladium black* is a material that will sorb many gases and vapors, and is particularly effective for hydrogen (better than activated charcoal). The sorption capacity is increased considerably by lowering the temperature. The material is prepared in the following manner: Dissolve palladium (sheet or wire) in aqua regia (in a beaker); evaporate on water bath (no more acid vapors). Dilute and warm the solution. Add concentrated sodium carbonate solution to neutralize the acid. Add a slight amount of acetic acid, warm the solution, and add a warm concentrated solution of sodium formate. The palladium comes down as a black flocculent precipitate. Decant the supernatant liquid and wash the precipitate with distilled water (until no trace of chlorides). Wash the palladium "black" with alcohol and transfer to a U-tube. Dry by blowing air over it and then evacuate U-tube on a roughing pump (with slight warming). Pass hydrogen over it for some time, and with gas still passing, seal off the tube. This leaves the palladium black in equilibrium with hydrogen, and it can be kept active for a long time. Palladium black is effective as a sorption material but is expensive and is prone to "poisoning." Care in preparation and in the method of usage is very important. *Platinum black* is a material having properties very similar to those of palladium black, with a high sorption capacity for hydrogen, and is prepared in much the same manner.

8.13   Temperature Data

It is often useful to have temperature data concerning metals and alloys readily available. Table 8.1 gives the melting points of a number of elements, and additional data are included in other sections of this chapter. However, for ready reference Fig. 8.5 is included here. This table shows the melting points of a number of metals and alloys commonly used in vacuum practice. The colors corresponding to various temperature ranges are also indicated in this figure. There are so many special purpose alloys available that it is not possible to include complete temperature data. The manufacturer (or supplier) should be consulted.

8.14   Alphabetical Listing

For ready reference, many of the materials used in vacuum practice are listed here. Where information concerning some of these materials is contained elsewhere in this book, reference is made to the appropriate section or chapter.

ACETONE. See Section 8.3.

ALCOHOLS. See Section 8.3.

ALLOYS. See Sections 8.7, 8.8, and 8.10.

ALUMINUM. This metal is available in pure form and in various alloys.

Properties. In pure form (Alcoa 2S, etc.) it is ductile and is a good electrical and thermal conductor. Mp = 660°C. Does not sputter

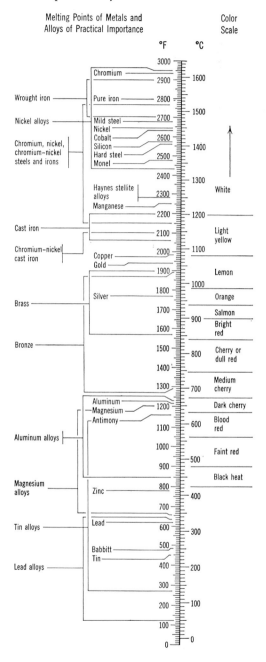

Fig. 8.5 Melting points and color scale.

easily. Readily evaporated to form thin films. Difficult to degas thoroughly. The alloys are generally readily worked in the shop: drilling, machining, etc. The workability depends on the composition. The alloys can be cast for metal vacuum parts. Sandcasting or the permanent mold process are generally used. The surface can be hardened easily by the "anodizing" and other processes. Parts can be joined together by using aluminum solder (see Chapter 10).

Uses. In the pure form, it is used to make aluminized mirrors (high reflectivity, particularly for the ultraviolet), as a getter (see Chapter 5), as electrodes in gas discharge tubes (low sputtering rate), to make thin windows (x-ray tubes, etc.), and to make bakable metal gaskets. Commercial tubing or piping can be used for various vacuum connections. Rolled plate is used for vacuum chambers and base plates. Cast parts are used for a variety of purposes—valves, diffusion pumps (particularly jet assemblies), grooveless flange gaskets, etc. The design of the dies is important in order to get vacuum-tight castings. Aluminum (usually in alloy form) is commonly used for vacuum parts where good heat and electrical conductivity is required.

APIEZON. See Section 8.11.

ASBESTOS

Properties. Heat resistant mineral fiber. Two natural forms are readily available, hornblende (mp = 1150°C) and serpentine (mp = 1550°C). Commercially they are combined with organic binders and sold as asbestos paper, textiles, millboard, etc. Transite = asbestos + Portland cement.

Uses. Not suitable for vacuum systems because of too much outgassing. Used as heat insulation in glass-blowing and brazing procedures, in ovens, and as lagging for pump heaters, etc.

BAKELITE. Trade name for various phenolic and other synthetic resins, usually combined with paper, textiles, etc.

Properties. Good electrical insulator. Easily cut, machined, etc. Its vapor pressure is high at temperatures above 30°C. Low heat resistance, chars at 280°C. Does not absorb water.

Uses. General purpose electrical insulator (outside vacuum system). For tube bases. Not used inside most vacuum systems (high vapor pressure, low heat resistance). Bakelite cements (mp = 80°C) are often used to seal tube caps.

BARIUM AND STRONTIUM. Barium is used as a getter (see Table 5.1). Carbonates of these elements are converted in vacuum to oxides, which are good electron emitters (oxide coated cathodes). Sometimes the carbonate of calcium is also used.

BEESWAX. See Section 8.11.

BERYLLIUM. A very hard metal which is used for windows in x-ray tubes. Films of the metal reflect strongly in the ultraviolet. It readily alloys to most hard metals. This metal is toxic, and an efficient hood (and respirator) should be used when working with it to avoid inhaling dust.

BERYLLIUM-COPPER. See Section 8.7. The comments in this section are generally applicable to beryllium-nickel alloys.

BERYLLIUM OXIDE. A refractory ceramic (mp = 2570°C) sometimes used in electronic tube structures. The powder is very poisonous, so that ceramics containing this material should be ground only under an efficient hood (preferably using an efficient respirator).

BRASS AND BRONZES. See Section 8.7.

CADMIUM. Used in special photocells and in cadmium lamps (reddish) which are used to give a standard wavelength in spectroscopy. It can also be used to trap mercury vapor (heat to between 300 and 400°C and water cool regions on both sides of trap). Cadium plated parts that must be baked should not be used in vacuum systems (particularly sealed-off systems).

CALCIUM. Keep in a nonoxidizing atmosphere or immersed in a liquid. Used as a getter where an almost invisible deposit is needed (low "noise").

CARBON

Properties. This material is obtainable as graphite, diamond, and carbon black. It sublimes, i.e., goes directly from solid to vapor without becoming a liquid. Carbon is a conductor of electricity, the value of conductivity depending on the form. It is difficult to degas (see Section 8.4). Aquadag is a colloidal graphite suspension in distilled water (or oil). It can be baked on glass at about 450°C. Sometimes 3 to 4% by volume of sodium silicate is added to the aquadag to improve its adhesion to glass. It can also be baked on metal at about 900°C. The metal should have a coefficient of expansion very near that of the carbon.

Uses. Graphite is used for anodes in small transmitting tubes and as a lubricant for moving parts in vacuum. Aquadag is used to provide electrical screening on walls of tubes, contacts to deposited metal coatings, etc. It is also used to improve the electrical conductivity of paper, asbestos, etc., by impregnation. Carbon is used in making jigs for glassblowing and air brazing, either with torch or induction heating. Diamond is being used in certain specialized applications in vacuum work.

CARBON DIOXIDE. See Section 8.4.

CARBON TETRACHLORIDE. See Section 8.3.

CEMENTS AND WAXES. See Section 8.11.

CERAMICS. See Section 8.6.

CESIUM. Oxidizes rapidly in air. Prepared in vacuum by heating a mixture of cesium bromide or chloride with calcium metal. Used in some special types of photocells.

CHROME IRON. General designation for chrome steels having carefully selected expansion coefficients to match those of glasses or ceramics to which direct fusion seals are to be made. See Section 8.6.

CONSTANTAN. A nickel-copper alloy with a very low temperature coefficient of electrical resistivity. See Section 8.7.

COPPER AND ITS ALLOYS. See Section 8.7.

EBONITE. Readily worked insulator. It should not be used in most vacuum systems because of its high vapor pressure. Its electrical and mechanical properties deteriorate rapidly above 40°C. It is used for switch panels, tube sockets, etc. In many cases, it should be cooled to avoid damage.

FERNICO. See Section 8.6.

FLUXES. See Chapter 10.

GLASSES. See Section 8.6.

GOLD

Properties. In the pure form (24 carat) it is malleable with a melting point of about 1063°C. It is readily evaporated and sputtered to form thin films. It is a good electrical conductor.

Uses. In the pure form it is used to prepare thin films for optical work, etc., and to make diffusion seals with copper and other metals. With this type of seal, the clean, accurately machined surfaces to be joined are gold plated, or separated by a thin gold foil, and pressed together at a moderate temperature well below the melting point of gold. Alloys of gold with copper, nickel, and/or other metals are used as brazing materials (see Chapter 10).

GREASES (VACUUM). See Section 8.11.

INCONEL. See Section 8.10.

INVAR. See Section 8.8.

IRON AND ALLOYS. See Section 8.8.

KOVAR. See Section 8.6.

LEAD. The vapor pressure of this metal is high at elevated temperatures (see Table 8.1). It is used as a gasket material for high speed dynamic systems (see Chapter 12). Systems using lead gaskets should be baked well below the melting point of lead.

MAGNESIUM. Oxidizes rapidly in air. Keep in a vacuum desiccator or under alcohol if it is to be used in vacuum. It is used as a getter and for highly reflecting front surface mirrors. In the latter case it is evaporated on glass with a covering layer of evaporated aluminum. The resulting layer is tougher than a pure aluminum layer.

MERCURY. See Section 8.2.

MICA

Properties. Good electrical properties. Shows good resistance to moderately high temperatures and to thermal and mechanical shock. It can be accurately cut, drilled, and slotted. Two forms: muscovite and biotite. Muscovite is nearly colorless in thin sheets and is preferred for tube and electrical work. Affected only by hydrofluoric acid. Can be split along cleavage planes into very thin sheets (0.0005 in. thick). Avoid heavily colored or striated mica as the inclusions are chiefly iron compounds which are likely to be gassy. Do not subject to temperatures in excess of 625°C because then the cleavage planes separate and its mechanical properties are partially or largely destroyed.

Uses. As an insulator, particularly to support electrodes in tubes, and for thin windows (nuclear physics and optical work). Pieces selected for tube work should be free of interlaminar defects, bubbles, inclusions, cracks or pinholes, and after cutting to shape should be handled carefully to avoid fraying the edges. For vacuum work the mica pieces, after machining, should be rinsed in two changes of acetone and then soaked in methyl alcohol for 30 min, followed by thorough drying in an air oven at 200°C. Aluminum oxide (alundum), the material used to coat heaters, may be used to coat the surface of mica to improve its ability to resist the effects of electrical leaks from evaporated metals in tubes. However, metal films evaporated on mica do not adhere very strongly and may be removed by rubbing. Low capacity condensers (less than 0.001 of a microfarad—μfd) are made by evaporating silver films on either side of mica sheet. Resistors are made by evaporating a layer of gold or platinum, of appropriate thickness, on mica. Mica sheet 0.001 in. thick transmits about 90% of normally incident white light.

MOLYBDENUM. See Section 8.9.

NICHROME. See Section 8.10.

NICKEL AND NICKEL ALLOYS. See Section 8.10.

NILVAR. A nickel-iron alloy similar to Invar (Driver-Harris Co., Harrison, New Jersey).

PALLADIUM

Properties. Similar to platinum. A closed-end tube, heated to around 700°C, passes hydrogen readily. At lower temperatures, it absorbs hydrogen, the absorption rate increasing rapidly as the temperature drops. *Palladium black* is used for this purpose (see Section 8.12).

Uses. Palladium barrier ionization gauges. Provides pure hydrogen (or deuterium).

PHOSPHORUS. A nonmetal that occurs in two forms—red and white. Can be used as a getter. A filament wire is dipped into a suspension of red phosphorus in alcohol with a little sodium silicate. When the wire is heated in vacuum, the phosphorus absorbs the common atmospheric gases. Phosphorus is not widely used as a getter since it has an adverse effect on thermionic and photocathodes.

PHOSPHORUS PENTOXIDE. See Section 8.12.

PHOSPHORS. These are chemical substances and mixtures that fluoresce in vacuum under the influence of a stream of electrons. Common types are: cadmium borate, cadmium sulfide, calcium tungstate, zinc orthosilicate (Willemite), and zinc sulfide. A Willemite screen can be prepared as follows. Mix a quantity of the dry powder, as a suspension, with distilled water and allow it to settle undisturbed on the backing (glass, nickel, etc.). The backing must be scrupulously chemically clean. When the settling is complete, partly decant the water carefully so as not to disturb the settled material. Allow the screen to dry in warm (not over 35°C) air, protected from drafts and dust, for several hours. Then put in an oven for 30 min at about 70°C. Detailed information regarding phosphors can be obtained from various tube manufacturers.

PLATINUM

Properties. Pure metal—very ductile and highly resistant to corrosion at high temperatures. Can be sealed directly to certain glasses (see Chapter 10). Easily spot welded. Passes hydrogen and deuterium when heated. Platinum black, at liquid nitrogen temperature, absorbs large quantities of hydrogen (see Section 8.12).

Uses. Platinum foil flux for spot welding tungsten or molybdenum. Pure platinum brazing material at its melting point (1770°C) for molybdenum, tungsten, and for joining tungsten metallized thorium oxide cathodes to their supports. High temperature resistance thermometers. Glass-platinum seals. Platinum and platinum-rhodium alloys—high temperature thermocouples. Provides pure hydrogen or deuterium.

POTASSIUM. Reacts strongly with air at normal temperatures. Used in photocells—high green and blue sensitivity. Can be used to trap mercury. Coat inside of trap with distilled film of potassium. Absorbs up to its own weight of mercury. Doesn't absorb other condensable vapors. A potassium trap is not convenient.

QUARTZ (silica). See Section 8.6 and Chapter 10.

RHODIUM. Not readily drawn into wire form. Used in electroplating. Gives best corrosion-free surface. Evaporated film on glass is hard and adhesive. Reflects all parts of visible spectrum uniformly.

RUBBERS. See Section 8.5.

RUBIDIUM. Similar to cesium. Used in rubidium oxide photocells—responds to white light in much the same way as the human eye. (Maximum sensitivity in the green.)

SILICA. See Section 8.6 and Chapter 10.

SILICA GEL. See Section 8.12.

SILVER AND ITS ALLOYS

Properties. Pure silver—best conductor of heat and electricity. Readily evaporated in vacuum to form metallic reflecting or conducting films (not strongly adherent). Can be chemically deposited on glass by the Rochelle salt, Brashear, or silver paste processes. Readily soft soldered or brazed but almost impossible to spot weld. Common alloys are: fine silver (99.9+% silver), sterling silver (92.5% silver, 7.5% copper), U.S. coin silver (90% silver, 10% copper), jewelry silver 80% silver, 28% copper).

Uses. Reflecting and conducting films. Brazing.

SODIUM. Soft, chemically active metal. Usually stored in kerosene. Oxidizes rapidly. Soluble in liquid ammonia. Used in photocells and sodium vapor lamps.

SOLDERING AND BRAZING ALLOYS AND FLUXES. See Chapter 10.

STEELS (stainless, etc.). See Section 8.8.

STRONTIUM. See BARIUM.

SULFUR. Good electrical insulator but little used in vacuum practice. Remove as an impurity, e.g., from rubber tubing, by dissolving in caustic soda solution.

TANTALUM. See Section 8.9.

TEFLON. See Section 8.5.

THERMOCOUPLES. Various combinations of metals can be used for thermocouples. The choice is chiefly determined by the range of tem-

Table 8.10   *Some Thermocouples*

Type of Thermocouple	Range for Continuous Service (°C)	Range for Short Service (°C)
Copper to Constantan (or Advance)	−190 to 350	to 600
Chromel to Constantan (or Advance)	0 to 900	to 1100
Iron to Constantan (or Advance)	0 to 900	to 1100
Chromel to Alumel	0 to 1100	to 1350
Platinum to platinum (13% rhodium)	0 to 1450	to 1700
Platinum to platinum (10% rhodium)	0 to 1450	to 1700
Tungsten to molybdenum	1000 to 2500	to 2500

perature to be measured. Some common types of thermocouples are shown in Table 8.10. Thermocouples for use in air must be protected against oxidation and mechanical injury. Thermocouples subjected to excessive temperatures or oxidizing or corrosive atmospheres cannot be depended on to hold their calibration. For heavy duty, use heavy gauge wire in making the couple. However, the thinner the wire, the more responsive the couple will be to small temperature changes. External connections from the couple can be made with copper or nickel wires, provided that both conductors are the same, and that the joints are made in the same manner, i.e., by soldering, brazing, welding, or clamping. For highest accuracy a cold junction should be used in series with the heated junction, i.e., an identical thermocouple immersed in melting ice.

THORIUM AND THORIUM OXIDE. Thoriated tungsten is made by combining a small amount (2% or so) of thorium oxide with tungsten. This material, used in the form of a wire filament, is a very good electron emitter when "activated," i.e., maintained at high temperature until the electron emission reaches the acceptable value. Rugged, directly heated cathodes have been made by using pressed and sintered mixtures of thorium oxide and tungsten or molybdenum. The mixture is formed into a hollow tube, current being applied from the ends through platinum brazed expansion-type ferrules of molybdenum. Such cathodes give good emission at fairly low temperatures (1650°C), do not have to be activated, and can be repeatedly exposed (cold) to the atmosphere without damage.

TIN. Pure tin has a low mechanical strength, low melting point (232°C), and low vapor pressure (see Table 8.1). In spite of its low mechanical strength, it is sometimes used as a soft solder for parts made of copper or other metals. Joints designed with pure tin should have no stresses on the solder. Tin is used for various static seals and bakable valves (see Chapter 11). It can also be used to join glass to metal tubing. There should be an overlap of about 20 mm, and a better seal is obtained by first sputtering the glass with a thin layer of platinum. Tin is an ingredient of many common soft solders.

TITANIUM. This metal can be evaporated in vacuum from directly heated (electrically) tungsten strips at about 2000°C, with pure films obtained at a pressure of 10^{-5} mm Hg or less. The adhesion to glass is good and such films are used as interference filters. Titanium is also widely used as a getter, particularly in getter-ion pumps.

TOMBAC. See Section 8.7.

TRANSITE. See asbestos. Obtainable in various forms—sheet, tubing, etc. Can be drilled, sawed, and machined. Useful for ovens, heat-resistant table tops, etc. (a Johns-Manville Company proprietary product).

TRICHLOROETHYLENE. See Section 8.3.

TUNGSTEN. See Section 8.9.

VITON. See Section 8.5.

WATER. See Section 8.2.

WILLEMITE. See PHOSPHORS.

WOOD'S METAL. See Chapter 10.

NINE

Cleaning Techniques

9.1 The Nature of Contaminants

Component parts intended for use in high vacuum systems must always be scrupulously clean. Visual examination is not adequate. Gases may be adsorbed on surfaces or absorbed in the interior of the vacuum components. As the pressure in a system is reduced the effect of these gases will become more important. At first the regular atmospheric gases will be pumped. When the pressure approaches the vapor pressures of contaminants in the system, then the pressure will drop more slowly, the rate of drop depending on the nature and extent of the contaminants. The most common contaminant is water vapor, and it must be kept in mind that it has a pressure of about 17 mm Hg above the liquid at room temperature. Other common contaminants are greases, oils, solder fluxes, etc., which are introduced during fabricating processes. The nature and extent of the contaminants will determine the time required to pump down to a particular pressure. The cleaning methods to be considered here should make it feasible to achieve operating pressures of 10^{-6} mm Hg or better in reasonable times. The lowest pressure obtainable will be determined by how carefully these cleaning procedures are followed.

One method of classifying the various types of contamination is the following:

a. Visible contaminants such as deposits of tapping compounds, cutting oil, polishing materials, varnish, etc.

b. Contaminants hidden in crevices and holes.

c. Gases and vapors adsorbed on surfaces.

d. Gases and vapors absorbed within vacuum materials.

e. Gases and vapors which are combined chemically with materials of the system (chemisorption).

As a very general rule it is advisable to eliminate contaminations of types *a* and *b*. How much effort should be expended in trying to eliminate contaminations of types *c*, *d*, and *e* depends to a great extent on the type of vacuum process involved.

9.2 General Methods of Reducing Contamination

The first step is to remove as much gross contamination as possible by mechanical means—blasting with abrasives, rubbing with emery cloth, filing, and scraping. Filings and small loose particles can be removed by use of an air hose. Clearly the use of mechanical methods will be limited by the nature of the vacuum components. They cannot be used on fragile parts, such as filaments. Also, ceramics and glasses should be wiped with a clean, lint-free cloth as a preliminary step. An air supply is very useful in various phases of the cleaning operation. However, to avoid oil contamination the supply should be equipped with a satisfactory oil filter between the compressor and the hose.

The following steps are then considered to be suitable for many purposes:

1. Clean with a soap solution or synthetic detergent—immerse part and agitate, brush on solution, etc.

2. Rinse with hot distilled water.

3. Rinse in a suitable solvent. Acetone is commonly used. This solvent is moderately effective as a cleaning agent, dissolves paints and varnishes, evaporates readily, and dissolves water to some extent. Other solvents which are used are MEK (methyl ethyl ketone) and methyl alcohol.

a. MEK—very effective for paints and varnishes. Moderately effective as a degreaser. Evaporates more readily than acetone.

b. Methyl alcohol—very effective for water. Doesn't evaporate as readily as acetone.

4. Use a vapor degreaser with a chlorinated solvent. Examples of this type of solvent are: methylene dichloride, chloroform, carbon tetrachloride, dichloroethane, and trichloroethylene. Trichloroethylene is generally used since it is considered somewhat less toxic than the other chlorinated solvents. It is a clear, colorless liquid with a boiling point of 86.7°C. It decomposes above 130°C. It is not flammable nor does it form combustible or explosive mixtures with air. However, it is harmful to most pump oils.

5. Rinse in two changes of methyl alcohol (or acetone).

6. Dry in warm air blast or oven at 70 to 110°C.

The above methods should be adequate to remove contaminants of types *a* and *b* (Section 9.1). These steps are carried out prior to assembly of the system (or incorporation of new components). The parts cleaned according to the above directions should not be left exposed longer than necessary to the atmosphere. This results in condensation of water and accumulation of dust particles on surfaces. When necessary, try to store clean components in a dry, dust-free area. The actual operating pressures that can be achieved by following the above procedures will be influenced by the constructional materials and types of traps used. Step 4 above is sometimes replaced by the use of rinses (three or more) of trichloroethylene. This is not considered to be as good as the use of a vapor degreaser, particularly for contaminants of type *b.*

Precautions:

1. Use hot, or preferably boiling, distilled water. Do not use tap water—this may result in contaminating surface layers.

2. Flushing with solvent (or distilled water) is preferable to the use of a bath. Dispersed particles in the bath may result in contamination. Several rinses in clean bath solution will minimize the problem. Sometimes a bath agitated by an ultrasonic generator is used.

3. The vapors of such solvents as trichloroethylene, methyl alcohol, and acetone are toxic. Use adequate ventilation. Avoid skin contact with trichloroethylene, particularly when using a vapor degreaser. If hot parts are handled, use clean, lint-free cotton gloves or equivalent.

To cut down the time required to reach a low operating pressure, some attention must be given to contaminants *c, d,* and *e.* Type *c* contaminant is primarily water and is caused by exposure to the atmosphere. To speed up the vaporization of the water, the procedure is to heat surfaces. This is best done in vacuum and can be carried out by, (1) incorporating a heating cycle into a preparatory phase of the pump-down, or (2) using a separate system. If method (2) has to be used, the transfer time to the main system should be as short as possible. If parts must be heated in air, then this should be done in an oven. The temperature should be raised above the boiling point of water (100°C). Higher temperatures will speed up the vaporization process. However, temperatures may be limited by construction materials, e.g., rubber gaskets. Heating in vacuum is usually done by means of resistance elements, induction heating or, ion bombardment. These methods are discussed in Section 9.7.

Removal of contaminants of type d again involves the use of heat but for longer times, in order that the absorbed gases and vapors can diffuse to the surface and be pumped away. Higher temperatures will speed up the process but, again, the temperature may be limited by construction materials. The choice of special materials and techniques to achieve very low pressures in complete vacuum systems is discussed in Chapter 14. Parts to be incorporated in a vacuum system can be heated up to temperatures where their vapor pressures become significant. This heating is usually done in vacuum or in a hydrogen furnace using resistor elements or induction heating. The hydrogen is helpful in removing oxide layers. Other methods for removing oxides are discussed in the following section. In most cases oxides and other chemical compounds (contaminants of type e) are removed by chemical or electrolytic means. It must be kept in mind that many vacuum construction materials can be used without removing oxide layers when the vapor pressure of the oxide is low. This is true of aluminum and stainless steel. On the other hand rusty iron contributes substantial amounts of vapor (water vapor) to the system. This is because rust takes up large quantities of water vapor whereas some other oxides do not. In many cases parts are chemically or electrolytically cleaned so that certain operations, such as brazing, sealing to glass, plating, etc., can be carried out. This is particularly true in the case of the production of sealed-off, evacuated devices, such as electronic tubes. The removal of gases and vapors from a material is often referred to as degassing.

Ultrasonic cleaning is an effective process that has come into general use, particularly for small parts. In this process, the parts to be cleaned are immersed in a fluid and high frequency sound waves (inaudible) are applied. The agitation produced by these waves will loosen contaminants. Details of the process can be obtained from the manufacturers of this type of equipment.

9.3 Descaling and Pickling Processes

The general term "scale" is applied to the coatings which develop on metals when they are exposed to high temperatures. Scale can form during many types of operations, such as casting, annealing, and forging. The physical structure and composition of the scale depend on how it was formed. It is generally necessary to remove scale in order to apply the usual cleaning procedures. Certain types of scale will tend to absorb gases and vapors. Also, the scale must be removed in order to apply various types of coatings, such as metal plating.

BLAST CLEANING

This method involves bombarding the part with high speed abrasive particles. The chemical nature of the scale has no effect in this case. However, it is sometimes necessary to use different sizes and types of abrasives to remove the more firmly bound scales. The method is particularly good for removing siliceous (sandy) scales produced during welding or sand casting. At one time, sand was most commonly used as an abrasive. Although sand is cheap the dust generated creates a health hazard. This applies to many other natural abrasives. Steel shot or grit are now commonly used, with the grit giving a better finish. The grit must be stored under dry conditions. To descale stainless steel it is necessary to use stainless steel shot, alumina, or iron-free sand. In the case of aluminum, abrasive materials include washed silica sand, alundum, emery, steel particles, and aluminum oxide (alumina). Silica sand is not recommended because it causes a hazardous environment for workers. In the case of any soft metal like aluminum it is best to use relatively low air pressures and a fine abrasive so as to avoid surface damage. Regardless of the type of abrasive used, the operator should wear protective equipment such as a face shield and a respirator. The hazard due to dust can be minimized by using vapor blasting. Here the blast stream is a slurry consisting of either abrasive and water or abrasive and solvent.

PICKLING

This process for cleaning metals is primarily used with iron and steel, usually as a preparatory step for plating or applying some other type of coating. An appropriate chemical bath is used, depending on the metal being cleaned. The part should first be cleaned by the usual methods using detergents, liquid solvents, or vapor degreaser, as required. No attempt is made here to list the numerous pickle baths that can be used. Only some typical baths that are generally useful are given. Iron and steel are usually cleaned in a solution of sulfuric acid or hydrochloric acid. Commercial sulfuric acid has a concentration of 95% or less and commercial hydrochloric acid consists of about 35% hydrochloric acid gas in water. The latter acid cannot be used at temperatures above about 40°C. In the case of castings, silicates and sand are often mixed with the scale and in this case hydrofluoric acid is added to one of the above acids to dissolve these materials. This mixture can also be used for welding scale. After the pickling process has been carried out the parts should be thoroughly washed. A recommended procedure is: (1) Wash in running cold water (preferably using high pressure jets). (2) Rinse in a running-water tank. (3) Rinse in very hot water.

A. *Cast Iron*

Sulfuric acid (conc.)	1 pint	in hard rubber
Hydrofluoric acid (conc.)	1 pint	or paraffined
Water	1 gal	glass container

Use at room temperature or higher.

B. *Iron and Steel*

(1) Sulfuric acid (conc.)   3 oz/gal
    Potassium nitrate   3 oz/gal
    Temperature, about 70°C.
(2) Phosphoric acid   10–15% solution
    Temperature, about 70°C.

C. *Monel*

Sulfuric acid (conc.)   1 pint/gal
Sodium nitrate   ¾ lb/gal
Sodium chloride   ¾ lb/gal
Temperature, 82–88°C.

D. *Inconel*

(1) Nitric acid (conc.)   1 gal
    Hydrofluoric acid (conc.)   1 pint
    Water   2 gal
    Temperature, 66–74°C.
(2) Sulfuric acid (conc.)   13 oz/gal
    Rochelle salt   13 oz/gal
    Temperature, 71–82°C.

E. *Stainless Steel*

With this type of material it is often advisable to pickle in two operations, the first to loosen the scale and the second to remove it. With light scales, the first operation can be omitted. A typical bath for loosening scale is concentrated sulfuric acid (12 oz/gal) at a temperature of 85°C. Some pickling baths are:

(1) Nitric acid (conc.)   1 gal
    Hydrofluoric acid (conc.)   1½ gal
    Water   2½ gal
    Temperature, 50–65°C.

(2) Hydrochloric acid (conc.)   5 gal
    Nitric acid (conc.)   1 gal
    Water   14 gal
    Temperature, 50–70°C

Any pickling operation on stainless steel should be followed by a passivation or immunization process. This is simply an additional chemical treatment which produces a very stable film on the part, leading to high corrosion resistance. A typical solution consists of 1 gal of concentrated nitric acid in 4 gal of water. Treatment is for 20 min at a temperature of 50°C.

In the pickling operations discussed above, inhibitors are added to the acids which are used. These materials minimize pitting of the metal. They are added in very small amounts, usually a fraction of a percent by volume. Examples of inhibitors are pyridine, quinoline, and aniline.

CAUTION: The acids used above are corrosive and should be handled with care. Hydrofluoric acid can be handled only in hard rubber, plastic, or paraffined containers. When mixing sulfuric or phosphoric acid with water, the acid should be poured into *cold* water slowly with stirring in order to avoid spattering. These acids are destructive of skin and clothing.

NOTE: Never pour water into concentrated acid.

9.4   Electrolytic Cleaning and Polishing

These methods again involve the use of a chemical bath (the electrolyte) but an electric current is passed through this bath between two electrodes. The work piece is generally the anode (positive electrode). The cathode (negative terminal) that is used depends on the electrolyte.

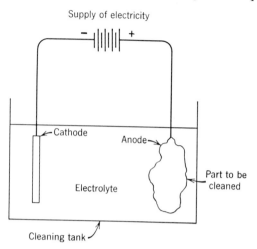

Fig. 9.1   Electrolytic cleaning and polishing.

The choice of the tank in which the process is carried out is based on its resistance to chemical action by the electrolyte. When the prime purpose is cleaning, including the removal of scale, the arrangement is such that the cleaning is done by the mechanical action of gases released at the work piece. In electrolytic polishing, material is dissolved from the surface of the work piece, resulting in a polished appearance. In a sense this can be considered to be a method for improving the appearance of the part which is competitive with mechanical methods. However, since cleaning does result from the application of this method, it is considered here. The parts to be electrolytically polished should first be cleaned by mechanical means. The basic setup for electrolytic cleaning and polishing is shown in Fig. 9.1. In some cleaning procedures the part to be cleaned is made the cathode.

ELECTROLYTIC CLEANING

A. *Nickel*

A suitable bath for the electrolytic cleaning of nickel is the following:

Sulfuric acid (conc.)	172 cc
Phosphoric acid (conc.)	546 cc
Water	308 cc

The work should be connected at the anode and sufficient current should be used to cause gassing of the work. Caution must be exercised since this solution rapidly attacks and dissolves nickel.

B. *Tungsten and Molybdenum*

These procedures should not be used for heaters and filaments since they should not be etched.

(1) Electrolyze for 30 sec or longer in a 20% solution of potassium hydroxide using a carbon electrode.

(2) Rinse in hot water.

(3) Rinse in cold distilled water.

(4) Dry in warm air blast.

The following method should be used only for molybdenum.

(1) Electrolyze as in (1) above.

(2) Rinse in water.

(3) Place in 50%, warm, concentrated and inhibited hydrochloric acid for 5 min.

(4) Rinse in running water to remove loose oxides.

(5) Repeat (3) and (4) until oxide removal is complete.

(6) Rinse thoroughly in water.

(7) Rinse in clean methyl alcohol.

(8) Dry in warm air blast.

ELECTROLYTIC POLISHING

A. *Copper and Brass*

These metals can be polished using orthophosphoric acid with concentrations from 25 to 60%. The bath can be operated at room temperature, and a voltage of 2 v is generally adequate. Several patented methods are available. An electrolyte used by Faust (U.S. Patent 2,347,039) is composed as follows:

Phosphoric acid (conc.)   59%
Sulfuric acid (conc.)   4%
Chromic acid (conc.)   0.5%
Water   36.5%

Currents around 3.5 amp/in.^{2} are used at temperatures of about 38°C. The time involved is generally 5 or 10 min. This solution can also be used to polish nickel, steel, and many alloys.

B. *Stainless Steel*

Since stainless steel is difficult to polish mechanically, many chemical baths have been used for electrolytic polishing. Several involve the use of phosphoric acid–sulfuric acid solutions. Sometimes chromic acid is added to this mixture. In the following baths concentrated acids are used:

(1) Phosphoric acid   60%
    Sulfuric acid   20%
    Water   20%

At a current density of 5 amp/in.^{2} the time involved is a few minutes. The temperature should not be above 80°C.

(2) Phosphoric acid   15%
    Sulfuric acid   60%
    Chromic acid   10%
    Water   15%

The current density to use is 4 amp/in.^{2} and the time necessary to complete the action is 30 min. The temperature should be 50°C.

The above electrolytes can also be used to polish mild steels. The surfaces should be ground or mechanically polished. Current densities of between 0.7 and 3.5 amp/in.^{2} and temperatures as high as 90°C are used with the phosphoric acid–sulfuric acid mixture.

9.5   Cleaning of Mercury and Glass

Steps to be taken in obtaining extremely clean mercury are:

1. Remove large particles by squeezing through chamois leather.

2. Shake thoroughly in a bottle with chromic acid and then with water.

3. Allow the mercury to fall in a fine spray through a long column of dilute nitric acid (25 parts acid to 75 parts distilled water). An arrangement for doing this is shown in Fig. 9.2a. The pinchcock shown should be closed before adding the acid and then opened when enough mercury has collected. The side tube should be a little more than 1/13.6 of the height of the acid column.

4. Pass the mercury through a column of distilled water. The arrangement of Fig. 9.2a can be used.

5. Dry the mercury by evaporation at around 350°C.

6. Distill the mercury in vacuum. A simple apparatus for doing this is shown in Fig. 9.2b. The process involved is simply vaporizing the mercury under vacuum, condensing it on cooled surfaces, and collecting it.

Original mercury used in vacuum pumps, gauges, valves, and cutoffs should be chemically pure (cp grade). When mercury is cleaned for these purposes, the steps outlined above should be followed. For mercury that is not too dirty, steps 1 and 2 can sometimes be eliminated.

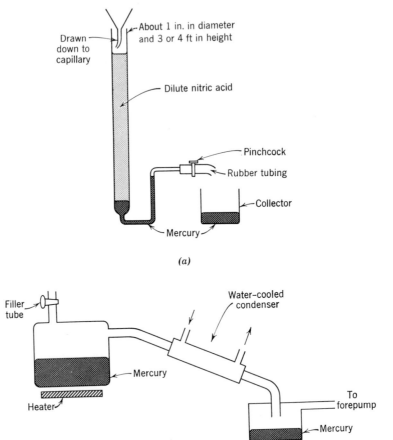

Fig. 9.2 (*a*) Spraying through dilute nitric acid. (*b*) Vacuum distillation apparatus.

A standard cleaning solution for glass consists of 35 parts of a saturated solution of potassium dichromate mixed with 1000 parts of concentrated sulfuric acid (chromic acid solution). The acid should be poured slowly into the dichromate solution, while stirring. The solution is most effective when hot and should be used at around 110°C. It should be red in color for best results. If the solution is muddy or greenish, it should be discarded. Often glass is first washed with a soap solution and rinsed in distilled water before using the chromic acid solution. Finally, the glass should be washed with warm distilled water and dried with hot, dust-free air. Additional precautions must be taken with glass which is to be coated with a metal film in vacuum, e.g., glass for aluminized mirrors. The treatment with chromic acid should be continued until a water film spreads uniformly over the glass surface (this indicates freedom from grease). Next dry the glass with cotton wool. Finally, polish with cotton wool and alcohol until a uniform water vapor film forms on the surface when breathed upon. It should be pointed out that often glass parts for vacuum systems can be adequately cleaned by following the general cleaning methods of Section 9.2. Chromic acid solution can also be used to clean tantalum.

9.6 Miscellaneous Cleaning Procedures

REMOVAL OF FLUX

All soft solder fluxes are more or less corrosive with time and should be thoroughly removed with appropriate solvents. The acid or chloride types of fluxes are removed by means of very hot water, preferably by boiling the part in several changes of water. The oil or resin types of fluxes are removed by degreasing agents while rosin can be dissolved with methyl alcohol. The grease paste type of flux contains chlorides. This type of flux can be removed by using a combination of degreasing agents and boiling water, in sequence.

Hard soldering fluxes generally contain borax, boric acid, and sometimes fluorides and other compounds. These should be removed after brazing by immersing the component in several changes of boiling water. In many cases it is also necessary to scrub with a wire brush or clean steel wool.

CLEANING OXIDIZED IRON

The cleaning of oxidized iron is in a sense a pickling process. A chemical bath that has been found to be effective is 50% inhibited hydrochloric acid. The use of inhibitors to avoid pitting has already been mentioned and many such materials are available on the market. The procedure in removing oxide from iron is:

1. Place oxidized iron in the acid solution for about 5 min.
2. Rinse thoroughly with water.
3. Dry.

REMOVING HEAVY OXIDE ON COPPER

In this case a warm, 75%, inhibited solution of hydrochloric acid is used. The procedure is:

1. Immerse part in acid solution for a few minutes.
2. Rinse thoroughly with water.
3. Dry.

TUNGSTEN AND MOLYBDENUM HEATERS AND FILAMENTS

In this process use a 20% potassium hydroxide solution, e.g., 300 g of potassium hydroxide in 1200 cc of water. This is the procedure:

1. Boil part in solution for 5 min.
2. Rinse thoroughly in tap water.
3. Rinse in cold distilled water.
4. Dry in warm air blast.

9.7   Cleaning by Degassing

Degassing is considered under cleaning techniques since the word clean can be used in vacuum practice to apply to systems which release very little gas or vapor. As has been indicated, the principal methods used to release gases and vapors are heating and discharge cleaning (ion or electron bombardment). To these methods must be added the use of getters. Gettering materials have already been considered as pumps. In a sense they can also be considered as cleaning agents since they result in "clean-up" of gas. Similarly ion or electron bombardment can be considered as either a pumping or a cleaning procedure.

DYNAMIC SYSTEMS

These are continuously pumped systems and most of them involve mechanical pumps or mechanical pumps in combination with diffusion and/or ejector pumps. Operating pressures are commonly greater than 10^{-6} mm Hg. The general cleaning methods already discussed, when carried through the vapor degreasing stage with subsequent drying, are generally adequate. It must always be kept in mind that, after cleaning and drying, parts should not be left exposed to the normal atmosphere any longer than is absolutely necessary. Degassing these systems by heating is usually limited by certain materials in the vacuum system, such as rubber gaskets and vacuum grease. As a general rule, systems involving the use of natural or synthetic rubber gaskets and vacuum grease are not designed for degassing by heating. However, in some cases such systems have been heated by the use of gas-air flames (torching), electrical resistance elements, or circulating hot fluids. Care must be taken to avoid temperatures high enough to damage critical components. Glass systems are often degassed by torching. A low heat should first be used and then gradually increased in intensity. Also the heat should be applied uniformly over the apparatus so as to avoid unequal expansions causing damage. Again, care must be taken to avoid damaging critical components such as stopcocks. Of course, glass apparatus can be baked in ovens prior to assembly as long as the temperature is kept below the softening point of the glass. Assembly of parts should be done as rapidly as possible after they have been baked and cooled down. A gas discharge is often used for degassing glass systems. Such a discharge can be produced by passing a Tesla coil or a high frequency oscillator over the outside of the system (see Chapter 15) or by applying high voltage to two electrodes inside the system. The use of high voltage is illustrated in Fig. 9.3. This method can also be used in metal systems. In this case the electrodes must be insulated from the vacuum walls. A system involving the use of a getter-ion (or sputter-ion) pump can be considered to be either a dynamic or a static vacuum system. Since many of these systems are used in a cycling operation they will be considered here as dynamic systems. Many of the remarks made above will also apply here. However, such systems are often used to achieve pressures in the ultra-high vacuum region. The special methods involved are discussed in Chapter 14.

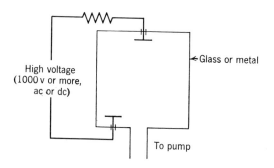

Fig. 9.3 Use of high voltage between electrodes.

Sealed-off or static systems require particular attention to degassing procedures. Examples of such systems are various types of electronic tubes such as radio tubes, cathode ray tubes, x-ray tubes, television tubes, etc. If such tubes were sealed off from the dynamic pumping system without going through an adequate degassing procedure, there would be a gradual evolution of gases and vapors, making the tube useless in a short time. The usual steps involved in preparing electron tubes are:

1. Clean the parts by chemical means.

2. Degas the metal parts by heating (firing) in: (*a*) a hydrogen atmosphere, or (*b*) a vacuum. Metals such as molybdenum, nickel, iron and platinum can be heated in hydrogen. However, tantalum and unplated stainless steel cannot be heated in hydrogen. Resistance or induction heating is commonly used. After heating, allow the parts to cool in vacuum.

3. Assemble the parts. The clean parts should only be handled with tweezers or forceps and the operator should wear clean, lint-free gloves. The operation should be carried out in a dust-free, preferably air-conditioned room.

4. Close up the tube.

5. Connect to the pumping system.

6. Pump down to the operating pressure, usually around 10^{-5} mm Hg.

7. Heat (fire) the tube up to a temperature of 500°C or more. Some kind of insulated oven, made of asbestos or other type of insulating compound, is used with resistance heating. Sometimes gas heating is used.

8. Flash the getter or activate by heating.

9. Seal off the tube.

The above steps are very general. The actual details involved in the manufacture of tubes are numerous and complex and cannot be considered here. Brief mention of the matter is made here to give some indication of the difficulties inherent in obtaining static vacuum systems that will be useful for many thousands of hours. Also, some of the procedures used in making tubes are applicable to other aspects of high vacuum work. Resistance heated ovens, that is, ovens incorporating electrical heating elements, can be used to degas various vacuum components in a hydrogen atmosphere. Such resistance heating can also

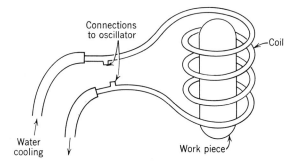

Fig. 9.4 Use of induction heating.

be used in vacuum. Similarly, induction heating can be used in either hydrogen or vacuum. This type of heating involves the use of a coil (usually water cooled) which is energized by an oscillator (radio-frequency generator). This coil is placed over the part to be degassed and is then energized. The part will heat up to a temperature determined primarily by its size and by the amount of energy being fed into

Table 9.1 Degassing Temperatures for Some Common Materials

Material	°C max
Tungsten	1800
Molybdenum *	950
Tantalum	1400
Platinum	1000
Copper and alloys †	500
Nickel and alloys (Monel, etc.)	750–950
Iron, steel, stainless steel	1000
Graphite	1500–1800
Lava No. 1137 (previously air or hydrogen fired) ‡	800

* Embrittlement takes place at higher temperatures. The maximum degassing temperature is 1760°C.

† Except zinc bearing alloys, which cannot be vacuum fired at high temperatures owing to excessive evaporation of zinc (see Chapter 8).

‡ American Lava Corporation.

the coil. The size of coil and number of turns will be determined by the size of the work piece. Usually hollow tubing is used so that cooling water can be passed directly through it. The arrangement of such a coil is shown in Fig. 9.4. The oscillators usually operate at radio-frequencies less than one million cycles (mc) per second. The range 500 to 600 kilocycles (kc) is commonly used.

To effectively degas a metal in a reasonable time, it must be heated to a suitably high temperature. The upper limit will be determined by the metal itself. The maximum degassing temperatures for several metals are listed in Table 9.1. Graphite and one grade of lava are also included in this table. The various glasses can be heated to temperatures just below their softening points (see Chapter 8). The properties of various getters are listed in Table 5.1. Other getters commonly used in the tube industry are discussed in several of the references in Appendix E. The article, "Getter Materials for Electron Tubes" by Espe, Knoll, and Wilder (*Electronics*, Oct. 1950, 80–86) is of particular interest. Cold traps and drying agents can be considered to be degassing agents although they have been treated as pumps in Chapter 5. The design and maintenance, including cleaning, of various types of traps are considered in Chapter 11. As has already been noted, some substances, such as molecular sieve material, will remove not only water vapor but also other vapors and gases.

TEN

Some Fabrication Techniques

10.1 Glass versus Metal

The trend in recent years has been to all-metal systems due to their greater ruggedness as compared to glass systems. However, many small systems are still constructed of glass, and glass components are commonly used in large systems, usually because of the electrical insulating properties of the glass or because it is necessary to observe some phenomenon in the vacuum system. In addition, glass can be fairly easily cleaned (see Chapter 9), it can be effectively baked, and Tesla coils or ordinary spark coils can be used to get an indication of leaks or of the pressure. The most commonly used glasses are the hard, borosilicate glasses (Pyrex, Hysil, etc.). Occasionally there is a need for a high silica glass, such as Vycor or quartz, and sometimes for a soft glass. Often combinations of these glasses occur in a single envelope, e.g., where seals are to be made with Kovar, tungsten, and molybdenum in the same tube, each metal requiring a glass with approximately matching coefficients of expansion if good seals are to be obtained (graded seals). This requires graded seals consisting of one, two, or more intermediate glasses, the number required depending on the difference in the thermal expansions of the two outside glasses. This matter is discussed further in Section 10.2.

No attempt will be made here to discuss glassblowing techniques. Skill in glassblowing is acquired primarily through experience. Several of the references listed in Appendix E contain material covering various glass-blowing techniques, e.g., Strong, *Procedures in Experimental Physics* (Prentice-Hall, 1938). Only a few points regarding glassblowing will be considered here. Soft glass (soda glass) is worked in a gas-air flame (about 900°C) while hard glass is worked in an oxygen-gas flame (about 1400°C). The general steps involved in working glass involve:

1. Clean the parts thoroughly according to the methods outlined in Chapter 9.

2. Start off by heating slowly, gradually increasing the temperature until the softening point of the glass is reached, at which time the actual operation can be carried out.

3. The parts should be heated uniformly.

4. After performing the working operation, the glass should be cooled gradually to room temperature (annealing).

The annealing operation is very important. The glass should be cooled slowly and uniformly to avoid strains and subsequent cracking. The actual annealing schedule will depend on the hardness of the glass, its thickness, and the complexity of the apparatus. Large apparatus is best annealed in a temperature controlled oven. In flame annealing in air, it is necessary to keep the temperature uniform while cooling by extra application of heat to those parts which tend to cool more rapidly, either because they are thinner or because they are subject to greater heat losses through radiation and convection. After the temperature is well below the strain point, set the work aside for final cooling in a place free of drafts. As has been noted above, to join two different types of glass, they must have nearly the same coefficient of expansion to avoid strains and subsequent cracking. If the two coefficients are too widely different then it is necessary to use intermediate glasses. In the case of Pyrex and soda glass four intermediate glasses must be used with progressive increases in expansion coefficients.

The metals commonly used in vacuum chamber construction are mild steel, stainless steel, copper, brass, and aluminum. Other components are made of these materials as well as various pure metals (gold, silver, etc.) and special alloys, depending on the vacuum requirements. Some indication of where various metals and alloys are used in vacuum practice has been given in Chapter 8. No attempt will be made here to go into detail regarding routine machine shop practices such as cutting by lathe, milling machine, and saw, grinding, drilling, polishing, etc. Such operations, together with preparation and maintenance of tools, are covered in various shop manuals. However, although such manuals are useful to the novice, actual machine shop experience is a necessary requirement. The more intricate machine-shop operations are best left to the expert machinist. In certain operations with metals, vacuum practice requires precautions not normally involved in other fields. The concern here is largely with these special requirements, particularly with regard to such matters as soldering, brazing, and welding. Soldering (or soft soldering) refers to the use of low melting point materials (below about 800°F) to join metals (or other materials) together. Brazing (or hard soldering) requires the use of "brazing filler metals" (or hard solders), which generally have melting points above 1000°F. In both soft soldering and brazing the metals being joined are not melted. In welding, the metals being joined are melted and filler metals may or may not be used.

10.2 Glass-to-Metal Seals

A wide variety of glass-to-metal seals are available commercially. Some of these are discussed in Chapter 12. In some cases where suitable commercial seals are not in stock or are not available in the form desired, it may be necessary to fabricate seals for laboratory vacuum use. Only a few of the simplest types are discussed.

As an approximate working rule, glasses and metals which are to be joined should not differ in their expansion coefficients by much more than 10×10^{-7} cm/°C. The properties of several glasses and of some metals commonly used in making seals have been included in Chapter 8. Some general rules which should be followed in order to obtain good seals are:

1. Clean the wire and the glass to ensure freedom from grease (see Chapter 9).
2. Be sure there are no surface marks on the wire (scratches, machining cuts, etc.).
3. Carefully anneal after making the seal.
4. Use fluxes on metals that oxidize.
5. With large seals, pre-anneal the metal.

Wires which are to be sealed into glass, such as lead wires in presses, must be beaded with the appropriate type of glass before sealing in. Metals which can be sealed directly to soft glass (soda lime or lead) are: platinum (wire or tube) and Dumet [wires—copperclad, nickel iron, and SA-50 alloy (Callite)]. Platinum is not commonly used be-

cause of its expense. Typical soft glasses which can be used are: Corning 0010, 0080, 0120; Wembley X8, L1; and Chance GW1, GW2. Commercial platinum is usually an alloy of platinum and iridium (90 to
10%). In making a platinum-glass seal, a bead of soft glass is first
fused to the wire. The wire and bead are then heated to about 1000°C
and the bead then sealed into the wall (glass) of the device being
used. No flux is needed on the platinum. Dumet metal is sometimes
used with a borated copper surface. In making a Dumet–soft glass
seal, borax is used as a flux for the metal if it has not been borated.
To apply the flux, the wire may be dipped in powdered borax or borax
solution. The borax is then fused on the wire in a soft flame to form
a thin skin. The end of a glass tube should be drawn down to fit the
seal. It is advisable to "work" the wire by pulling it slightly while the
glass is soft. This procedure is also advisable with other types of seals.
The final seal is rich red in color.

Metals which are used to make seals with hard glasses (borosilicate
glasses) are: tungsten, molybdenum, Kovar, Fernico, and certain types
of glass-sealing ferrous alloys. The general procedure for making a
seal with these metals is as follows. First, the wire should be clean and
free of scratches, etc. Next a sleeve of glass is put on the wire and a
bead formed. This is done by heating the wire with a soft flame just
beyond the sleeve over a length equal to the sleeve length until it is
oxidized. It is heated to a bright red heat for 10 or 15 sec. It is
important not to overheat the wire. The bead is then drawn down
rapidly to the oxidized portion of the wire. Heating is done with a
hotter flame, starting from one end so the glass is melted progressively
on the wire. Characteristic colors of seals are: tungsten, straw yellow;
molybdenum, chocolate brown; Fernico or Kovar, rich brown. The
resulting seal is then sealed directly to the main hard glass apparatus
or through a graded seal to soft glass. Specific steps in making a tungsten wire to glass seal are:

1. Heat the wire to a white heat in a gas flame to prevent bubbles
from forming at the surface of the seal.
2. Clean the surface of the tungsten by heating and touching with a
piece of potassium or sodium nitrite.
3. Wash the tungsten.
4. Fuse a short sleeve of hard glass of the proper type to the tungsten.

The heating should be started at one end so that shrinking progresses
from that end, thus avoiding the trapping of air bubbles between the
metal and the glass. The interface between glass and tungsten is light
straw colored to light brown due to dissolved oxide. After sealing the

glass sleeve and tungsten, the sleeve is welded into the apparatus. Since tungsten wire is often fibrous, to avoid leaks through its length the tip of the tungsten should always be closed by fusing nickel or Advance wire over it. Copper wires can also be attached to these tips.

It is possible to seal copper to hard or soft glass (feather-edge-seal) by a process developed by W. G. Housekeeper. Only high grade OFHC copper should be used. This contains a minimum copper content of 99.92%. A "certified" grade of OFHC is available that contains a minimum copper content of 99.96% and is preferred over the ordinary grade. Very high purity copper is obtained by vacuum casting. The taper of the feather edge is very critical, the thickness of the edge being 1.5 ± 0.5 mils and the taper, about 5°, extending back from the edge until the wall is 40 mils thick. The surface of the feather edge should be smooth and polished. After degreasing and cleaning, the part is frequently borated by heating to redness and immersing in a concentrated solution of sodium borate. With this procedure, any impurities or contaminations will show up as dark spots. A properly treated part will have a uniform deep red to purple sheen. In making the seal, standard glass working procedures are followed with the copper barely heated to an orange heat. With larger seals it is usually necessary to use a glass lathe and anneal carefully in a furnace. Housekeeper seals can be made in diameters up to several inches. Simple types of tungsten and Housekeeper seals are discussed by Strong (*Procedures in Experimental Physics*, Prentice-Hall, New York, 1938, pp. 24, 26).

It is possible to seal windows of certain glasses, such as Corning 7052, into Kovar. Glass tubing is cut in short lengths and in half lengthwise, then flattened between carbon plates in a furnace. The pieces are then ground and polished on both sides to the required thickness by a lens grinding process. They are then sealed to the Kovar on a vertical glass sealing machine. The use of induction heating, in which the metal is heated directly and the glass only by induction, tends to avoid overheating, which causes bubbles. Such windows must be thoroughly oven heated. In the case of assembled tubes, the annealing is accomplished by the baking-out process at the start of the evacuation. The elements inside the tube are then protected from oxidation.

10.3   Fabrication of Some Other Types of Seals

In bonding metals to ceramics, proper matching of expansion coefficients is important. Various methods have been developed for bonding ceramics to metals. One method involves the use of zirconium hydride. The general procedure involved in making this type of joint involves coating the surfaces to be brazed with thin films of zirconium hydride. A water paste or nitrocellulose solution binder works quite well in producing thin, uniform coatings of the hydride. A piece of hard solder (silver-copper eutectic, BT or pure silver) is then placed in contact with the hydride-coated surface. The material is heated to a temperature at which the solder flows readily (800 to 1000°C) in a vacuum, in an atmosphere of pure, dry hydrogen, or in commercial (tank) dry nitrogen. When the proper temperature is reached, the brazing alloy will melt and flow over the hydride coating. Titanium and other hydrides are sometimes used. Alloys or intermetallic compounds of silver or zirconium, i.e., 85% silver and 15% zirconium, will wet and bond ceramics, diamond, sapphire, carbides, etc., in much the same manner as zirconium hydride. Good results can be obtained by placing pieces of zirconium metal and silver wire on the surface to be brazed and heating to the flow point. The alloy which is formed wets and bonds nonmetallic as well as metallic materials in tank dry nitrogen. Good bonds are obtained on stainless steels in either forevacuum pressures (a few microns) or dry nitrogen. The brazing process is not affected by oxide films on the metal since the alloy readily wets and bonds oxides. Good brazes with molybdenum can also be obtained with this alloy. The zirconium brazing process can probably best be used in vacuum. However, good seals can be obtained in tank dry nitrogen, in commercial tank helium, and in argon. As in the case of glass-to-metal seals, proper matching of expansion coefficients is extremely important. Certain remarks regarding the use of titanium hydride may be pertinent. Around 300 mesh hydride suspended in cellulose lacquer is brushed on the ceramic to be bonded. Copper, silver-copper alloys, and silver are suitable brazing compounds. When copper is used it is applied as a powder (72 parts copper, 28 parts hydride). Strong seals of tubes, rods, or wires to ceramic can be made.

"Liquid" gold or platinum can be used to make ceramic (or glass)-to-metal seals. A colloidal solution of either metal in an essential oil is painted on the ceramic, which is then baked at about 500°C or just below the fusion point of the soft glass. Oil is driven off, leaving an opaque film of metal fused in the ceramic. Careful annealing must be done. The seal can then be soldered to stronger metals. Silica-metal seals are considered here although sometimes they are treated in connection with glass-metal seals. Difficulties are experienced in sealing electrodes into silica because of the low thermal expansion of silica. Invar has a low coefficient of thermal expansion and it is possible to make seals using metallized silica and Invar. The platinum layer, deposited as above and heated in a Bunsen flame or furnace, is built up by electrolytically depositing copper on it, the joint being completed with soft solder. Seals have also been made with molybdenum foil. A strip of molybdenum foil about ½ mil thick and about ⅜ in. long is spot welded at each end to tungsten or molybdenum wires. This composite wire is positioned at the sealed end of the silica tube, which is pumped continuously with a mechanical pump. The tube is then collapsed along the length of the molybdenum strip. The silica should be melted slightly past the two spot welds to give mechanical strength to the seal, although the seal is not vacuum tight. Molybdenum wire can be sealed to silica by a method described by Sands [*Proc. Phys. Soc. (London)*, **26,** 127, 1914].

10.4   Soldering

In general, soft soldering should be avoided in vacuum practice. This is due to the fact that soldered joints are not too strong and are likely to break due to some physical strain or too high a temperature. However, in many cases it is necessary to use soldered joints, e.g., when it is not possible to raise the temperature to the point where brazing or welding can be carried out. In some cases it is possible to use strong adhesives such as the epoxy resins rather than solders. The properties of such adhesives have been discussed in Chapter 8. In any case, soldering is widely enough used in vacuum practice to merit giving some rules to be followed. Several of the references in Appendix E provide detailed information concerning soldering procedures.

Great caution must be used in selecting solders for use inside vacuum systems. Solders consist of various components of relatively low melting point. Some of these components can contribute substantially to the pressure in the system at fairly low temperatures. A list of the more common solder constituents is given in Table 10.1. It is clear that cadmium and zinc can contribute substantially to the pressure at temperatures around 298°F and 412°F, respectively. The first of these temperatures is not a great deal above the boiling point of water (212°F). Water cooling solder joints can sometimes avoid considerable difficulty.

A number of points to be followed in obtaining a good vacuum solder joint can be summarized as follows:

1. Design parts so that the solder joint proper does not provide all the mechanical strength.

Table 10.1   Vapor Pressure of Solder Components (mm Hg) at Temperatures Shown (°F)

Metal	mp	10^{-7}	10^{-6}	10^{-5}	10^{-4}	10^{-3}	10^{-2}	10^{-1}	10^0
Bismuth	519	662	752	886	997	1127	1189	1473	1712
Cadmium	611	196	238	298	356	428	508	611	—
Phosphorus	—	—	—	—	—	—	384	428	517
Lead	622	680	788	902	1019	1157	1324	1530	1785
Tin	450	1183	1345	1512	1692	1907	2172	2502	2932
Indium	312	968	1093	1232	1372	1542	1742	1992	2292
Antimony	1175	644	743	871	977	1102	1271	1452	1658
Zinc	788	284	347	412	479	558	649	761	—

2. Clean the surfaces to be soldered as thoroughly as possible. Do not depend on the flux to do all the cleaning.

3. Choose the appropriate solder on the basis of melting point, strength, and vapor pressure.

4. Choose the right flux for the solder.

5. Do not overheat parts.

6. Do not use excess solder.

7. Make joints as rapidly as possible and at the proper temperature.

8. Space adjacent solder joints well apart so that previously made joints are not softened.

9. Keep low melting materials such as plastics at a sufficient distance from the joint.

10. Use simple jigs and fixtures to hold parts firmly in position while soldering.

11. Pre-tin surfaces wherever possible before actually making the joint.

12. Clean flux from the pre-tinned surfaces.

13. Clean the joint thoroughly before installing in the vacuum system.

Some comments regarding these various points may be pertinent. To provide mechanical strength, overlapping parts and reasonably large soldered surfaces help. Cleaning of surfaces is usually done by mechanical means, such as grinding, filing, or the use of emery cloth, followed by the use of some form of solvent. Vapor degreasing is quite effective. More detail regarding cleaning procedure is given in Chapter 9.

Various sources of heat can be used in soldering. Often a soldering iron, heating coil, or gas flame is used. The size of iron, coil, or flame should be adequate for the sizes of parts involved. The use of wet cloths adjacent to the joint to be made often makes it possible to position joints closer together. The usual steps are:

1. Clean surfaces.
2. Start heating parts.
3. Apply flux.
4. Apply solder while heating.

The solder should flow smoothly over the surfaces to be joined. If it is necessary to rub the solder on the surface a poor joint will result. Although the appropriate flux, properly applied, is usually adequate to remove oxides, a completely nonporous joint is usually obtained by soldering in an inert atmosphere furnace or a hydrogen furnace. Such furnaces are readily obtainable commercially although they are not, at present, a commonly accepted part of every vacuum setup. In using a furnace the procedures recommended by the manufacturer should be followed in detail. It is clear that the procedure necessary in using a furnace is different from the procedure for soldering by hand. All of the parts, plus the flux and the solder, have to be set up ahead of time. Everything has to be arranged so that a good solder joint is obtained by surrounding the joint with an inert atmosphere (dry nitrogen, argon, hydrogen, etc.) and applying heat. It should be pointed out that hydrogen is not truly an inert gas; it works by combining with oxygen to prevent oxides from forming. The term inert gas is properly applied to helium, argon, etc., which simply keep oxygen away from the joint so that oxides cannot form.

In using a solder the usual practice is to employ a flux to remove oxides so as to obtain a good bond. It is possible to achieve the same result by preparing clean surfaces and soldering in an inert atmosphere or in a hydrogen atmosphere. In this case no further cleaning is needed after soldering and the parts can be used directly in the vacuum system. However, the practice of soldering in air is so common that it is worthwhile saying a word about some common fluxes. Fluxes may be classified as corrosive fluxes, e.g., zinc chloride, ammonium chloride, hydrochloric acid, and phosphoric acid, and noncorrosive or resin-type fluxes. The most commonly used flux is zinc chloride dissolved in water. The water, or other solvent used, quickly evaporates when the flux is applied to the hot joint and the zinc chloride then melts and combines with surface oxides. To reduce the melting point of zinc chloride (689°F) ammonium chloride is usually added in such proportion as to form a eutectic mixture of the lowest melting point. To perform its function, the flux should melt before the solder. The correct selection of flux composition as well as that of the solder will have a distinct bearing on the strength and reliability of the joint.

Resin or rosin is a gum exuded from cuts in the bark of pine trees. At ordinary temperatures it is solid and does not cause corrosion, but at or near the soldering temperature it reacts mildly. It melts readily at 257°F. Usual solvents for resin are methyl or ethyl alcohol, propyl or butyl alcohol, or turpentine. "Activated" resin fluxes contain additions which decompose at the temperature of the soldering operation and become noncorrosive. Flux pastes of either the corrosive or noncorrosive type are often a convenience when liquid fluxes are liable to run off the work. When they are mixed with powdered solder, solder paints are obtained. Flux core solder contains the paste flux in a tube of solder. Common paste-forming ingredients are petroleum jelly (vaseline), tallow, and lanoline, with glycerine as the moisture retaining substance. The selection of fluxes is controlled by the design of the joint, the materials involved, and the degree of possible removal of corrosive fluxes after the joint is made. Flux should be removed from pre-tinned parts before the actual joint is made.

Table 10.2 lists some solders and fluxes which are commonly used in the United States. Many of these materials are available in other countries. Also, solders and fluxes comparable to these are produced in other countries. The list in Table 10.2 is not intended to be complete. New solders are continually being developed and each new product should be carefully tested unless it is sponsored by reputable suppliers. Soft solders containing antimony should not be used since the vapors are almost impossible to remove. Soft solder joints should not be used where the part involved is sealed off from the pumps or where baking is required. One special method of soldering which can be used with metals or nonmetals may be worthy of mention. This involves tinning the surfaces by using a grinding wheel loaded with a low melting point solder such as Wood's metal. The heat of grinding flows the solder on the surfaces. The actual soldering is then carried out, using an ordinary lead-tin solder such as 50/50 lead-tin (see McGuire, *Rev. Sci. Instr.*, **26,** 893, 1955). Ultrasonic techniques have also come into use for cleaning surfaces.

10.5   Brazing

Brazing takes place at temperatures above about 1000°F using a filler material which is nonferrous. The physical mechanism of bonding is essentially the same as in soft soldering, except that it takes place at a higher temperature. The bond is achieved below the melting point of the joined metals by the penetration of the filler material. The various brazing processes have been tabulated in the following manner by the American Welding Society in their *Brazing Manual* (Reinhold Publishing Corporation, New York, 1955):

1. Torch brazing.
2. Twin-carbon arc brazing.
3. Furnace brazing.
4. Induction brazing.

5. Resistance brazing.
6. Dip brazing.
7. Block brazing.
8. Flow brazing.

Each of these methods will be discussed briefly here.

TORCH BRAZING

This is done with a gas torch or torches. Depending on the temperature and amount of heat required, the fuel gas (city gas, propane, acetylene, etc.) may be burned with air, compressed air, or oxygen. The brazing filler material is often placed in position in the forms of rings, washers, strips, slugs, powder, etc., or it may be hand fed, usually in the form of a wire or rod. Proper cleaning and fluxing are essential. Manual torch brazing is particularly useful on assemblies involving sections of unequal mass.

TWIN-CARBON ARC BRAZING

This method is not commonly used except where extremely rapid heating is required. The heating is done by a flaming arc between two carbon electrodes.

FURNACE BRAZING

This method is widely used where the parts to be brazed can be assembled with the filler already in place in or near the joint. The filler may be in the form of wire, foil, slugs, powder, paste, etc. Fluxing is usually employed except when an atmosphere, such as hydrogen, is introduced into the furnace to perform the same function. Inert gases, such as argon or helium, are often used in brazing furnaces to prevent oxidation of parts. Particular care must be taken in brazing vacuum parts to avoid porosity. Hydrogen brazing is discussed in

Table 10.2   Some Useful Solders and Fluxes

Ser. No.	mp (°F)	Flow Pt. (°F)	Material and Supplier	Composition (%)	Flux	Comments
Fusible Alloys						
1	100	110	Cerrolow-105, C.D.P.	Bi: 42.91, Pb: 21.70, Sn: 7.97, Cd: 5.09, In: 18.33 Hg: 4		Pre-tinning required on all Cerro alloys. Acid flux gives best results with Bi alloys.
2	117	117	Cerrolow-117, C.D.P.	Bi: 44.70, Pb: 22.60, Sn: 8.30, Cd: 5.30, In: 19.10		
3	134	149	Cerrolow-140, C.D.P.	Bi: 47.50, Pb: 25.40, Sn: 12.60, Cd: 9.50, In: 5.0		
4	136	136	Cerrolow-136, C.D.P.	Bi: 49, Pb: 18, Sn: 12, In: 21		
5	142	149	Cerrolow-147, C.D.P.	Bi: 48, Pb: 25.63, Sn: 12.77, Cd: 9.60, In: 4		
6	158	165	Wood's metal, C.D.P.	Bi: 50, Pb: 25, Sn: 12.50, Cd: 12.50		
7	174	174	Cerrolow-174, C.D.P.	Bi: 57, Sn: 17, In: 26		
8	203	203	Newton's alloy	Bi: 52.5, Pb: 32, Sn: 15.5, or Bi: 50, Pb: 31.25, Sn: 18.75		
9	203	239	Darcet's alloy	Bi: 50, Pb: 25, Sn: 25		
10	212	212	Rose's alloy	Bi: 50, Pb: 28, Sn: 22 or Bi: 46, Pb: 20, Sn: 34		
11	240	260	Cerroseal-35, C.D.P.	Sn: 50, In: 50	No flux on non-metals.	Low vapor pressure. Adheres to glass, metal, mica, quartz, glazed ceramic. Strong and corrosion resistant.
12	274	358	Alkali-resistant solder	Sn: 37.5, Pb: 37.5, In: 25	A	
13	291	291	Bi-Cd eutectic	Bi: 60, Cd: 40		
14	351	351	Sn-Cd eutectic	Sn: 67.75, Cd: 32.25		
Soft Solders						
1	313.5	313.5	Pure indium, I.C.A., C.D.P.	In: 100		Expensive. Rarely used.
2	360	—	Alkali resistant solder	In: 50, Pb: 50	A	Strong and corrosion resistant.

				Composition	Flux	Remarks
3	361	361	Eutectic soft solder	Sn: 61.9, Pb: 38.1	A	Works easily. Strong. Adheres to and covers metals readily.
4	361	374	"Fine" solder	Sn: 60, Pb: 40	A	
5	361	421	50-50 soft solder	Sn: 50, Pb: 50		
6	361	460	—	Sn: 40, Pb: 60	A	Weak.
7	392	500	20-80 soft solder	Sn: 20, Pb: 80	A	Electrical work and copper tubing.
8	450	464	—	Sn: 95, Sb: 5	A	For joints up to 406°F.
9	554	595	Lead-tin	Sn: 5, Pb: 95		
10	363	504	Lead-tin-antimony solders	Sn: 25, Pb: 73.7, Sn: 0.96	A	E-wiping solder (Bell System).
11	363	—	}	Sn: 34.5, Pb: 64.1, Sb: 1.25		
12	425	450	Eutec Rod 199B, E.W.A.	Sn: 90-92, Zn: 8-9, Ni: 0.75–1.25	E 199B	Soft solder for aluminum, ferrous, and nonferrous alloys.
13	446	—	Alkali resistant solder	Pb: 75, In: 25	A	Strong and corrosion resistant.
14	450	450	Pure tin	Sn: 100	A	Shrinks. Cu-Sn alloys brittle. Low strength. Rarely used.
15	575	580	Eutec Rod 153, E.W.A.	Pb: 93-95, Ag: 5-6, Sn: 1-2	E 153	Ferrous and nonferrous alloys.
16	579	579	Lead-silver eutectic	Pb: 97.5, Ag: 2.5	A	
17	579	715	—	Pb: 94-95, Ag: 6-5	A	
18	598.5	598.5	C.D.P.	Pb: 95, In: 5	B	
19	621	621	Lead	Pb: 100	E 155	
20	725	735	Eutec Rod 155, E.W.A.	Cd: 94-95, Sn: 1-2, Ag: 5-6		Ferrous and nonferrous metals. High strength.
21	752	—	Intermediate solder, I.C.A.	Sn: 75, Ag: 20, Cu: 3, Zn: 2	A	Allow time for silver to diffuse.

Fluxes

A (1) Liquid: 40 parts zinc chloride + 20 parts ammonium chloride + 40 parts water

   (2) Paste: 90 parts petroleum + 10 parts ammonium chloride

   (3) Solution of resin in alcohol

B Lloyd's No. 6

E Eutectic fluxes, E.W.A.

Suppliers

C.D.P.   Cerro de Pasco Copper Corp., New York, N.Y.
E.W.A.   Eutectic Welding Alloys Corp., Flushing, New York, N.Y.
I.C.A.   Indium Corporation of America, New York, N. Y.
Lloyd's   Lloyd S. Johnson, Chicago, Ill.

Symbols

Ag	silver	Hg	mercury	Sb	antimony
Bi	bismuth	In	indium	Sn	tin
Cd	cadmium	Ni	nickel	Zn	zinc
Cu	copper	Pb	lead		

Sections 10.6 and 10.7. Vacuum furnaces are also used extensively in vacuum work.

INDUCTION BRAZING

Induction heating was discussed briefly in Chapter 9 in connection with degassing. In brazing, the work to be heated is surrounded by a water-cooled coil of copper tubing which is connected by special water carrying cables to the vacuum tube oscillator (source of radiofrequency power). For maximum energy transfer, the coil (inductor) should be as close to the work as possible but not in contact with it. When work is heated in a vacuum with the inductor outside, the vacuum vessel walls should be of nonconducting material such as glass or quartz. If the work is to be heated to a high temperature, e.g., tantalum to 1800°F or higher, the vessel must be of such size and the work so centered or positioned as not to cause softening and collapse of the walls by radiated heat. The heating of the work is greatest when the coil is small. Therefore, a coil appropriate to the work must be chosen. Where the work is not to be raised to a high temperature, and where close control of heating is required, a coil of larger diameter may be used. A rule of thumb for estimating the amount of power required to heat a given piece of work is 25 kw for each square inch of surface to be heated. This will vary considerably with the size and shape of the work. For surface heating of metals and carbon, radiofrequencies of 200 to 500 kc/sec are commonly used. Since the voltages employed in induction heating are high enough to cause serious or fatal injury, adequate provision should be made for protection of personnel. Metal tables or frames which are part of or close to an induction heating setup should be thoroughly grounded to a cold water pipe. The rubber covering on the cables used on the small and medium heaters is not sufficient insulation and contact with them should be avoided. Condensation of water sometimes occurs in humid weather on the work coils and can cause breakdown troubles. This type of trouble can be reduced by increasing the spacing of the turns or by covering the coil with unsaturated fiberglass sleeving.

RESISTANCE BRAZING

The heat necessary for resistance brazing is obtained from the resistance to the flow of an electric current through the electrodes and the joint to be brazed. A common method is to put the work pieces in a vessel or tube with parallel sides which, in turn, is placed between special graphite pieces machined to fit. By suitable electrical connections, a low voltage high current is made to flow between the graphite pieces and through the work piece so as to heat it to the brazing temperature. The vessel or tube is designed so that carbon dioxide bubbling through isopropyl alcohol can be introduced inside and carried off at a moderate flow rate. At brazing temperatures this gas mixture prevents oxidation and is not flammable. This gas mixture does no harm to unactivated vacuum tube cathodes, although the outside of the metal will be oxidized. This latter effect can be removed later by mechanical and chemical cleaning, taking precautions against the entrance of water or chemicals into the tube by plugging up the openings.

DIP BRAZING

This method involves using a bath to provide the heat required. Two types of baths can be used—a chemical bath or a molten metal bath. The chemical bath is usually a molten salt and the work piece is dipped into it. Often the bath provides necessary protection against oxidation. If not, a suitable flux must be used. The metal bath is the molten brazing filler metal. The parts must be cleaned and fluxed, if necessary, with a cover of flux maintained over the molten bath. This method is usually confined to small parts.

BLOCK AND FLOW BRAZING

In block brazing the heat is obtained from large metal blocks, which are separately heated and applied to the work. The brazing filler metal is usually put in position first. In flow brazing, the molten brazing filler metal is poured onto the joint, preferably from one side. This process is not widely used.

Of the above methods, torch brazing, furnace brazing, induction brazing, and resistance brazing are usually used in vacuum work. Hydrogen, inert gas, and vacuum furnaces are commonly used. As has been noted, the filler metals contain no iron (nonferrous) and are generally silver- or copper-base alloys. The former melt between about 1100 and 1650°F while the latter have a melting range between 1300 and 2200°F. Pure metals such as silver and copper are sometimes used. The American Welding Society has made a general classification of brazing filler metals (*Brazing Manual*, Reinhold Publishing Corporation, New York, 1955).

The purpose of a brazing flux is to promote the formation of a brazed joint. In this sense a gas that is used to surround the work and provide an active or protective atmosphere is a flux. The main purpose of a flux is to prevent the air from combining chemically with the work parts. Oxidation is the prime offender although nitrides and even carbides can be formed. Reliance should not be placed on fluxes to remove oxides, oil, grease, dirt, and other foreign materials. Proper cleaning techniques should be used for this purpose. The flux must remain in contact with the braze area to prevent further formation of oxides. Also, the flux must be readily displaced by the molten brazing filler metal. Many chemical compounds are used in making up fluxes. Some common ingredients of fluxes are:

Borates (sodium, potassium, lithium, etc.).
Fused borax.
Fluoborates (potassium, sodium, etc.).
Fluorides (sodium, potassium, lithium, etc.).
Chlorides (sodium, potassium, lithium).
Acids (boric, calcined boric).
Alkalies (potassium hydroxide, sodium hydroxide).
Wetting agents.
Water.

These fluxes differ in their operating temperature. The choice of flux depends on the metals to be joined and on the brazing filler metal. They are available in the form of powder, paste, or liquid, the form selected depending on the individual work requirements, the brazing process, and the brazing procedure used. Often, powdered flux is mixed with water or alcohol to make a paste. Dry powder can also be used. Paste is probably the most commonly used form for applying brazing flux. The flux residue can usually be removed by washing in hot water, followed by thorough drying. Difficult residues can be removed by a chemical dip, the particular dip depending on the base metals. Mechanical methods, such as scrubbing with a fiber brush or a wire brush, sandblasting, shotblasting, applying a steam jet, chipping with hammer and chisel, etc., are sometimes necessary. When the flux has become saturated with oxide it will no longer act as a flux and also will be difficult to remove during clean-up. Very active, free-flowing fluxes are generally short-lived and sluggish fluxes are long-lived. The latter have a higher melting point and are preferable during prolonged brazing operations at high temperatures. Various types of controlled atmospheres are used to serve the function of a flux. They are not intended to serve the purpose of a cleaning agent. Such atmospheres are primarily used in furnace brazing although they are also used in induction and resistance brazing in special cases. These atmospheres prevent the formation of oxides and scale over the whole part and permit machining to be done before brazing in many applications. Elimination of ordinary chemical fluxes is very important in vacuum operations. As has been indicated, hydrogen, inert gases, and vacuum are commonly used as controlled atmospheres in vacuum practice.

In brazing, the acceptable tolerance for fit of parts to be joined is much smaller and more critical than with soft-soldered joints. No hard-and-fast rule can be given for the tolerance in fit since it depends on several factors, particularly the metals to be joined, the filler material, and the geometry of the parts. However, a few thousandths of an inch covers most circumstances. Large fillets are objectionable. The brazing alloy is usually located at the joint to be made in the form of rings, disks, or foil, or applied in paste form with provisions to prevent the alloy from "running off" before entering the joint. When the heat is controlled manually, the brazing alloy or a sample thereof, located near the joint, should be visible to the operator so that overheating or inadequate heating can be avoided. With massive parts in the furnace the temperature distribution at the joint to be made is often difficult to determine, even in controlled furnaces, unless carefully calibrated thermocouples are directly attached to the joint. Basic joints are of the lap or butt types.

During the processing of complex assemblies it is often necessary to use step brazing, in which the first joint is made at the highest temperature permissible and subsequent brazes are carried out at progressively lower temperatures after intervening assembly operations. Of course, it is necessary that the previous joint not be weakened during the following brazing operation. Table 10.3 shows a representative list of brazing filler metals. This table is not intended to be complete and only shows some materials commonly used in the United States. Comparable materials are available in other countries.

The melting points of gold (1945°F) and copper (1980°F) are rather too close for comfort when a copper assembly is brazed with gold. Gold tends to erode copper, as does silver. In using a gold-copper brazing alloy, it is best to start with one containing less than 40% gold by weight. A useful alloy for brazing vacuum parts has 37.5% gold and 62.5% copper. This alloy melts at 1742°F and gives satisfactory joints between copper and copper, copper and steel, and copper and Fernico. The brazing is carried out in hydrogen. A number of other gold-copper and some silver-copper brazing alloys are included in Table 10.3. In some cases metals are electroplated for the purpose

Table 10.3   Some Brazing Alloys and Fluxes

Ser. No.	mp (°C)	mp (°F)	Material and Supplier	Composition (%)	Flux	Comments
1	600	1112	Low melting hard solder	Ag: 46.5, Cu: 32.5, Sn: 21	H_2	Very hard. Brittle. Low vapor pressure.
2	620	1148	Easy Flo, H. & H.	Ag: 50, Cu: 15.5, Zn: 16.5, Cd: 18	G	PDS 7452 for ferrous and non-ferrous use. Yellow. Mechanically strong.
3	641	1186	Sil Fos, H. & H.	Ag: 15, Cu: 80, P: 5	None, C	For nonferrous metals. Self-fluxing. Yellow. Mechanically strong.
4	646	1195	Easy Flo #3, H. & H.	Ag: 50, Cu: 15.5, Zn: 15.5, Cd: 16, Ni: 3	A, B, C	For ferrous and nonferrous metals where fillets are required. Yellow. Strong.
5	677	1250	ET, H. & H.	Ag: 50, Cu: 28, Zn: 22	C, F, G	Yellow-white.
6	707	1304	Phos-Copper, Westinghouse	Cu: 93, P: 7	None, C	For nonferrous metals. Hard and strong. Very free flowing.
7	721	1330	DT, H. & H.	Ag: 40, Cu: 30, Zn: 28, Ni: 2	C, F, G	Pale yellow.
8	745	1370	SI-1, G. P.	Ag: 68, Cu: 26.6, Sn: 5	C, F, G	Suitable for vacuum work.
9	752	1385	RE-MN, H. & H.	Ag: 65, Cu: 28, Mn: 5, Ni: 2	C, F, G	For stellites, carbides, and refractory alloys containing tungsten.
10	780	1435	Silver-copper eutectic, W. G. P., BT, H. & H.	Ag: 72, Cu: 28	G, H_2	Excellent for copper. White. High electrical conductivity.
11	875	1607	Brazing compound	Cu: 54, Zn: 46	C, D E 181	Common brazing alloy for metal work shops.
12	950	1742	Gold-copper alloy	Au: 37.5, Cu: 62.5	D, H_2	For vacuum components.
13	950	1742	Gold-nickel eutectic	Au: 82.5, Ni: 17.5	D, H_2	Good wetting on tungsten. For vacuum components.
14	960	1760	Silver	Ag: 100	D, H_2	For vacuum components.
15	970	1778	Gold-copper alloy, W. G. P.	Au: 35, Cu: 65	H_2	For vacuum components.
16	980	1796	Nicoro, W. G. P.	Au: 35, Cu: 62, Ni: 3	H_2	For vacuum components.
17	1063	1945	Pure gold	Au: 100	D, H_2	For vacuum components.

18	1083	Pure copper (OFHC)	Cu: 100	H₂	Wets tungsten.
19	1160	Platinum solder	Ag: 73, Pt: 27	H₂	Wets tungsten.
20	1205	Nickel coinage (prewar)	Cu: 75, Ni: 25	H₂	Wets tungsten and molybdenum
21	1450	Nickel	Ni + Co: 99–99.5; C, Mn, Si: traces	H₂	

Fluxes

A Low temperature flux No. 1100, A.P.W.
B All-purpose flux No. 1200, A.P.W.
C (1) Handy flux (fluid at 1100–1600°F). 5–10% KOH facilitates fluxing with certain refractory oxides.
   (2) Special handy flux type LT. Contains less fluorine. Higher mp, more viscous, less active, and longer lived.
   (3) Special handy flux type H. Still less fluorine. More difficult to wash off. For brazing above 1600°F.
D (1) 10 parts powdered borax + 1 part boracic acid.
   (2) Borax applied dry
E 181 Eutectic flux, E.W.A.
F Flotectic flux 1100 (1100–1500°F), E.W.A.
G Handy flux, Lloyd's No. 7
H₂ Hydrogen

Suppliers

E.W.A.   Eutectic Welding Alloys Corp., Flushing, New York, N.Y.
G.P.   General Plate Division, Metals and Control Corp., Attleboro, Mass.
H.&H.   Handy and Harmon, New York, N.Y.
Lloyd's Lloyd S. Johnson Co., Chicago, Ill.
A.P.W.   American Platinum Works, Newark, N.J.
W.G.P.   Western Gold and Platinum Works, San Francisco, Calif.

Symbols

Ag	silver	Ni	nickel
Au	gold	P	phosphorus
Cd	cadmium	Pt	platinum
Co	cobalt	Sn	tin
Cu	copper	Zn	zinc
Mn	manganese		

of obtaining a stronger joint. Tungsten is frequently silver-plated although successful brazes of unplated tungsten filaments to copper with BT brazing alloy (Table 10.3) have been made. These brazes have been successful only because of thorough cleaning, e.g., a rinse in a 50/50 by volume mixture of concentrated nitric and hydrofluoric acid, with subsequent rinses in distilled water and methyl alcohol. Platinum can be used to braze tungsten to molybdenum, and molybdenum to braze tungsten to tungsten.

10.6 Hydrogen Brazing

Brazing and heat treating in an atmosphere such as hydrogen offer definite advantages over performing these operations in air. Copper, nickel, Monel, Kovar, silver, and other metal and alloy parts generally come out bright and clean and are usually stress relieved due to uniform heating. No particular skill is required to produce clean brazed joints with good vacuum and mechanical properties and a high degree of uniformity. Consideration must be given to: design of surfaces at joints, dimensional tolerance of fits, disposition of brazing filler, jigs or other means of holding parts together, time and temperature cycles. Parts to be brazed must be degreased and chemically clean and should not be handled before firing.

Comments regarding hydrogen brazing of several materials follow.

ORDINARY ELECTROLYTIC COPPER

Hydrogen brazing of this material results in porosity and lowered mechanical strength. Parts treated in this manner are not suitable for vacuum work. To obtain nonporous ductile parts, use OFHC copper. The brazing filler should have no ingredients which vaporize (see Table 10.1) and should have a melting point below that of copper. Suitable fillers (hard solders) are: silver-copper eutectic (72% silver, 28% copper, mp 1434°F—Handy and Harmon BT), other alloys of silver and copper, gold-copper alloy (80% gold, 20% copper, mp 1634°F), gold-nickel eutectic (82.5% gold, 17.5% nickel, mp 1742°F), pure silver (mp 1760°F), and several alloys containing varying amounts of gold and copper. Pure gold can be used, sometimes with the "diffusion" process, in which the parts are accurately machined and held together under pressure with the gold as foil, wire, or electroplate, making very good contact.

Electrodeposited copper can be used very effectively to join copper parts. The parts are held together by press or close mechanical fit, with due allowance for plating thickness. The plating can be limited to definite areas by the use of a stop-off lacquer, e.g., aluminum oxide (alundum) in a nitrocellulose binder, during the plating process. The binder is burned off harmlessly during the firing and the alundum can be brushed off after the work has cooled. In a hydrogen atmosphere, materials such as copper present a very "clean," i.e., wettable, surface so that the solder flows up into small gaps and cracks and even upward along exposed surfaces. Precautions to be taken are: (1) do not use excessive solder or silver paste, (2) stop-off and position work in furnace to take advantage of the pull of gravity, (3) make parts of jigs in contact with or close to work and solder of a material which will not be brazed fast to the work (SS 303, Nichrome), (4) prefire the parts of the jig in tank hydrogen to get protective coating of green chromium oxide.

COLD ROLLED, HOT ROLLED, AND CARBON STEELS

Many of the brazing steps for steels are similar to those for copper, particularly with regard to design, machining, and fit of parts to be joined. Chrome-free steels can be brazed with the fillers noted for copper. Steel can be brazed to steel by using pure copper as a brazing material, with the steel previously nickel and/or copper plated. Chrome- and aluminum-free steels can be hydrogen brazed with silver solders by partial purification of the hydrogen. Pass the gas through one or more catalytic-type purifiers which convert any oxygen present to water vapor. The water vapor is removed by drying agents or a liquid nitrogen trap. Copper brazing is not suitable for joining steel to copper.

NICKEL, MONEL, AND RELATED ALLOYS (chrome- and aluminum-free)

Braze with silver-copper eutectic in a hydrogen furnace. Nickel can be copper or silver brazed without special treatment. Phosphor bronze is handled like copper, with due regard to the subsequent heat treating (if required). Beryllium copper should be copper or nickel plated before brazing. Special precautions must be taken with brass and zinc-bearing alloys since the zinc vaporizes. *A hydrogen furnace used with brasses and bronzes should not be used with other materials* (*particularly tube parts*). The temperature schedule should ensure fast heating and melting of the solder to avoid excessive evaporation of volatile parts. The vapors are noxious and contaminate the furnace. The "bottle" process can be used for materials containing volatizable ingredients (brass, etc.). Kovar is handled much like nickel and its alloys. Copper brazing is preferred for joining Kovar to Kovar, nickel, or steel. Gold-copper and gold-nickel alloys, as well as pure silver and gold, may be used to braze Kovar to copper. Silver-copper eutectic can be used if the Kovar has been stress relieved above 1650°F and copper plated. Always braze Kovar with quick melting solder, i.e., pure metals or simple alloys such as eutectics. In general, copper or nickel plate Kovar for brazing.

STAINLESS STEELS

Generally, these materials cannot be brazed in tank hydrogen due to the formation of a protective film of chromic oxide. This difficulty can be overcome by properly purifying the hydrogen, by passing it through suitable chemicals to avoid formation of this oxide. A low rate of flow through the chemicals should be used. A second method for handling stainless steel and other chrome alloys involves carefully cleaning the metal parts and then nickel plating them. The well-bonded nickel plate protects the chromium from oxidation so that brazing can be accomplished in the furnace with regular tank hydrogen.

COMMENTS

A number of procedures are generally applicable to hydrogen brazing. These are: (1) Parts should be clean and firmly held by jigs, clamps, screws, wires, weights, etc., with allowance made for differences in expansion of the work pieces, jigs, etc. (*Example:* stainless steel jig and copper piece. The jig expands less than the copper. Take care that the jig does not bite into, crush, or squeeze the work as it heats.) (2) Weights are sometimes preferred to clamps and screws for metals which become soft at brazing temperatures (copper, nickel, etc.). (3) Exercise care so nothing can loosen, fall, or strike walls of the furnace as the work is pushed through. (4) Load pieces into a nickel or stainless steel box or tray. (5) If a cover is used, use vents large enough for free circulation of gases. (6) Preheat massive pieces with small orifices in front part of furnace so (*a*) work and muffle are protected from thermal shock, and (*b*) the hydrogen has time to displace the air in interior spaces. (7) Purge large cavities with dry nitrogen previous to preheating to avoid possibility of explosion. (8) The brazing temperature depends on the solder used and the time is found by experiment. Temperature can be determined by putting a small piece of solder where it can be seen while heating and where it will not drip on the muffle or work (sometimes solder on the work can be observed).

10.7 Welding

The usual method of joining steel parts is by welding, gas or arc techniques being the most common. In welding parts for vacuum service particular care must be exercised in order to avoid porosity in the joint.

A technique which minimizes oxide formation, porosity, and other structural flaws is that of inert gas shielded arc welding. This method involves doing the welding in an inert atmosphere, often argon. The equipment in which such welding is carried out consists of an enclosure for the inert gas (usually steel), means for removing the air (usually a mechanical vacuum pump), means for filling the enclosure with inert gas to atmospheric pressure, insulators for introducing electrical leads, and means for manipulating the welding rods in the inert atmosphere. In some cases both mechanical and diffusion pumps are used in order to pump down to a lower pressure. Often viewing and lighting ports are included. The usual method for manipulating the welding rods and the work is by rubber gloves attached to openings in the enclosure. Clearly, these openings must be covered by plates during pump-down. Figure 10.1 shows a schematic layout of a typical inert gas shielded arc welding arrangement.

Some precautions to be followed in making welds are listed below. These are intended to cover arc or gas welding in air but many of the points are applicable to inert gas shielded arc welding.

1. Whenever possible, use single pass welds. When multiple pass welds are necessary, remove all slag from the preceding pass by chipping and brushing with a wire brush. Do not peen or hammer the preceding pass excessively.

2. Use the longest practical welding rod so as to permit the longest possible uninterrupted weld. With an interrupted weld make the overlap as generous as practicable to eliminate voids.

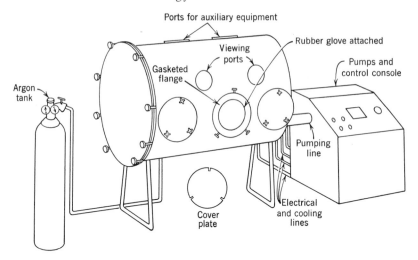

Fig. 10.1   Typical setup for inert gas shielded arc welding.

3. Avoid an undue amount of machining, grinding, hammering, or peening since cracks or pinholes may thus be temporarily closed and open up later under stress or corrosion. Where machining is necessary, e.g., with mating flanged surfaces, extra precautions should be taken to make good welds.

4. Whenever possible, welds should be made from the outside of the vessel only. With double welds, gas is trapped and there is a possibility of virtual leaks.

5. Welds of a structural nature should never cross sealing joint welds continuously. Here again gas may be trapped and a virtual leak develop.

6. Structural welds inside the vessel should be discontinuous or just tack welds to allow easy flow of gas from any channels or pockets.

The possibility of virtual leaks, i.e., trapped pockets of gas exposed to the vacuum through cracks, pinholes, or a porous weld, is an important consideration in welding for vacuum service. Figure 10.2 shows good and poor welds for a number of common situations. In all examples shown the good welds avoid the formation of gas pockets with the possibility of virtual leaks. Particularly serious are the cases where the channel between welds has a connecting leak to both the inside and outside of the vessel. Here the high flow impedance of the leak, coupled with the storage capacity of the pocket or channel, makes it almost impossible to locate the leak. An example of this type of trouble, together with a method of avoiding it, is illustrated in Fig. 10.3*b*. Figure 10.3*a* shows a method of avoiding virtual leaks where a structural weld crosses a sealing weld. Where double welds are absolutely necessary for structural reasons, a definite channel between the welds should be provided in the joint. Small pipe taps, located so that the channel can be pressurized, should be provided. The openness of the channel is determined by a flow test from one tap to the next; then one is sealed off and the channel pressurized. The soap bubble technique, or other leak detection methods (see Chapter 15) are then used on the welds inside and out.

Spot welding with an electric current is often used to join metals to be used in vacuum systems. Usually spot welders are designed to

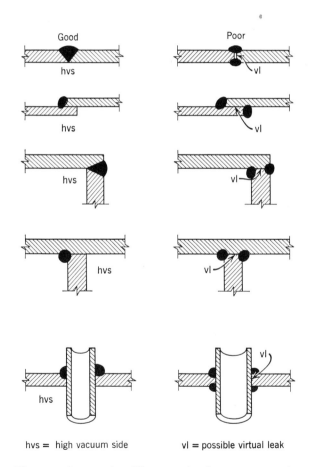

hvs = high vacuum side   vl = possible virtual leak

Fig. 10.2 Suggested welding practice for vacuum service.

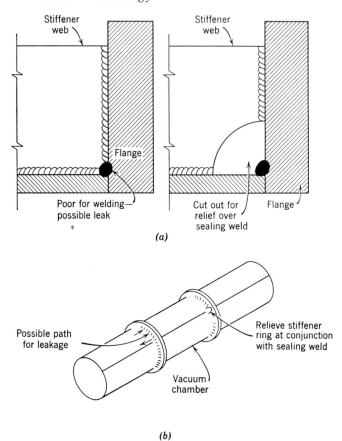

Fig. 10.3 (*a*) Structural weld crossing sealing weld. (*b*) Avoiding virtual leak by means of relieving stiffener ring.

operate at low voltage (often about 6 v) and high current (up to several thousand amperes). The required voltage and current are obtained from the secondary of a transformer, the primary of which is connected to the line voltage. Some means, such as a rheostat, is provided to regulate the welding current. Many designs of various power capacities are available on the market. Units designed to handle very thin wires (a thousandth of an inch or less) are very useful. In carrying out the welding, the electrodes are brought in contact with the work, and a definite pressure is applied, usually by means of a foot pedal. The pressure and duration of the current are important. Inadequate pressure results in burning and "spitting" at the joint, while too much pressure decreases the joint resistance and consequently the heating action. Metals which weld together best are those of similar melting temperature and heat conduction. Table 10.4 shows the relative spot-welding characteristics of various laboratory metals.

In many cases, special welding techniques are used. Heliarc welding is common, in which the work is covered by helium to prevent the formation of oxides. Atomic hydrogen welding is also common. In this method the welding is carried out in an atmosphere of hydrogen which is at a high temperature. At such high temperatures as these hydrogen is very effective in reducing oxides or in preventing them from forming, thus eliminating porosity. To obtain particularly pure metals or alloys, the material is melted in vacuum by an electron beam. Welding by this method prevents the formation of oxides. The method involves heating the metals to be joined by an electron beam in vacuum.

10.8   Electroplating

The process of electroplating involves depositing metal on a material (usually another metal) by means of a chemical bath and an electric current. The metal to be plated is made the cathode while the anode depends on the metal being plated. Electroplating metals is carried out for various purposes. Iron is often plated with nickel to provide

Table 10.4   Spot Welding of Some Metals

Best	Good	With Difficulty
Nickel to iron	Nickel to $\begin{cases} \text{tungsten} \\ \text{molybdenum} \\ \text{tantalum} \end{cases}$	Nickel to $\begin{cases} \text{copper} \\ \text{aluminum} \end{cases}$
		Molybdenum to $\begin{cases} \text{molybdenum} \\ \text{tungsten} \end{cases}$
	Iron to $\begin{cases} \text{copper} \\ \text{constantan} \end{cases}$	Iron to $\begin{cases} \text{tungsten} \\ \text{molybdenum} \\ \text{tantalum} \end{cases}$
		Tungsten to tungsten
		Aluminum to aluminum

a clean surface and to prevent the formation of oxides and other compounds on the iron surface which would give rise to vapors. In a sense, electroplating for this purpose could be considered to be a cleaning technique. In other cases electroplating is used to prevent metals from being corroded by some particular vapor with a resultant loss in structural strength. As an example, such metals as copper and aluminum can be plated with nickel so that they will not be attacked by mercury vapor. Some metals are difficult to braze by the general methods outlined previously. In such cases the use of electroplating facilitates the brazing process.

The examples of electroplating procedures given below are readily adaptable to laboratory scale. Large parts or those requiring heavy deposits should be handled by commercial plating companies. Details regarding various baths and special techniques are to be found in publications of the plating industry, such as *The Plating and Finishing Guidebook,* published by the Metal Industry Publishing Company (*The Journal of Metal Finishing*). Other information is in several of the references listed in Appendix E. The cleaning process and preparation of metal surfaces for plating are extremely important for proper adhesion and a good plate. In general the work piece and the anode should be of about equal size. To protect certain parts of the work from being plated, a stop-off lacquer can be used. This is a nitrocellulose vehicle, usually with a red dye added for identification. It is applied to the cleaned work piece by dipping, spraying, or brushing. The lacquer is removed after plating by immersion in ethyl acetate or acetone. To avoid thin tarnish films on plated surfaces, as soon as the work is plated it should be rinsed in hot running water, dipped in clean alcohol, and dried in hot air. It is generally considered desirable when applying more than one deposit on a given material to electroplate in a specified order.

The electroplating procedures for only a few materials will be considered here.

COPPER

May be directly plated with gold, silver, or nickel.

KOVAR

May be silver plated directly, although for best results it should have an initial thin layer of copper (copper flash or strike).

BRASS

May be silver, gold, or nickel plated directly but it is best to use an initial copper strike.

CHROMIUM

Plating is carried out using copper first, then nickel, and finally chromium.

COPPER PLATING IRON AND STEEL

Carried out so that other metals can then be electroplated on them, e.g., silver, nickel, or gold.

SILVER PLATING

Normally carried out after first depositing a thin layer of silver (silver strike). This is done with the following bath:

Silver cyanide	6.5 g/l
Potassium cyanide	68.0 g/l

The current density is 15 amp/ft^2 and the work should be connected to the current source before immersion in the bath. The strike solution has a low silver content and a high cyanide concentration. The strike solution will prevent the deposition of silver by immersion, which results in poor adhesion of the ultimate deposit. The work is kept in the strike just long enough to coat the surface.

The actual silver plating can be carried out with the following bath:

Silver cyanide	41 g/l
Potassium cyanide	40 g/l
Potassium hydroxide	11 g/l
Potassium carbonate	62 g/l

The current density is 5 to 15 amp/ft^2. The yield is 0.001 in./hr at a bath temperature of 48°C but operation can be carried out at room temperature with reduced thickness of deposition. The anode should be 0.9995 fine (pure) silver. Sterling is *not* suitable. The anode should be removed from the bath when not in use to prevent excessive solution of the metals.

GOLD PLATING

Can be carried out in the following type of bath:

Potassium gold cyanide	3.9 g/l
Potassium carbonate	7.9 g/l
Potassium cyanide	3.9 g/l

The current density is 5 to 15 amp/ft^2 and, as with silver plating, the anode should be removed from the bath when not in use. The yield is about 0.001 in./hr at a temperature of 50 to 70°C. Operation at room temperature results in reduced yields.

NICKEL PLATING

Commonly used in vacuum practice. The only methods to be discussed here are concerned with the plating of difficult metals such as stainless steel, molybdenum, and tungsten. These methods involve the use of a Watts Nickel Bath. To obtain a matte finish, which is preferred for the brazing and heat treating of chromium alloys, the following bath can be used:

Nickel sulfate	200 g/l
Nickel chloride	45 g/l
Boric acid	30 g/l

The current density is about 25 amp/ft^2 at a temperature of 38 to 42°C. At room temperature the deposition is slowed up. A Watts-type bright nickel bath is described in *Metal Finishing* (Finishing Publications, Inc., New York, 1949, p. 274). The characteristics of this bath are as follows:

Nickel sulfate hexahydrate	225–375 g/l
Boric acid	30–45 g/l
Nickel chloride hexahydrate	30–60 g/l
Temperature	49–71°C
Current density	20–60 amp/ft^2

The proper brighteners can be specified by suppliers.

Some of the nonmagnetic stainless steel alloys, e.g., types 304, 347, etc., are very useful in making various vacuum components, such as electronic tube parts. The presence of chromium in these alloys makes ordinary hydrogen brazing difficult owing to the moisture content of tank hydrogen. To braze below 1740°F, an electrodeposited coating such as nickel has been found very satisfactory. Such a coating is necessary when brazing types 304, 307, etc., stainless steel alloys with silver and gold brazing alloys such as silver-copper eutectic (BT), Silfos, and gold-copper materials. Probably the most difficult metals to plate are those containing amounts of chromium and/or nickel or cobalt. Small amounts of these materials can result in loose bonding of the plate, which is likely to peel and blister. The procedure for electroplating such alloys with nickel is a fairly complex process and will not be discussed here. A method using the Watts Nickel Bath which has proved quite satisfactory has been developed by Westinghouse Electric Corporation, Bloomfield, New Jersey. Other methods are discussed in the various publications of the plating industry.

10.9   Gasketed Joints

There are two major types of gasketed joints. Fixed joints are normally found in pipe and flanged connections, which are usually set up once for emergency reasons and then disassembled. The second type of joint involves seals which provide for motion into a vacuum system. These are dynamic devices with the sealing material constantly being exposed to motion; some designs are discussed in Chapter 12. The emphasis here is on fixed joints gasketed with natural or synthetic rubbers, such as Hycar, neoprene, butyl, and silicone, although some attention is given to the fluorocarbons (Teflon, Viton A, Fluorel, etc.) and soft metals. Properties of these materials are covered in Chapter 8.

Fixed rubber gasket joints can be made with simple flat gaskets and no grooves, although this is not too common a practice. If this method is used, the surfaces to be sealed by the gasket need not be particularly smooth although they should be machined so as to ensure a uniform loading of the gasket. Quite often the surfaces are rough machined, leaving nearly concentric rings, which are forced into the gasket material, ensuring a better seal due to the higher local load. A more realistic approach to the problem of providing a gasket between two flanged surfaces is to make use of flange gaskets (see Chapter 12). The rubber cord, having a relatively small area, can be uniformly loaded without undue bolting stresses.

It is common practice to use extruded rubber shapes in gasket grooves to obtain a seal between two flanged or fixed joints. In general extruded rubber cords with square, rectangular, or circular cross sections

(O-rings) are used. Gasket grooves are usually designed so that the gasket is compressed about 30%. Since rubber is essentially incompressible the groove must provide enough space for deformation of the gasket. The groove width for square or rectangular gaskets is greater than for O-rings. Commercial rubbers can be obtained in a range of hardnesses (durometer values) and other values can be produced by special compounding through local fabricators. A durometer value between 50 and 65 is commonly used so that a tight seal can be obtained using reasonable pressure values. The choice of rubber depends on a number of factors, including expected exposure to oil or other deteriorating substances, the operating temperature range, etc. (see Chapter 8).

The most common fixed gasket seal employs a single rubber gasket. In some cases a double gasket system is used which involves the use of two concentric gaskets with a pump-out between the gaskets. The pump-out connects to the outer atmosphere and in case of a leak in the inner gasket it can be sealed off or pumped on with an auxiliary mechanical pump so as to keep the vacuum system in operation. A double gasket system can be vacuum tested without pumping down the system by pumping on the space between the gaskets. With a double gasket system, a pump-out is essential. Single gasket grooves are cheaper to machine and they occupy less space. For most purposes, the single gasket joints have been found to be entirely satisfactory. Two general types of gasket grooves are used. The first type (regular or spaced) involves compressing the gasket, but not enough to obtain a metal-to-metal contact between the surfaces being sealed. In the second type, the design is such that the compression will bring about a metal-to-metal contact. Grooves of the first type are cheaper to machine and the gaskets may be further compressed if found leaking. The metal-to-metal type of gasket is standard for O-rings and is used where alignment of parts is essential, where a minimum exposure of rubber to vacuum is desired, or where electrical contact is required.

No hard-and-fast rules can be given for the choice of gasket arrangement. O-rings are most commonly used since many stock sizes can be obtained commercially and the grooves are cheaper to machine than those required for square or rectangular gaskets. As has been noted, such joints normally involve the use of a single gasket with metal-to-metal seal. Grooves are generally machined on a lathe or milling machine. Where substantial numbers of grooves of the same size are required it is worthwhile making up special cutting tools. The finish of the surface is not too critical, the main thing being to avoid any scratches across the groove. Concentric scratches or cuts (such as from the cutting tool) around the groove are not too serious. Rather than rely on experience one can specify a finish of 63 microfinish or better, which covers most situations. The usual procedure is to machine the groove and then polish it out with fine emery cloth until no scratches or flaws are evident.

Commercial O-rings are readily available with rubber diameters up to $\frac{1}{4}$ in. and ring diameters up to 16 in. (length about 50 in.). However, in many cases O-rings are required which must be specially fabricated, either by local fabricators or in the laboratory. For example, a cover plate 2 ft by 4 ft would require a gasket about 136 in. long (about 43 in. in diameter). As long as the groove is of uniform dimensions and the gasket is compressed uniformly, the diameter of the rubber need not be larger than $\frac{3}{16}$ in. to get good sealing. This requires making up the gasket from standard stock. This can be done by use of a mold and heat (vulcanizing) or by the use of an appropriate cement. As a rule, care must be taken in vulcanizing because there is a good chance that the cross section of the rubber will not be uniform unless considerable care is taken in the construction of the mold. Quite satisfactory joints can be made with an appropriate cement if certain simple precautions are taken. These precautions include:

1. Use a simple fixture to cut the ends as smoothly and reproducibly as possible.

2. Apply the cement as soon as possible after cutting the ends which are to be joined.

3. Use appropriate supports or fixtures to ensure a smooth joint.

4. Smooth the joint with emery cloth and check its strength by pulling back and forth on it (after being sure that the cement has dried).

A simple fixture which works well for cutting the ends of O-ring stock is shown in Fig. 10.4. The rubber stock is placed in the groove and then cut by a razor blade, using the slot provided. By this method the rubber stock is cut at right angles to its length. The surfaces to be joined are then minimum. Various cements specifically formulated for joining rubber parts are available. These are usually solvent-type cements, which are solutions of rubber in organic solvents such as gasolene, benzene, etc. Many new high strength cements are continually being developed that are very suitable for joining rubber to rubber. Using the fixture shown in Fig. 10.4, very good bonds are achieved by using a fast drying, high strength cement such as Eastman Kodak No. 910. Many laboratories use diagonal cuts so as to get larger surface contact. This method may be advisable when lower strength cements are used. The fixture shown in Fig. 10.4 could be

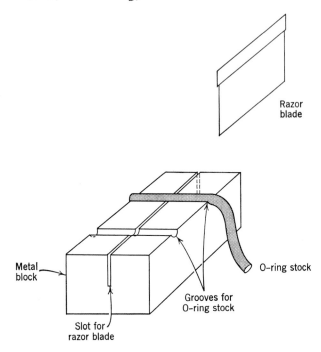

Fig. 10.4   Method of cutting O-ring stock.

used to make diagonal cuts by changing the direction of the razor slot with respect to the groove for the rubber stock. Rubber parts that have been lying around for some time should be cleaned with a suitable solvent such as acetone, benzene, or gasolene. Otherwise, new surfaces should be cut.

The grooves for O-rings are usually of rectangular cross section, although other shapes, such as those with semicircular cross section, are used for special situations. The actual dimensions of rectangular grooves for various O-rings may differ somewhat from one organization to another. In general, the cross section of the groove must be sufficient to take up the deformed rubber after it has been compressed about 30%. Various manufacturers of O-rings list recommended values of groove dimensions. Table 10.5 lists the groove dimensions which have been used by the Lawrence Radiation Laboratory, University of California, Berkeley, California, for several years with quite satisfactory results. The dimensions shown are based on Hycar rubber of durometer value 55 to 65 and the depths of the grooves allow about $35 \pm 5\%$ compression of the gasket, which gives a pressure of about

Table 10.5 Suggested Dimensions for O-Ring Grooves *

O-Ring Dimensions

Nominal Size			Actual Size		Groove Dimensions		
W	ID	OD	W	ID	X Diam	Y	Z
1/16	1/8	1/4	0.070	0.114 ± 0.005	0.128 ± 0.004	0.113	0.044
	3/16	5/16		0.176	0.191		
	1/4	3/8	±0.003	0.239	0.254	+0.015	+0.004
	5/16	7/16		0.301	0.317	−0.000	−0.000
	3/8	1/2		0.364	0.380		
3/32	7/16	5/8	0.103	0.424 ± 0.005	0.447 ± 0.010	0.151	0.064
	1/2	11/16		0.487	0.510		
	9/16	3/4	±0.003	0.549	0.573	+0.015	+0.006
	5/8	13/16		0.612	0.637	−0.000	−0.000
	11/16	7/8		0.674	0.699		
	3/4	15/16		0.737	0.763		
1/8	3/4	1	0.139	0.734 ± 0.006	0.761 ± 0.010	0.205	0.086
	7/8	1 1/8		0.859	0.887		
	1	1 1/4	±0.004	0.984	1.013	+0.015	+0.009
	1 1/8	1 3/8		1.109	1.140	−0.000	−0.000
	1 1/4	1 1/2		1.234	1.266		
	1 3/8	1 5/8		1.359	1.392		
	1 1/2	1 3/4		1.484	1.518		
1/8	1 3/4	2	0.139	1.734 ± 0.010	1.772 ± 0.010	0.205	0.086
	2	2 1/4		1.984	2.027		
	2 1/4	2 1/2	±0.004	2.234	2.280	+0.015	+0.009
	2 1/2	2 3/4		2.484	2.532	−0.000	−0.000
	2 3/4	3		2.734 ± 0.015	2.790		
	3	3 1/4		2.984	3.042		
	3 1/4	3 1/2		3.234	3.294		
	3 1/2	3 3/4		3.484	3.547		
	3 3/4	4		3.734	3.799		
	4	4 1/4		3.984	4.051		
	4 1/4	4 1/2		4.234	4.304		
	4 1/2	4 3/4		4.484	4.556		
	4 3/4	5		4.734	4.809		
	5	5 1/4		4.984	5.061		
	5 1/4	5 1/2		5.234 ± 0.023	5.321		
	5 1/2	5 3/4		5.484	5.574		
	5 3/4	6		5.734	5.826		
	6	6 1/4		5.984	6.079		

Table 10.5 (Continued)

O-Ring Dimensions							
Nominal Size			Actual Size		Groove Dimensions		
W	ID	OD	*W*	ID	*X* Diam	*Y*	*Z*
1/8	6 1/2	6 3/4		6.484	6.583		
	7	7 1/4		6.984	7.088		
	7 1/2	7 3/4		7.484 ± 0.30	7.601		
	8	8 1/4		7.984	8.105		
	8 1/2	8 3/4		8.484	8.610		
	9	9 1/4		8.984	9.115		
	9 1/2	9 3/4		9.484	9.620		
	10	10 1/4		9.984	10.125		
3/16	1 5/8	2	0.210	1.600 ± 0.010	1.639 ± 0.010	0.297	0.134
	1 7/8	2 1/4		1.850	1.892		
	2 1/8	2 1/2	±0.005	2.100	2.144	+0.015	+0.010
	2 1/4	2 5/8		2.225	2.270	−0.000	−0.000
	2 3/8	2 3/4		2.350	2.397		
	2 5/8	3		2.600	2.649		
	2 7/8	3 1/4		2.850 ± 0.015	2.906		
	3 1/8	3 1/2		3.100	3.159		
	3 3/8	3 3/4		3.350	3.411		
	3 5/8	4		3.600	3.664		
	3 7/8	4 1/4		3.850	3.916		
	4 1/8	4 1/2		4.100	4.168		
	4 1/2	4 7/8		4.475	4.547		
1/4	5	5 1/2	0.275	4.975 ± 0.015	5.052 ± 0.010	0.385	0.177
	5 1/8	5 5/8		5.100 ± 0.023	5.186		
	5 1/2	6	±0.006	5.475	5.565	+0.015	+0.012
	6	6 1/2		5.975	6.070	−0.000	−0.000
	6 1/2	7		6.475	6.575		
	7	7 1/2		6.975	7.080		
	7 1/2	8		7.475 ± 0.030	7.591		
	8	8 1/2		7.975	8.096		
	9	9 1/2		8.975	9.106		
	10	10 1/2		9.975	10.116		
	11	11 1/2		10.975	11.125		
	12	12 1/2		11.975	12.135		
	13	13 1/2		12.975	13.145		
	14	14 1/2		13.975	14.155		
	15	15 1/2		14.975	15.165		

* All dimensions are in inches.

400 psi based on the projected area. The same dimensions should also be satisfactory for any other elastomer, such as neoprene, natural rubber, silicone rubber, etc., as long as the durometer value is about the same. Of course, as the durometer value increases, the pressure required to deform the gasket also increases. As a result the depth of the groove must be increased as the durometer value increases if

additional pressure is undesirable. In all cases the cross-sectional area of the groove must be such as to accept the deformed rubber when metal-to-metal contact is achieved. When placing a stretched O-ring in the groove the ring should be given a twist between thumb and fore-finger in such a way that the ring will not roll out of the groove. The grooves for circular rings, such as are used on flanges, are rather readily made on a lathe. However, in the case of grooves for various types of cover plates it is necessary to use a milling machine or some special type of machine. This results in grooves which intersect at right angles, as shown in Fig. 10.5*a*. This leaves some unfilled parts of grooves. To avoid having the O-ring creep out into this space, several methods can be used. Figure 10.5*b* shows one method of doing this, by using a solder or brazing fillet to fill in the unfilled parts of the groove. Figure 10.5*c* shows a method using a piece of O-ring gasket to fill in the space so that the regular O-ring cannot expand. The corner should be rounded to avoid damage to the O-ring. Figure 10.5*d* shows another arrangement, whereby the corners of the gasket groove are milled out on relatively sharp curves. The trouble with this method is the cost involved. Commercial O-rings or O-ring stock should be examined carefully for flaws before using.

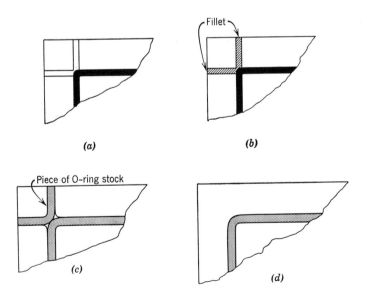

Fig. 10.5 (*a*) Unfilled corner grooves. (*b*) Use of solder or brazing fillet. (*c*) Use of O-ring to fill corner grooves. (*d*) Milled corner groove.

Suggested dimensions for square and rectangular gasket grooves are shown in Table 10.6. These dimensions have been used with satisfactory results at the Lawrence Radiation Laboratory, University of California, Berkeley, California, and are fairly typical. Rectangular metal-to-metal gaskets are easier to install and are better retained in the grooves than the square metal-to-metal gaskets. Square metal-to-metal gaskets are used *only* where required by space limitations. The dimensions in Table 10.6 are based on Hycar rubber of durometer

Table 10.6   Suggested Dimensions for Square and Rectangular Gasket Grooves *

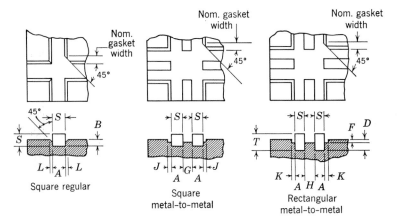

Gasket Dimensions				Groove Dimensions											
Nominal		Actual													
Sq.	Rect., Width × Ht.	S	T	A	B	C	D	E	F	G†	H†	J†	K†	L‡	
$\frac{3}{16}$	$\frac{3}{16} \times \frac{9}{32}$	0.208	0.301	0.188	0.108	0.140	0.233	0.071	0.103	$\frac{1}{4}$	$\frac{3}{16}$	$\frac{1}{8}$	$\frac{3}{32}$	$\frac{3}{64}$	
		0.193	0.286	0.193	0.113	0.146	0.239	0.077	0.109						
$\frac{1}{4}$	$\frac{1}{4} \times \frac{3}{8}$	0.270	0.395	0.250	0.146	0.186	0.311	0.093	0.137	$\frac{5}{16}$	$\frac{1}{4}$	$\frac{5}{32}$	$\frac{1}{8}$	$\frac{1}{16}$	
		0.255	0.380	0.255	0.151	0.193	0.318	0.100	0.144						
$\frac{3}{8}$	$\frac{3}{8} \times \frac{9}{16}$	0.395	0.582	0.375	0.221	0.279	0.466	0.140	0.199	$\frac{3}{8}$	$\frac{5}{16}$	$\frac{3}{16}$	$\frac{5}{32}$	$\frac{3}{32}$	
		0.380	0.567	0.380	0.226	0.287	0.474	0.148	0.207						

* All dimensions are in inches.
† Tolerance: $+\frac{1}{64}$, -0.
‡ Tolerance: $\pm\frac{1}{64}$.

value 55 to 65. In general, such grooves are satisfactory for other rubbers with the same durometer values. The groove depths for metal-to-metal contact allow about 34 ± 4% compression of the unrestricted portion of gasket for rectangular gaskets and about 42 ± 3% for the square gaskets. Based on the area of the gasket face before clamping, this results in about 300 psi for the rectangular gasket and about 500 psi for the square gasket. The metal-to-metal types of grooves shown in Tables 10.5 and 10.6 should not be used with elastomers which "cold flow," i.e., flow under pressure through any available openings. A material such as Teflon has a strong tendency to cold flow and, therefore, the grooves must be designed so that the gasket has "no place to go" after being compressed. Special arrangements are often used to maintain a constant pressure on the Teflon as it cold flows. These comments also apply to some formulations of fluorocarbons and special gasketing materials, such as Kel-F, Viton A, and Fluorel. However, new compounding techniques have resulted in gaskets of lower durometer values and subsequently lower cold flow tendencies so that they can be treated much as natural or synthetic rubbers are treated. The principal advantages of the fluorocarbons are their low vapor pressures and their ability to stand rather wide temperature ranges. Silicone rubbers also have good temperature tolerance and can be obtained in a wide range of hardnesses so that the grooves can be the same as shown in Tables 10.5 and 10.6 for the appropriate durometer values. Butyl rubber is claimed by some observers to be superior to other rubbers in vapor pressure properties. Grooves for gaskets of this material can be designed in the same manner as for other rubbers.

Rubbers and other elastomers are usually cleaned with an appropriate solvent such as acetone, alcohol, or trichloroethylene and then dried in air at room temperature. J. R. Young (*Rev. Sci. Instr.*, **30,** 291, 1959) claims that, if commercial O-rings are baked for several hours at about 100°C after cleaning as above, the result will be considerably lower vapor pressures and, therefore, a lower ultimate pressure. An alternative method suggested by Young is to install commercial gaskets after unpacking without cleaning. Of course, care must be taken to avoid touching the gasket unnecessarily with bare hands since various secretions from the hands will result in high vapor pressure. Sulfur can be removed from rubber by boiling in caustic soda. Usually gaskets are coated with a thin layer of a good vacuum grease. Only a very thin layer is necessary. Too much grease simply results in additional outgassing. Vacuum grease should never be left exposed to the air for any length of time since it will absorb gases and vapors from the atmosphere. In some cases it is necessary to install gaskets dry. Dry gaskets generally require considerably higher forces to compress than gaskets which have been coated with vacuum grease because there is little slippage on surfaces. The increase in force will depend on the gasket material, on the surfaces involved, and on the roughness of the surfaces.

Metal gaskets are used when it is desired to reduce the vapor pressure due to the gasket or for operation at high temperatures. Usually such seals are arranged so they can be baked at a high temperature. Both soft and hard metals have been used as gasket materials although soft metals such as copper, lead, aluminum, gold, indium, and tin are most common. OFHC copper should be used. Both thin metal stock and stock of circular cross section are commonly used. In general if a soft gasket material is used, the parts being sealed should be made of a hard material (steel or stainless steel). This is not a hard-and-fast rule. Gaskets of thin stock material can be used between two smooth surfaces while gaskets of circular cross section can be used in O-ring grooves of the type discussed above for rubber. In each case the gasket and the surfaces to be sealed must be highly polished. Also, provision must be made to obtain high pressures (up to many thousands of pounds per square inch) on the gasket. Metal gaskets have come into more general use with the advent of ultra-high vacuum technology. Consequently, many different designs of joints have been reported in the last few years. A number of these are discussed in Chapter 12.

10.10   Construction of Vacuum Vessels

The brazing, welding, and gasketing techniques discussed above are commonly used for making up vacuum vessels. Of course, often commercial systems come provided with vacuum vessels (bell jars, furnace chambers, etc.) and several sizes of bell jars (glass, steel, and stainless steel) are available. In fabricating metal vessels, the proper choice of metal should be made (see Chapter 8) together with the proper fabrication technique. It should always be kept in mind that the atmosphere exerts a force of about 1 ton on every square foot of vacuum wall. Consequently, if the walls are made too thin there is a danger of the vessel "oil canning," i.e., simply collapsing. The problem becomes more severe for large process equipment where it is undesirable to use extra heavy walls to get a large safety factor because of considerations of economy. For this large equipment it is necessary to make as accurate a compromise as possible between economy, safety, and special requirements of the process. This means making the best possible use of existing engineering codes.

For the smaller types of experimental systems, using walls somewhat heavier than necessary is not of prime concern since the cost is relatively minor. Particularly for the larger size of system there is one rule that should always be kept in mind. Because of the force exerted by the atmosphere, it is best to use curved walls, particularly cylinders and spheres, rather than flat surfaces. This makes it possible to use thinner walls. Cylindrical sections are often used which are sealed at the ends by dished (convex outward) sections. In addition, strengthening rings and rods are often welded on the chamber so that thinner walls can be used, thus resulting in a saving in cost. The exact geometry of the vacuum vessel is often dictated by the function of the vacuum system. A simple type of small vacuum system is shown in Fig. 10.6. This type of chamber has been found relatively easy to construct and is useful for many purposes.

Referring to Fig. 10.6, typical dimensions of one chamber are 18 in. OD and 12 in. in height for the cylinder with a wall thickness of ½ in. The lid is also ½ in. thick and is not dished since this thickness of metal for the diameter used results in a very small deflection due to the atmosphere. Aluminum, mild steel, stainless steel, or brass can be

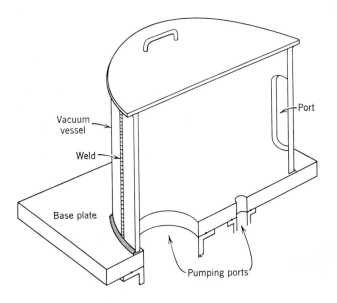

Fig. 10.6 Simple type of small vacuum vessel.

used as construction materials. The cylinder was rolled from mild steel sheet and the ends were welded. Either brazing or welding can be employed depending on the metal used. Standard pipe or tubing can be used for smaller chambers. The ends of the cylinder were machined to get flat surfaces. Sealing of the lid to the cylinder and the cylinder to the base plate can be done with standard bell jar gaskets or O-rings. The atmosphere will exert a large enough force to produce good seals. The ½ in. wall thickness is sufficient to permit cutting O-ring grooves on the ends and milling ports in the sides (for viewing windows, etc.). The design also permits attaching various vacuum components, such as electrical lead-throughs, liquid nitrogen traps, gauges, etc. to the lid, all of which can be readily removed for maintenance or changes. The base plate can be made of aluminum, brass, mild steel, or stainless steel with the top surface machined.

For large systems such as those used with vacuum furnaces, dehydration plants, etc., the design of the vacuum vessel must be done carefully in order to keep costs down. Various sections of the American Society of Mechanical Engineers (ASME) code for boilers and pressure vessels may be used to determine wall thicknesses for various standard designs. This code, Section VIII, for unfired pressure vessels, was revised and re-issued in 1952. In addition to this standard reference a number of manuals on pressure vessel design are available. The design of dished heads and spheres is also covered by the above-mentioned code.

In general, the mechanical loads imposed on a vacuum vessel are those contributed by the weight of the structure itself, those imposed by the process within the vacuum chamber, and the principal load, which is the external atmospheric pressure loading. The atmospheric loading usually controls the design. Usually heavier sections than are required by the code are used in the design of vacuum vessels. There are two reasons for this. First, corrosion must always be considered in the design of process equipment. Second, damage or bending must be minimized for those movable structural elements which are being handled daily. Consider the following example. If one has a cylindrical vessel in need of a head, and a flat steel circular plate is compared with a standard steel dished head using code design technique, one finds that for the same deflection the flat plate would have a thickness of approximately 1 in. for a 60 in. diameter head, while the dished head would have a thickness of approximately ⅛ in. for the same diameter. These data are based on a spherical shaped head. The weight saving would be considerable. However, a steel dished head ⅛ in. thick would be susceptible to damage and might fail from additional stresses caused by the damage. A spherical head of ¼ in. steel or even ⅜ in.

steel would still permit a substantial saving in weight and would be much safer to handle without fear of damage. For cylinders, *minimum* wall thicknesses (mild steel) are given in the following list (*ASME Boiler Code*):

Cylinder Diameter	Wall Thickness
36 in. or less	$\frac{1}{4}$ in.
37–54 in.	$\frac{5}{16}$ in.
55–72 in.	$\frac{3}{8}$ in.
over 72 in.	$\frac{1}{2}$ in.

In view of the wide use of glass bell jars it may be advisable to consider some factors involved in the use of such vessels. Various sizes of glass bell jars are available commercially. Popular heights of such jars are 18 to 20 in. In using such jars it is important not to use any handy glass jar. The jar must stand both atmospheric pressure and *heat*—due to the evaporating dish or a high voltage discharge. A good quality, heat-resisting glass should be used. The walls should be as uniform as possible and at least $\frac{1}{4}$ in. thick. The top should be domed so as to withstand atmospheric pressure. There has been talk about "implosion" of glass bell jars, i.e., failure of the glass under vacuum with complete collapse due to atmospheric pressure. It is often said that an implosion of this type should not harm anyone in the vicinity because the air pressure should force glass fragments toward the center of the bell jar. Unfortunately, although glass fragments will be forced to the center of the jar they will not stop there but will continue on through. With commercial glass bell jars such implosions are very unlikely. However, to avoid any possibility of implosion certain simple precautions can readily be taken. The jar should be examined for thin areas and flaws and it should "ring" properly. Any jar with large or numerous air bubbles should be rejected. Also, jars departing considerably from the cylindrical should be rejected. A tolerance of $\frac{1}{4}$ in. variation per foot diameter is just tolerable. In some cases the jar is covered with a wire mesh, perforated metal, or expanded metal screen to provide protection in case of implosion. Too fine a mesh will obstruct the view and too coarse a mesh will afford no protection. Blown or molded covers of plastic, such as methyl methacrylate, are recommended (about 0.2 in. thick). This cover should fit loosely over the bell jar with a central top hole large enough to prevent temporary exhaust if the jar does break. The plastic cover can be reinforced with a coarse wire cover. Generally larger glass bell jars are lifted by a mechanical hoist with appropriate mechanical drives and cushioning features.

ELEVEN

Baffles, Traps, and Valves

11.1 The Uses of Baffles and Traps

The primary purpose of a baffle is to prevent pump fluid from getting into the vacuum chamber. Such fluids will limit the ultimate pressure and may interfere with the vacuum process. *Backstreaming* and *back migration* of vapors have been defined in Chapter 4. Most backstreaming can be stopped with a water-cooled baffle. However, to stop most back migration it is necessary to use a refrigerated baffle or to follow up the water-cooled baffle with a cold trap (or refrigerated baffle). A water-cooled baffle will condense the pump fluid and allow it to flow back into the pump. If a refrigerated baffle (or cold trap) is used above the pump, vapor will be kept out of the system but also may be prevented from returning to the pump. This will depend on the type of pump fluid. In the case of mercury, the freezing point is $-40°C$. Consequently a baffle can be cooled to near this temperature. As a result, refrigerated baffles are commonly used with mercury pumps. On the other hand, going to a temperature this low will freeze oil vapors. This represents a loss of oil from the pump. Any significant backstreaming and back migration means more maintenance on the pumps. Consequently, most oil diffusion pump systems use a water-cooled baffle, often followed by a liquid nitrogen cold trap. In some cases heated baffles have been used. In this case the oil vapor is decomposed by the hot surfaces of the baffle, producing volatile gases and impure carbon. The gases are pumped from the system and the carbon has a low vapor pressure. However, cooled baffles are by far the most common type.

It should be pointed out that, although they are effective in keeping pump fluid out of the vacuum system, cold traps are also very effective pumps for various vapors (see Chapter 5). In particular, they are very effective for water vapor, the chief offender in most systems, as well as for grease vapors, mercury vapor, etc. Some types will also pump carbon dioxide. The principal methods routinely used to obtain low temperatures for cold traps are:

1. Liquid nitrogen.
2. Dry ice.
3. Mechanical refrigeration.

At the temperature of dry ice ($-78°C$) the vapor pressure of water is about 0.6 μ. It is necessary to go to temperatures well below $-100°C$ to make the vapor pressure of water negligible. However, at $-78°C$ the vapor pressure of mercury is around 3×10^{-9} mm Hg (Torr) so dry ice is effective in trapping mercury. Freon 12 or 22 is commonly used in mechanical refrigeration systems. With Freon 22, a single stage system will give temperatures around $-40°C$ while a compound system will give values around $-100°C$. Consequently, a suitable Freon refrigeration system can be used to effectively trap mercury and also water, if the ultimate pressure desired is not too low. Mechanical refrigeration, where it can be used effectively, has an advantage over dry ice or liquid nitrogen since the refrigerant does not have to be replenished frequently. Refrigerated baffles act as traps as well as baffles. However, their design will be considered under baffles since they do not differ substantially from ordinary water-cooled baffles.

Several points regarding the general design of baffles and traps should be made:

1. They should be effective in trapping vapors but should not introduce an undue amount of impedance to gas flow. This is a matter of compromise.

2. They should be readily accessible or easily removed from the system for cleaning.

3. In the case of baffles operated at temperatures above the freezing point of the pump fluid, such as water-cooled baffles, provision should

be made so condensed fluid is returned to the pump without striking the hot jet assembly. Otherwise, it is simply re-evaporated.

The emphasis above has been on baffles and traps for use between the pumps and the vacuum chamber. Often baffles or traps are used to prevent pump fluid from reaching some specific part of the system. Traps or refrigerated baffles are commonly used between pumps where several pumps are operated in series, as in the case of a diffusion pump followed by a booster pump, in order to prevent fluid from one pump from getting into another. It is common practice to use a cold trap or refrigerated baffle in the foreline before the mechanical pump with mercury pumps so as to prevent mercury from getting into the mechanical pump (and into the air), as well as to prevent oil from getting into the mercury pump. This practice is sometimes followed with a diffusion pump–mechanical pump system, particularly in larger systems. Small liquid nitrogen traps are often used to prevent vapors from getting into a vacuum gauge.

Only cold traps have been considered above. Various sorption materials are also used to trap vapors (and some gases). The effectiveness of such a material depends on its nature and on whether it is used at room temperature or is refrigerated (see Chapter 8). Whether room temperature or refrigerated sorption materials are used, the design of the trap is guided by the same considerations, viz., effective trapping without too large a loss in pumping speed and ease of servicing. Copper foil traps are used fairly commonly with small ultra-high vacuum systems and will be considered briefly. Special "sand" or "dirt" traps are sometimes used with very large systems to prevent loose particles of matter from getting into the mechanical pumps. Regardless of the type of baffle or trap being considered, it must always be kept in mind that using these units results in a substantial loss of pumping speed. Baffles and traps should only be used when absolutely essential to the vacuum process involved.

11.2   Some Designs of Baffles

Many designs of baffles have been reported in the literature. Emphasis will be placed here on some more common designs. A baffle is simply a surface (usually cooled) which is put in the path of the gas flow. The surface is provided by some metal, and to keep the surface uniformly cool (when cooled) a metal with good heat conductivity should be used. Copper is commonly used. The refrigerant used is usually passed through copper tubing which is brazed to copper plates. In many cases the housing is also cooled. The tubing can be attached to the housing by brazing or welding, depending on the materials involved. In this case the inside surface of the housing acts as part of the cold surface. In some cases the designs are such that refrigerant can be circulated through volumes that are sealed off from the vacuum.

The baffles considered here are of the optically dense type, i.e., molecules entering the baffle cannot see through the baffle. Any refrigerant could be used with these baffles, although as the temperature of the refrigerant is lowered more attention must be given to fabrication techniques and insulation. Contraction of parts due to low temperatures must be considered and lack of adequate insulation can make it impossible to attain the desired low temperatures in the baffle. Glass wool, asbestos, and some of the newer plastic materials are commonly used. Figure 11.1a shows a simple form of baffle which is in fairly common use. This simply consists of a disk inside a housing. Cooling tubes are brazed to the disk and the outside of the housing, somewhat as shown in Figs. 11.1b and 11.1c. These figures show just one typical arrangement of tubing. At least with the smaller units, the housing is often made of brass for simplicity in machining. The tubing can be

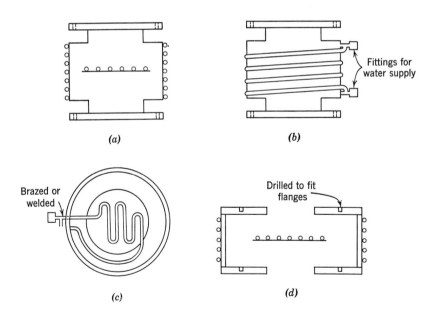

Fig. 11.1 (a) Baffle with flanged ends. (b) Cooling of housing. (c) Cooling of disk. (d) Alternative arrangement.

used to support the disk in place. Where the disk is too heavy, supports can be attached to the inside walls of the housing. The baffle shown in Fig. 11.1a includes flanges at both ends. Actually such units can have the ends drilled to fit other flanges as is shown in Fig. 11.1d. The thickness of the disk is not too important except that it should be sufficient to get a fairly uniform temperature distribution. This remark also applies to the housing, although mechanical strength considerations are more important there. If the ends are to be drilled and tapped so other flanges can be attached, then these ends must be of sufficient thickness.

Figure 11.2a shows another form of the baffle discussed above. Here, a central disk of the type shown in Fig. 11.1a is again used. However, in this case the disks with the holes in them serve the function of the necks at top and bottom of Fig. 11.1a or the end plates in Fig. 11.1d. Often this type of baffle is made as a unit separate from the housing (as shown). The disks can be brazed inside a cylinder and the whole unit can be inserted in the housing. Cooling of the baffle arrangement is done by cooling the outer wall of the housing and relying on good metallic contact between the inner cylinder and the inside wall of the housing. This does not lead to as low temperatures as can be obtained by directly cooling the baffle plates. On the other hand, the entire baffle assembly can be removed for cleaning. The construction becomes much more difficult if the baffle disks are to be directly cooled. It means mounting the middle disk (with cooling coils) before brazing the upper and lower disks (also with cooling coils). It should be pointed out that the holes in the disks are somewhat smaller than the middle disk so as to have an optical baffle. Figure 11.2b shows an arrangement using three baffle plates of the shape shown in the insert. These plates are designed so that the upper and lower ones overlap the middle one to some extent so as to have an optical baffle. The comments concerning the baffle of Fig. 11.2a with regard to cooling are applicable here. In some cases the plates are brazed or welded directly to the inside of the housing with only the housing cooled. Sometimes this arrangement, particularly with small pumps, is incorporated in the upper part of the pump barrel.

Figure 11.2c shows a multiple-ring type of baffle. A series of metal rings (usually copper) of increasing diameter from top to bottom are arranged as shown. These rings overlap in such a manner that an optical baffle results. The gas molecules pass through the spaces between the rings. The arrangement of Fig. 11.2c shows cooling tubes on housing and rings. Sometimes the ring assembly is arranged so that it is cooled by conduction to the housing walls. One such arrangement

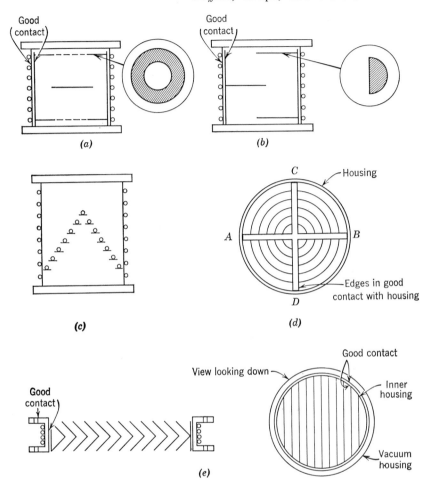

Fig. 11.2 (*a*) Use of three disks. (*b*) Use of half-disks. (*c*) Multiple ring baffle. (*d*) One form of multiple ring arrangement. (*e*) Chevron type of trap.

is shown in Fig. 11.2*d*. Two heavy metal plates, *AB* and *CD*, are used. These plates make good contact with the inner wall of the housing, which is cooled. The segments of rings are brazed to the metal plates. Of course, the whole assembly could be brazed to a cylinder which fits closely inside the housing. Figure 11.2*e* shows a chevron-type trap. Such traps are often made with the chevron arrangement brazed to an inner housing which makes good metallic contact with the vacuum housing of the baffle. The metal chevron strips overlap sufficiently to provide an optical baffle. Of course, the strips could be directly cooled using tubing but this leads to more difficulties in fabrication. Another arrangement available commercially, e.g., from Consolidated Vacuum Corp., Rochester, New York, has the chevron elements brazed or welded to a housing through which refrigerant (water, Freon, or liquid nitrogen) can be circulated. Some designs make it possible to remove the chevron baffle arrangement from the vacuum housing for ease of cleaning without breaking vacuum line connections. One big advantage of the chevron type of baffle is that its thickness can be kept to a minimum. This leads to fairly high efficiency in baffling with not too much loss of pumping speed.

The baffle arrangements shown in Figs 11.1 and 11.2 are commonly used with water cooling. However, other cooling methods can be used, including mechanical refrigeration and liquid nitrogen. In the case of liquid nitrogen suitable means for circulating the liquid nitrogen through tubing must be provided. Also, the tubing should be well insulated so as to reach the lowest temperatures and cut down on loss of liquid nitrogen. Many other baffle arrangements have been used and are reported in the literature. However, most of them are simply variations of the designs shown in Figs. 11.1 and 11.2.

11.3   Some Designs of Traps

Some common forms of traps are shown in Figs. 11.3a, b, and c. The traps shown are of the thimble type. The container A is filled with liquid nitrogen through the filler tube E. Using liquid nitrogen means that the trap must be kept in essentially a vertical position. In the case of Fig. 11.3a the container is supported by the filler tube, which is brazed or welded to a flange. This flange is sealed to the housing by an O-ring. Naturally the hole in the housing must be large enough so that the thimble can be removed for cleaning. This means that the trap must be provided with a flat surface to accommodate the flange. Of course, the filler tube could be brazed or welded to the wall of the housing. In this case any shape of housing, including cylindrical, could be used. This will make it more difficult to clean the trap, making it necessary to disconnect the trap from the system. Often the container A is made in a spherical shape to keep the size of flange to a minimum. The same basic arrangement is used in Figs. 11.3b and c, the changes being in the position of inlets and outlets. The traps shown in Figs. 11.3a, b, and c are of the optical type. The design of the container or thimble must be such as to provide adequate mechanical strength. A leak in the thimble could lead to disastrous results in cases where there are hot parts in the vacuum system. Therefore, the larger thimbles are usually made of stainless steel and are sometimes dished at the ends as is shown in Fig. 11.3*d*. There is not the same danger with smaller traps and these are often made of brass or

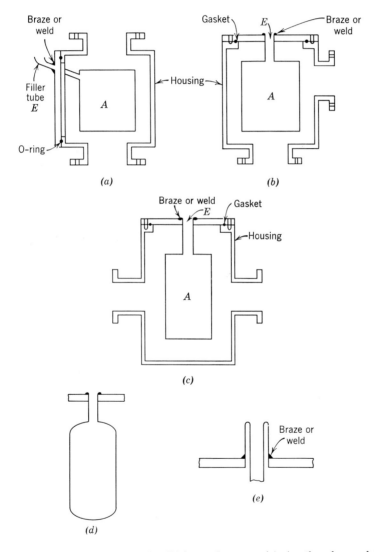

Fig. 11.3 (*a*) "In-line" trap. (*b*) Right angle trap. (*c*) Another form of "in-line" trap. (*d*) Dished ends. (*e*) Method to reduce heat loss.

copper. These metals offer an advantage over stainless steel since they are better heat conductors. This helps keep the whole trap cold in spite of a drop in the level of the refrigerant. If the top of the trap warms up, the frozen vapors will be released into the system.

Heat gets to the trap in two ways—by radiation from the housing and by conduction through the filler tube. The first type of heating can be minimized by making the thimble surface shiny. Stainless steel traps are usually polished while brass or copper traps are plated (ordinarily with nickel). Heating through the filler tube can only be avoided by using a material that is a poor conductor of heat. Practically, stainless steel tubing is most commonly used. Heating of the thimble can be minimized by using tubing with thin walls. The thickness will be determined by the weight of thimble and refrigerant to be supported. Sometimes additional supports are provided inside the vacuum system. In this case the filler tube could be made thinner. However, the additional supports must be poor heat conductors. Figure 11.3e shows an arrangement often used to minimize heat conduction through the filler tube. The tube has simply been made longer. The inside walls of the housing should be shiny as in the case of the thimble itself. After a trap has been constructed it should be thoroughly tested for leaks (see Chapter 15). Welds or brazes may open up due to contraction of parts when the trap is cooled with refrigerant. Figures 11.3a, b, and c show the inlets and outlets of the traps with flanges. With the smaller sizes of traps the inlet and outlet tubes are often just brazed or welded to the walls of the housing.

Glass cold traps are often used as part of small vacuum systems, or to protect vacuum gauges. Several forms of glass traps are shown in Figs. 11.4a through e. The trap in Fig. 11.4a is commonly used to protect vacuum gauges. It is more efficient than the trap of Fig. 11.4b but it also offers more resistance to gas flow. It will be noticed that both these traps are immersed in the refrigerant, which is held in a suitably insulated container, usually a Dewar flask. Figures 11.4c, d, and e show three forms of a trap into which the refrigerant is poured. They differ from each other only in the positions of inlet and outlet tubes. The outside walls of these traps are exposed to the atmosphere but suitable insulation can be used to cut down on loss of liquid nitrogen. The "spherical" trap shown in Fig. 11.4e is fairly efficient in trapping and offers a fairly low impedance to gas flow. The positions of inlet and outlet can be varied. Figure 11.4f shows a form of glass water-cooled trap.

Many special designs of traps have appeared in the literature. Particular attention has been given to traps which are cooled by conduc-

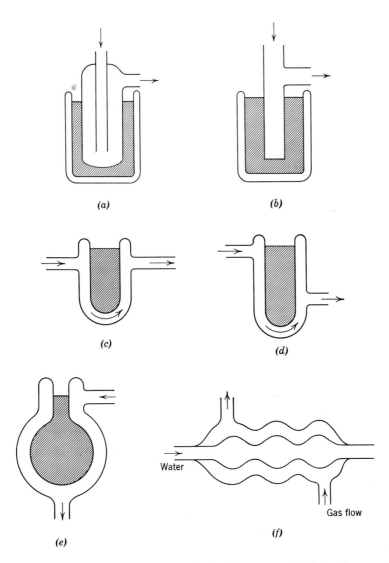

Fig. 11.4 (*a*) Re-entrant trap. (*b*) Simple glass trap. (*c*) Internal reservoir trap No. 1. (*d*) Internal reservoir trap No. 2. (*e*) Form of spherical trap. (*f*) Special water-cooled trap.

tion from an external reservoir. This general method keeps the entire trap cold as long as there is refrigerant in the reservoir. Of course, any baffle that is cooled by refrigerant being circulated through it serves the same purpose. Some means must be provided for circulating the refrigerant and this is sometimes a problem with liquid nitrogen. Self-contained Freon systems are convenient although the temperatures reached are not as low as with liquid nitrogen. Baffles of the type shown in Figs. 11.1a and e and Figs. 11.2c and e have been used with liquid nitrogen or Freon circulated through tubing brazed to the plates. The refrigerant is generally not used to cool the housing because of the resulting high rate of consumption. Some trap arrangements making use of an external reservoir are shown in Fig. 11.5. Figure 11.5a shows an arrangement whereby the cold surface S in the vacuum chamber is cooled by conduction through a rod R which is screwed into a massive piece of metal M (Conn and Daglish, *J. Sci. Instr.*, **34**, 245, 1956). Parts R, M, and S are made of polished copper. The outer tube T is made of poor conducting metal such as nickel-silver. An O-ring arrangement is shown for detaching the bottom part of the trap. This feature could be eliminated with the trap attached directly to the system. The arrangement shown serves to trap pump fluid vapor as well as vapors being released in the vacuum chamber. The trap is about 12 to 14 in. in total length with a diameter of about 2 in. and a pumping line 1 in. in diameter. The rod R could be removed and the outer tube T replaced by glass tubing. This trap would then become a re-entrant trap of the type shown in Fig. 11.4a.

In the trap of Fig. 11.5a the temperature of the cold surface does not get anywhere near that of liquid nitrogen. To reach lower temperatures it is necessary to get the liquid nitrogen as near the surfaces to be cooled as possible. A general arrangement for doing this is shown in Fig. 11.5b. The hollow connector tube C is filled with liquid nitrogen from the reservoir R. The structure holding the baffle plates could also be hollow so as to get liquid nitrogen as near as possible to the baffle plates. The arrangement shown uses semicircular plates of the type illustrated in Fig. 11.2b but other types could be used. By using a large enough reservoir it is possible to use this kind of trap overnight on systems that have reached their operating pressure and contain no great amount of water vapor. Martin (*J. Sci. Instr.*, **32**, 400, 1955) has described a trap of this type in some detail. He claims that a maximum temperature of $-150°C$ is achieved in the baffle system. A final example of a trap using an external reservoir is shown in Fig. 11.5c (Holland, *Vac. Symp. Trans.*, **168**, 1960). This arrangement is similar to that shown in Fig. 11.5a. No particular baffle system is

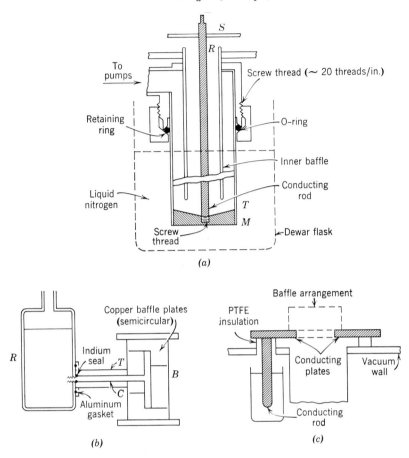

Fig. 11.5   (*a*) Special form of re-entrant trap.   (*b*) External reservoir trap No. 1.
(*c*) External reservoir trap No. 2.

shown since this is optional depending on the efficiency required and
what loss of pumping speed can be tolerated. It is necessary to avoid
heat getting through the support for the conducting plate. Also,
the support must be connected to the vacuum wall by a vacuum-tight
joint. A PTFE (polytetrafluorethylene) support is shown but other
materials can be used as long as they are vacuum tight.

   In trying to obtain very low pressure (very high and ultra-high
vacuums) it is important to keep traces of pump oil out of the system.
With liquid nitrogen traps, the housing is usually not cooled. Consequently, unless the design is right it is possible for oil to creep up the

inside wall of the housing and into the system.   In general, thimble-type traps suffer from creeping of oil since the housing is not cooled and there are no baffles near the connection to the system to stop the oil.   Of course, a baffle could be added, although this would cause an additional drop in pumping speed.   However, this solution is generally necessary to stop oil creep.   The baffles shown in Fig. 11.1 could be used as no-creep baffles by refrigerating the baffle plate and water-cooling the housing.   The baffle in Fig. 11.2a could be used as a no-creep trap by refrigerating the center baffle and water-cooling the housing.   In most cases operating the housing at room temperature is adequate.   The key to a no-creep trap is to have a baffle at the top of the trap so oil can't creep around this baffle but has to flow back again into the trap.   Cooling the baffle always helps.

In the very high and ultra-high vacuum regions, particular care must be taken to avoid oil creep and also re-evaporation of vapors if the parts of the trap warm up due to a drop in the liquid nitrogen level. This implies the use of a no-creep trap with continuous circulation of refrigerant or an external reservoir with suitable conduction paths. Instead of more or less standard cold traps, it is possible to use special types of traps which may or may not be refrigerated.   Various versions of the copper foil trap developed by Alpert (*Rev. Sci. Instr.*, **24**, 1004, 1953) have been used with systems operated at very low pressures and even with systems operated in the more conventional range, say around 10^{-6} mm Hg (Torr).   The original Alpert trap was used to prevent vapors from getting into the system even when a liquid nitrogen trap was used.   These vapors could get by the trap or could be re-evaporated from the trap as the liquid nitrogen level dropped.   The general nature of this type of trap is shown in Figs. 11.6a, b, and c.   Figure 11.6a shows a re-entrant type of trap in which the space inside the trap is filled with a continuous spiral of corrugated copper foil placed between the inner and outer tubing.   In effect, this arrangement forms a system of small diameter "straws" or "tubes."   The corrugated foil was fabricated from OFHC copper sheet 0.003 in. thick and 6 in. wide by means of gears.   Figure 11.6b shows the general appearance of the "straws" or "tubes."   The impedance of such a trap is not particularly high.   A cylindrical tubing 2.8 cm in diameter completely filled with copper straws has approximately the same conductance as an unfilled tube of the same length 1 cm in diameter.   This type of trap has been found to be effective for oil vapors at room temperature as well as at low temperatures (liquid nitrogen).   To attain very low pressures—10^{-9} mm Hg (Torr) or less—it is general practice to bake out these traps and then operate them at room temperature or at liquid nitrogen temperature. For small systems (say with oil diffusion pumps of 2 in. diameter or less) these traps are quite efficient. However, with larger systems they tend to saturate and need to be baked out too often to be very effective. Furthermore, they do not trap mercury effectively. Figure 11.6c shows a form of trap used by Alpert which makes use of straight tubing with corrugated copper foil.

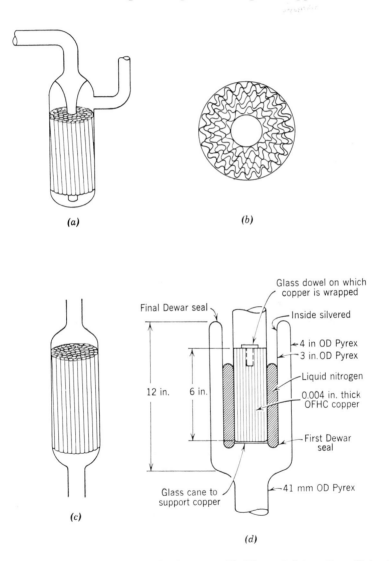

Fig. 11.6 (*a*) Re-entrant type of foil trap. (*b*) View of "straws" or "tubes." (*c*) Trap with straight tubing. (*d*) Foil trap with Dewar.

Various versions of the original Alpert trap have been reported. One of these is shown in Fig. 11.6*d*. This figure shows a form using copper foil which is cooled by liquid nitrogen in a Dewar flask (Burns, *Rev. Sci. Instr.*, **28**, 469, 1957). The pertinent features are shown in the figure. Grade A nickel foil was tried instead of OFHC foil. The nickel foil was hydrogen fired at 500°C before being put into the trap and appeared to function about as well as the copper. This type of trap is claimed to reduce the pumping speed much less than the original Alpert-type trap with the same nominal tube diameter. It is also claimed that liquid nitrogen lasts overnight with such a trap. A foil trap with a central cold thimble has been described by Thomas, Destappes, and Dupont (*Vacuum*, Vol. III, No. 4, Oct. 1953). This trap made use of stainless steel sheet 0.1 mm thick which was corrugated in the manner shown in Fig. 11.6*b*. The foil was cooled by the liquid nitrogen thimble. The housing was also made of stainless steel since a mercury diffusion pump was used.

The use of certain sorption materials for pumping was discussed in Chapter 5. Such materials include activated charcoal, activated alumina, and artificial zeolite. The latter two materials have found considerable use in traps, particularly to eliminate oil vapor from vacuum systems. Some traps using these materials are shown in Figs. 11.7*a*, *b*, *c*, and *d*. Figure 11.7*a* shows a trap for use with small ultra-high vacuum systems, say with a volume of around 1 l and an internal surface area of about 10^4 cm² (Biondi, *Rev. Sci. Instr.*, **30**, 831, 1959). The trap was designed to have a greater conductance than the tubing to and from the trap (14 mm diameter Pyrex). Either artificial zeolite or activated alumina could be used, it was found. With the first material, ⅛ in. pellets were used. The whole trap could be baked for use in ultra-high vacuum systems (also to activate the alumina and zeolite). Part *A* of the trap is to prevent creep of oil. Figure 11.7*b* shows a trap designed for larger systems (Biondi—see above). The design is intended to provide high trapping efficiency with high conductance. When this trap and the system were baked for 8 hr at about 430°C the pressure dropped to around 3×10^{-10} mm Hg (Torr) and then rose to 6×10^{-10} Torr in 60 days. The pressure remained below 10^{-9} Torr for 70 days. The particular trap reported by Biondi was about 8 in. in diameter and was intended for use with a 4 or 5 in. diffusion pump. Biondi noted that the effectiveness of his trap depended on the pump fluid being used. Figure 11.7*c* shows an alumina trap designed to minimize contamination of cathodes (Harris, *Rev. Sci. Instr.*, **31**, 903, 1960). His trap consisted of a 1 in. length of 3½ in. diameter tubing closed on each end by wire mesh. A heater

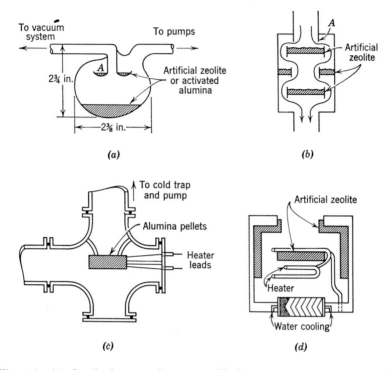

Fig. 11.7 (*a*) Small glass sorption trap. (*b*) Large metal sorption trap. (*c*) Alumina trap. (*d*) Zeolite trap with chevron baffle.

was imbedded in the alumina pellets for outgassing at about 500°C. Figure 11.7*d* shows a combination of a water-cooled chevron baffle and an artificial zeolite trap (Goerz, *Vac. Symp. Trans.*, **65,** 1960). The zeolite pellets ($\frac{1}{8}$ in. by $\frac{1}{4}$ in.) were held to the walls by $\frac{3}{32}$ in. stainless steel mesh. Heater elements are shown for outgassing the zeolite. With bake-out at 350°C and pumping for several hours a pressure of 1 to 5×10^{-9} Torr was reached. New designs of non-refrigerated traps using artificial zeolite or activated alumina are reported continually in the literature. Although they appear to be highly effective in trapping oil vapors, it must be kept in mind that such traps do not trap mercury. The effectiveness of alumina or zeolite is increased by refrigerating with liquid nitrogen. Activated charcoal, refrigerated or nonrefrigerated, can also be used to trap various vapors and gases (see Chapter 8). Often the charcoal is simply contained in a Pyrex or metal tube which is attached to the system and can be heated. The charcoal is activated by being heated to about 440°C (not over 700°C) for a matter of an hour or more while being pumped and then allowed to cool to room temperature. Cooling to liquid nitrogen temperature increases the trapping efficiency considerably.

In many cases dehydrating (drying) agents (see Chapter 8) are used in vacuum systems to absorb water vapor. Traps using such materials are designed for maximum trapping efficiency and conductance. Drying agents are also used in letting a vacuum system, or part of a system, down to air. In this case the design of the container is not too critical in view of the fact that the only requirement is to effectively remove water vapor and still let air through. In very large systems, "sand" or "dirt" traps are sometimes used before the mechanical pumps to prevent loose particles from damaging these pumps. Screens can be used for this purpose although a fine mesh will seriously affect the pumping speed. Another method is to use a deflector to deflect particles into an oil trap, where they are held by the oil.

11.4   Operation and Maintenance of Baffles and Traps

Water-cooled baffles present few problems in the way of operation or maintenance. Usually the water flows through the baffle first and then through the diffusion pump. It is turned on whenever the diffusion pump is in operation. The temperature of tap water is such that "ice" does not form. Consequently there is essentially no cleaning problem associated with water-cooled baffles. However, it is a good idea to set up a regular cleaning schedule to take care of foreign material from the vacuum system that collects on the baffle. The cleaning schedule will depend on the vacuum process.

When it comes to refrigerated baffles or traps the problems associated with operation and maintenance become more serious. These problems stem primarily from the formation of "ice." Most of this ice is ordinary frozen water. Frozen carbon dioxide is also present when liquid nitrogen is used as the refrigerant. The ice that forms on the trap (or baffle) is a poor conductor of heat. Consequently as the ice layer builds up the outer surface will reach a higher temperature than that of the refrigerant. The difference in temperature between the surface of the ice (which does the trapping) and the refrigerant will increase with the thickness of ice. No set rule can be given for the maximum thickness of ice that is permissible. This depends entirely on the nature of the particular vacuum process. When a system is operated continuously it is necessary to set up a trap cleaning schedule. Such a schedule has to be set up on the basis of experience. When a system is not shut down at night it is often possible to let the refrigerant evaporate and then refill the traps in the morning. The vapors that evaporate from the traps will be removed by the pumps during the night. This is true for any reasonably clean system.

The problem of ice formation exists for all traps using a refrigerant that is at a low enough temperature to freeze water vapor. In general the build-up of ice is greatest when the refrigerant is first applied since this is when there is the greatest amount of water vapor (and other vapors) present. With refrigerated baffles using an external reservoir or circulated refrigerant the entire baffle becomes cold on application of refrigeration. Consequently, ice will begin to accumulate all over the baffle. With traps (or baffles) using a reservoir inside the vacuum system, e.g., thimble traps, when the refrigerant is applied ice starts to form over the outside surface of the trap. As the level of the refrigerant drops, the top part of the trap becomes warmer and the ice there can evaporate. To avoid this, a thimble-type trap should never be filled completely at first. Filling about one-quarter full and waiting a few minutes before filling completely will cause most of the ice to collect on the lower part of the trap. As long as the refrigerant is not allowed to get below the one-quarter point the ice on the lower part will not evaporate. With this method of filling, there is less ice evaporated as the level of the refrigerant drops. Traps of the type shown in Fig. 11.3 and Fig. 11.4 should be filled in this manner. The two traps shown in Figs. 11.4a and b are often said to have external reservoirs. However, the distinction here is between traps or baffles which develop differences in temperature over their parts as the level of refrigerant falls and those that maintain a fairly constant low temperature regardless of the level of refrigerant (as long as there is some left).

With traps or baffles using liquid nitrogen or dry ice there is always a concern with loss of refrigerant. Rapid loss of refrigerant means filling more often, which may result is added labor costs. Also, the cost of additional refrigerant must be considered. The actual design of trap or baffle will largely determine the rate of loss of refrigerant. However, even with a given design certain steps can be taken to reduce the loss of refrigerant. If the inside of the housing is not shiny (polished or plated), thin, shiny foil, such as aluminum, can be applied to cut down losses. There is always heating of the refrigerant through the filler tube (or the external reservoir). Proper insulation can cut down losses considerably. In the case of Dewar flasks, glass or cotton wool can be used to cover the tops. Another arrangement that has been used to cover and insulate the top of a Dewar flask is described by Svec (*Rev. Sci. Instr.*, **27,** 969, 1956). His arrangement makes use of a foamed polystyrene shield. The shield described could be used with small glass traps of the type shown in Figs. 11.4*a* and *b*. Such a shield could be modified to fit other types of traps. The shield has a hole somewhat smaller than the diameter of the cold finger and is forced over it by a slight rotary motion. The compressed foam is sufficiently resilient to keep the shield in place. A cork or plug made of foamed polystyrene is used to close the filling port when not in use. Shields of this type have been made of metal. However, in this case ice forms on the shield (from moisture in the surrounding air) and the shield is not as efficient.

In cleaning a trap (or baffle) it is first removed from the system and allowed to stand until it reaches room temperature (or near it). The ice will then have disappeared. The removal of ice can be speeded up by scraping and/or by the application of heat. Often a special hot air blower is used. An electric fan with a heater in front of it (or incorporated in it) can be used. The trap should then be wiped with a clean, lint-free cloth or paper that leaves no fragments on the surface. If the cloth or paper shows evidence of "dirt" (which is usually pump oil), the trap should be cleaned further with a solvent, such as acetone. The trap should then be allowed to dry thoroughly before being installed in the system. During installation, care should be taken not to touch any surfaces that will be inside the system with bare hands. The usual care required with gaskets when used on flanges should be taken. The above steps apply primarily to thimble-type traps. Modification of the procedure is necessary for other types of traps. With many glass traps, e.g., that in Fig. 11.4*a*, it is impossible to physically clean the inside. In such cases, after the ice has melted, it should be shaken out. Further cleaning can be done by shaking a solvent such as acetone in the trap and draining thoroughly. The trap should be completely dry before being installed in a system. Sometimes glass traps are cleaned by simply heating them to near the softening point. A gas and oxygen flame or oven might be used. Having two traps saves a lot of time. The clean one can be used while the other one is being cleaned. Dry ice traps are generally of the thimble type and are cleaned according to the procedures outlined above. Usually acetone or trichloroethylene is used as the heat transfer liquid in such traps. The vapors of these liquids are toxic (also, acetone vapor is flammable) so adequate ventilation should be provided.

11.5 Liquid Nitrogen Filling Systems

The inconvenience of filling traps by hand has led to the development of various types of automatic liquid nitrogen filling systems. Only a few will be discussed here. Most of these systems operate on a change in temperature resulting from a drop in the level of the liquid nitrogen. The systems differ in the methods used to take advantage of this change in temperature. Systems depending on mechanical action are also used. A device making use of a thermistor is shown in Fig. 11.8*a* (Flinn, *J. Sci. Instr.*, **36**, 374, 1959). The electrical resistance of a thermistor decreases as the temperature rises. It is this change in resistance that is utilized to automatically control the filling of the trap. The thermistor *T* is used to control the solenoid *S*. When immersed in the liquid nitrogen the thermistor has a high resistance and the electrical circuit is so arranged that the solenoid *S* is not energized. This leaves the tube *C* open so evaporating nitrogen escapes to the atmosphere. When the liquid nitrogen level falls below the thermistor it is now exposed to the nitrogen vapor rather than the liquid and its temperature begins to rise. The lowered resistance of the thermistor is used to pass current through *S* and energize it. This moves the disk *D* down to seal off tube *C*. Nitrogen vapor pressure rapidly builds up in the reservoir and forces liquid nitrogen into the trap. When the liquid nitrogen touches the thermistor, the temperature of the thermistor drops, causing an increase in electrical resistance. This cuts off the current to the solenoid and *D* rises, allowing nitrogen vapor to escape from the reservoir. In the arrangement of Fig. 11.8*a*, *D* is a neoprene disk and the spring *A* limits the upward movement of the disk. Instead of a thermistor it is possible to use a diode. A diode is a transistor which also has the property that its resistance decreases as the temperature rises (like the thermistor). The operation with a diode is essentially the same as with a thermistor. Miller and Petersen (*Vacuum*, **9**, 231, July/Sept 1959) have described one arrangement using a diode. Instead of thermistors or diodes, many arrangements have been described that use a resistance element (tungsten, Nichrome, etc.). However, such elements are not as sensitive as thermistors or diodes.

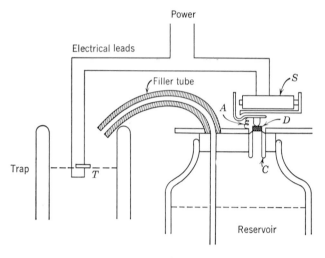

(a)

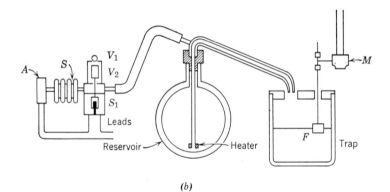

(b)

Fig. 11.8   (*a*) System using thermistor. (*b*) System using a float.

An arrangement depending on a mechanical action is shown in Fig. 11.8*b*. (Henshaw, *J. Sci. Instr.*, **34,** 207, 1957). The action here depends on a float *F* that rises and falls with the liquid nitrogen level. This float is used to operate a mercury switch *M*, which is used in turn to control the solenoid valve S_1. Suppose the trap is filled to the desired level. Then switch *M* will be open. As the nitrogen level drops, the float *F* drops with it and closes *M*. The movement required may be of the order of ½ in. although this depends on the specific design. When *M* is closed current passes through *S* and S_1

and closes valve V_2. At the same time current is supplied to the heater at the bottom of the filler tube, thus increasing the pressure above the liquid nitrogen more rapidly. Liquid nitrogen flows into the trap and when the level is high enough float F opens M. This stops current to S_1 and valve V_2 opens, allowing valve V_1 to vent the reservoir. The arrangement could be used without the heater although the build-up of pressure would then be slower. As long as there is some liquid nitrogen in the reservoir the vapor pressure is sufficient to open valve V_1 and at the same time keep microswitch A closed through the pressure exerted on the sylphon S. When the liquid nitrogen is exhausted switch A opens and the system will no longer operate. This condition is indicated by a light. Many other examples of mechanical arrangements have been reported in the literature.

Some of the more recent articles concerning liquid nitrogen filler systems are listed in Appendix E. These articles give references to earlier reports. In any type of liquid nitrogen filling system it is important to insulate all parts handling liquid nitrogen so as to avoid losses. Synthetic rubber, plastic, and metal tubing have been used for the filling line.

11.6 Pumping Losses in Baffles and Traps

Although baffles and traps are necessary for the purposes outlined previously, they do introduce impedance to gas flow. This means that a diffusion pump of larger capacity is required than in the case of no baffles or traps. Of course, the pump needed will be determined by all impedances between the pump and the vacuum chamber. A baffle, a liquid nitrogen trap, a valve, and piping are common. Methods of combining conductances have been considered in Chapter 2. Application to a system is covered in Chapter 13. Only some commonly used types of baffles and traps will be considered here. The mathematical treatment becomes quite involved for some designs. In most cases it is best to measure the conductance of the baffle or trap. This is readily done by measuring the speed of the pump with and without the baffle or trap (or combination). The baffles and traps considered here are of the optical type, in which molecules entering cannot see through them.

The first type of baffle to be considered is the type shown in Fig. 11.1a, or d. The conductance of such a baffle is fairly readily calculated. Figure 11.9a shows the pertinent geometry and dimensions of the type of baffle shown in Figs. 11.1*a* and *d*. The diameter *B* of the baffle plate must be somewhat greater than *C*, the dimension of the outlet (or inlet), in order to have an optical baffle. The resistance to gas flow becomes less as the diameter of the housing *A* becomes larger. However, the resistance of the outlet to gas flow then increases to some extent, although not too rapidly. There is a limit to how large *A* can be made due to considerations of space and cost. Dimension *D* is not extremely critical but should be kept to a reasonable minimum. Table 11.1A shows the dimensions and conductances of this type of baffle. These are inside dimensions. The nominal size shown is the size of pump with which the baffle can be used. The dimensions used are such as will give a drop in pumping speed of 30% or less. This is shown in Table 11.1B. The pumping speeds without baffle are the speeds of typical commercial pumps. Actually there can be considerable variation in speeds between different commercial pumps.

Table 11.1A   *Dimensions and Conductances of Baffle Plates*

Nominal Size (in.)	Dimensions (in.)				Conductance (l/sec)
	A	*B*	*C*	*D*	
2	4	3	2	3	120
4	8	6	4	4	750
6	10	8	6	6	1400
10	14	12	10	8	2700
16	24	18	16	12	9900

Table 11.1B   *Loss of Pumping Speed Due to Baffle Plates*

Nominal Size (in.)	Pumping Speed (l/sec)		Percent Drop in Pumping Speed
	Without Baffle	With Baffle	
2	50	35	30
4	250	185	25
6	500	365	27
10	1200	890	25
16	4000	2850	29

This will affect the speed with baffle since this is obtained by the formula

$$\text{Speed with baffle} = \frac{\text{pump speed} \times \text{conductance of baffle}}{\text{pump speed} + \text{conductance of baffle}}$$

The results shown in Tables 11.1A and B are fairly typical of this kind of baffle. Taking a pumping speed loss of 30% is usually fairly safe. It is more difficult to calculate pumping speed losses for multiple ring and chevron baffles. In most cases the conductances of such baffles are measured. The specific design will have a considerable effect on the conductance. Any attempt to increase trapping efficiency will result in an increased resistance to gas flow (smaller conductance). Fair trapping efficiency can be obtained without much loss in pumping speed by using a "one-bounce" baffle. This is a baffle in which the rings or chevron plates overlap sufficiently so that molecules must hit these rings or plates once in passing through. Of course, in many cases the molecules will make more than one hit. The average value of pumping loss is around 30% for chevron baffles of this type. The value for the ring baffle is dependent on the spacing between the rings. However, with proper design the above value of pumping loss can be approached. In setting up a vacuum system pumping losses due to baffle, liquid nitrogen trap (when used), valves, and pumping lines must all be considered. It is usual to use an average value for the overall pumping speed loss, assuming common types of baffles, traps, and valves and the shortest possible pumping lines. This is discussed in Chapter 13. The above discussion is in connection with the region of molecular flow which is where such baffles are most commonly used.

Figure 11.9b shows a commonly used type of liquid nitrogen trap. Variations of this design which involve only changes in positions of the inlet and outlet tubes are shown in Figs. 11.3b and 11.3c. The values of conductance quoted here are for the trap of Fig. 11.9b. However, the values for the other traps will not differ much from these. The dimensions used here were chosen on the basis of about 50% or less loss in pumping speed. Also, the trap itself and the outer housing are made of standard tubing (usually brass or copper) with a wall thickness of $\frac{1}{8}$ in. for the housing. This thickness could be varied somewhat without changing the conductance very much. For example, one might want to use a $\frac{1}{16}$ in. thickness for the smaller traps. This would give a little higher conductance than shown in Table 11.2A. On the other hand, if a $\frac{1}{4}$ in. thickness were used for the larger traps then the conductance would drop a few percent. The

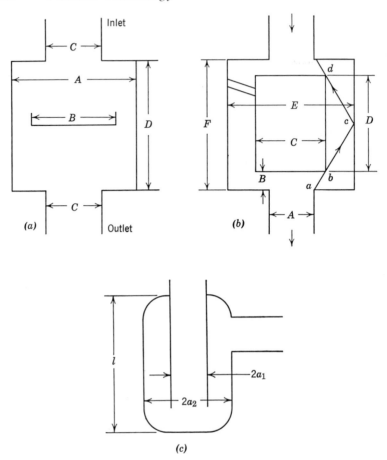

Fig. 11.9 (*a*) Form of optically dense baffle. (*b*) Thimble trap. (*c*) Re-entrant trap.

thickness of the trap itself has no effect on the conductance since only the outside diameter is used in the calculation. The effect of the filler tube on conductance is considered to be negligible. Conductances are given only for air. The conductances for other gases depend on their molecular weights. The conductance increases as the molecular weight decreases. The following relationship is adequate for estimating the conductance of any gas:

$$C_G = \sqrt{\frac{5.4}{M_G}} \, C_A$$

where C_G = conductance of the gas
M_G = molecular weight of the gas
C_A = conductance of air

Table 11.2A shows the pertinent dimensions and conductances of the trap outlined in Fig. 11.9*b*. The trap size is simply the nominal ID of the diffusion pump with which the trap is to be used. In many cases, of course, the trap might be preceded by a water-cooled baffle. Also, the actual ID of a commercial diffusion pump may differ somewhat from the nominal values used here. The dimension A and the size of the annulus between the trap (thimble) and the housing are the most critical dimensions. Dimension A often cannot be increased in those cases where the trap is attached directly to the pump. Also, it is fairly general practice to use tubing or piping of the same ID as the ID of the pump barrel. To get a higher trap conductance with a given diffusion pump it is probably best to use a larger size of trap rather than to increase A on a given trap. As a matter of fact, A

Table 11.2A Dimensions and Conductances of Thimble Traps

	Dimensions (in.)				Conductance
A (trap size)	B	C	D	E	(l/sec)
1	½	2	4	2⅞	32
2	1	3	6	4¾	90
4	1½	5	10	7¾	430
6	1½	8	12	11¾	1080

Table 11.2B Loss of Pumping Speed Due to Thimble Traps

	Pumping Speed (l/sec)		
Trap Size, A (in.)	Without Trap	With Trap	Percent Drop in Pumping Speed
1	10	7.5	25
2	50	30	40
4	250	155	38
6	500	315	37

cannot be increased too much or molecules will get through the trap without striking the thimble. These are optical traps and the design is such that no molecule can get through on one bounce from the housing. This is shown in Fig. 11.9b by the path of a molecule, labeled *abcd*. For the same reason the annulus cannot be made too wide. Furthermore, the cost is increased as the housing gets larger. As a rough rule, the area of the annulus should be two or three times the area of the inlet or outlet of the trap. Table 11.2B shows the effect of this type of trap on the pumping speed. It will be noted that the loss in pumping speed is around 50% for the 2, 4, and 6 in. traps. The lower loss for the 1 in. pump is due to the fact that the size of this trap can be made larger in a relative sense without its becoming cumbersome to handle. Increasing the length of the trap will make it possible to increase the total amount of liquid nitrogen available. However, this also reduces the conductance. The only way to increase the supply of liquid nitrogen is to increase the overall size of the trap. The pumping losses shown in Table 11.2B are calculated for "average" commercial diffusion pumps. Commercial pumps differ in pumping speed and this will show up in different pumping speed losses. However, the values in Table 11.2B can be considered to be fairly representative of this type of trap. It should be noted that the percent loss of pumping speed increases as the speed of the pump is increased. Therefore, in some cases, for a given trap (or combination of conductances) sometimes nothing is to be gained by going to a larger pump. The values of conductance in Table 11.2A are for molecular conductance.

Figure 11.9c shows a type of glass trap (re-entrant) that is commonly used with various types of vacuum gauges, although it has many other uses (Dushman, *Scientific Foundations of Vacuum Technique*, 2nd ed., J. M. Lafferty, editor, John Wiley and Sons, New York, 1962, pp. 101–103). One of these is trapping vapors which might be harmful to the system, as in the case of mercury devices. Only the case of maximum conductance is considered here. The maximum conductance for air at 25°C is given by

$$C = 189 \frac{a_2^3}{l} \, 1/\text{sec}$$

where $a_1 = 0.618a_2$ and l is greater than $5a_2$. All of the quantities a_1, a_2, and l are in inches. The above relationship is strictly true only when the wall thickness is negligible compared with the inner diameter of the tube. As an example of the application of the above relationship, consider a trap with $a_1 = 0.5$ in., $a_2 = 0.8$ in., and $l = 6$ in. The conditions for the above relationship are then satisfied ($a_1/0.618 = 0.5/0.618 = 0.81$, which is very near 0.8, and $5 \times a_2 = 5 \times 0.8 = 4$ in. whereas $l = 6$ in., which is larger). The conductance is then

$$189 \times \frac{(0.8)^3}{5} = 189 \times \frac{0.5}{5} = 19 \text{ l/sec}$$

This is actually a high conductance for use in connection with a vacuum gauge. However, such a trap could also be used as the main liquid nitrogen trap in small systems. In using this type of trap with mercury, care must be taken that mercury does not accumulate at the bottom to the point where the re-entrant tube is blocked off. With many special forms of traps it is usually necessary to measure the conductance.

11.7   The Functions of Vacuum Valves

Vacuum valves must possess the following characteristics:

1. Freedom from leakage.
2. Minimum flow resistance (maximum conductance).
3. Absence of outgassing.

Freedom from leakage means not only no leaks in the body of the valve (exposed to atmospheric pressure) but also no leaks through the sealing part of the valve when closed. A low flow resistance is needed so that there is not too great a loss of pumping speed. The pressure drop across a valve is usually considered adequate if it presents "full area," i.e., the cross-sectional area of the path through the valve is never less than the port area. Also, there should be no obstructions in the line of flow and the valve port should be completely uncovered. Outgassing is generally reduced by using a minimum of gasketing material and vacuum grease and by proper choice of construction materials. The degree to which a valve must meet the above criteria depends on the application in which it is being used.

Vacuum valves can be divided into a number of categories in several ways. One method is to classify valves in the following manner:

1. Those valves that will withstand atmospheric pressure on either side (vacuum on other side) without leaking.
2. Those valves that hold only when the excess pressure is exerted on one particular side.

This type of classification is not entirely satisfactory since the size of a particular design of valve may determine the category into which it falls. Another possible method of classification is by function, which could be done as follows:

1. High vacuum valves.
2. Foreline or roughing line valves.
3. Air admittance or release valves.
4. Isolation and cut-off valves.
5. Throttling valves.
6. Leak valves.

High vacuum valves are used between the pumps and the vacuum chamber and are the largest valves used in any system. Of course, when only mechanical pumps are used these become foreline or roughing line valves. Air admittance valves are used to let a system down to air (or to some other gas). Isolation valves are used to cut off one part of a vacuum system from the rest of the system. Often they need stand only a small pressure differential. Throttling valves are used to control the pressure in a system, usually by cutting down the flow of gas to the pumps. Leak valves are used to admit some desired gas into a system at a controlled rate. One common example is their use to admit helium to a helium leak detector at a known rate for calibration purposes. Other methods are used to admit gas to a system, which cannot strictly be called valves. Such methods are discussed in Chapter 12. Often a valve of particular design can serve several of the functions outlined above.

From the viewpoint of design, vacuum valves can be divided into several categories, such as:

1. Gate valves.
2. Disk valves.
3. Flap valves.
4. Stopcock valves.
5. Ball valves.

6. Plug valves.
7. Globe valves.
8. Needle valves.
9. Diaphragm valves.
10. Special valves.

Valves in some of these classes have many features in common. For example, gate, disk, and flap valves all make use of a disk which seals against a seat. The main difference is the manner in which the disk is mounted and seated. Also, ball and plug valves are, in a sense, versions of stopcock valves. Except for certain special types of valves, the operation is done through the wall of the valve. This means some type of seal must be used. Synthetic rubbers, such as neoprene and Buna N, are most commonly used. However, to minimize outgassing, special elastomers such as Viton A or Teflon are sometimes used. In the case of ultra-high vacuum systems, where bake-out at high temperatures is required, metal gaskets are commonly used. Instead of a seal, it is possible to use a bellows, which is usually brass or stainless steel. Also, a rubber or metal diaphragm can be used although the movement possible with this arrangement is quite limited. Bellows or diaphragms can be used with several of the categories of valves listed above. Care must be taken to avoid damage to the bellows or diaphragm. The manufacturer's recommendations for allowable movement should be followed.

11.8   Gate, Disk, and Flap Valves

These types of valves are normally considered to be high conductance valves, i.e., valves that offer low flow resistance, and are suitable for use between the vacuum chamber and the final pumps in the system. The gate and disk types of valves can assume many forms. Their main features are that the disk or gate is used to close off the flow of gas with a leak-free seal and, when open, there is a minimum resistance to flow. Commercial gate valves for fluid flow usually involve two gates which are forced apart by a suitable mechanism when being closed. A schematic drawing showing the principle of operation of a slide gate (or simply gate) valve is shown in Figs. 11.10a and b. These figures illustrate the basic features of some commercial designs. The disk D seals against a flange F to which piping or some vacuum component can be sealed. The mechanism A is designed so that the disk D seals directly against the inside of the housing without scuffing the sealing gasket G. When the valve is opened, the mechanism and disk are withdrawn up into the housing so as to leave a completely unobstructed opening. This results in a high conductance. This type of valve can be made quite thin. The design of the actuating mechanism will vary from manufacturer to manufacturer. In some cases the actuation is carried out through a rotary seal (Fig. 11.10a) while in others it is done by means of a sliding seal (Fig. 11.10b). The rotary method of operation has some advantages over the sliding method because less gas gets into the system during operation. The seal for the operating shaft can be an O-ring (or two), a Wilson seal, or a chevron seal. Commercial valves also include a body gasket. All gaskets are often made of Buna N. However, other gasket materials, such as neoprene, butyl, silicone and Viton A, can be used. Even

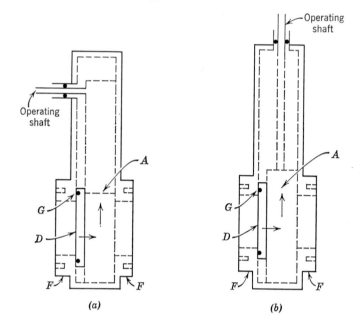

Fig. 11.10   (*a*) Gate valve with rotary seal actuator.   (*b*) Gate valve with sliding seal actuator.

metal (aluminum, indium, etc.) could be used for the body gasket. At this writing, valves of this basic design are manufactured in sizes ranging from a 1 in. opening to a 24 in. opening (1, 2, 3, 4, 6, 8, 12, 16, 20, and 24). These are nominal sizes since the openings on some valves may be larger than these sizes in order to fit "oversize" pumps. There is no reason why such gate valves cannot be made in considerably larger sizes.

The bodies and most internal parts of valves from 1 in. to 12 in. are made of aluminum alloy, usually by casting (sand or permanent mold). Such valves can be operated manually or pneumatically. Larger sizes (16 in., 20 in., and larger) are operated pneumatically and the bodies are made from mild steel plate. The pneumatic system differs between sliding seal and rotary seal types of valves. In the former type, an ordinary straight pneumatic cylinder can be used while in the latter case it is necessary to incorporate a mechanism for converting a linear motion (of the cylinder) into a rotary motion. The smaller sizes of valves (1 in. through 6 in.) are also made in a "necked" style. In this case the flanges for connecting to the system are connected to the body by necks. This allows bolting from under the flange. Some designs make use of a neck on one side, the other side being as shown in Fig. 11.10a. As has been noted, gate valves can be made quite thin. Commercial valves range in thickness from about 3 in. for a 1 in. valve to about 8 in. for a 20 in. valve. It is evident that such small thicknesses together with the straight-through feature result in a small loss of pumping speed. The conductance can be estimated closely enough by simply adding the thickness of the valve (flange to flange) to the length of piping used, if the piping has the same ID as the port opening of the valve. These valves can be used as throttling valves. Also, because of the fact that they can be closed rapidly (a small fraction of a second for the smaller sizes to 1 or 2 sec for the larger sizes) they can be used to protect pumps and other devices. Gate valves can be obtained from all major vacuum equipment manufacturers in the U.S.A.

Right-angle disk valves of the type shown in Fig. 11.11a are commonly used above diffusion pumps. This figure is schematic and no attempt is made to show flanging arrangements. The disk can be raised and lowered manually or pneumatically. A back seat is shown against which disk A seats when the valve is in the open position. This prevents any leakage through the Wilson seal into the system. Alternative seals, such as chevron seals, could be used instead of Wilson seals. Also many commercial disk valves do not include a back seat. These larger disk valves are connected into the system by welding or by means of flanges. They are usually made of mild steel or stainless steel. Sometimes the valve body is cooled with water by means of tubing brazed or welded to the body. The valve then acts as a combined valve and water-cooled baffle. These valves offer more resistance to gas flow than gate valves because of the longer path. However, where the connection to the chamber goes through a right angle, they are quite suitable. This basic design of valve is also made in small sizes for a variety of uses, such as in the foreline or roughing line and as an air admittance, isolation, or throttling valve. Some typical types of small disk valves are shown in Figs. 11.11b, c, and d. Figure 11.11b shows a right-angle valve with a stem gasket, cover or bonnet gasket, and disk gasket. The first two gaskets are usually O-rings while the disk gasket could be either an O-ring or an insert. A seat instead of disk gasket could be used. Synthetic rubbers or special elastomers such as Viton A, Kel-F, Teflon (with suitable design) are used. The body is usually brass or stainless steel. The dashed lines show the change in arrangement needed to have an in-line valve. To avoid any leakage through the stem seals, bellows or diaphragms are used. A bellows sealed valve is shown in Fig. 11.11c. The bellows is

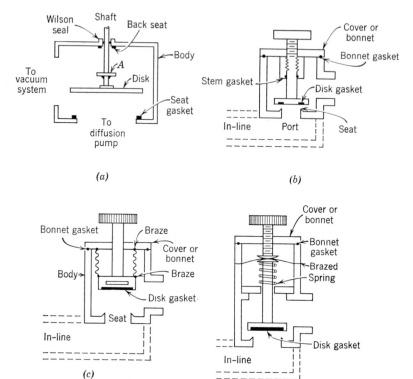

Fig. 11.11   (*a*) Large disk valve.   (*b*) Small disk valve with stem O-ring.   (*c*) Small disk valve with bellows.   (*d*) Small disk valve using diaphragm.

usually brass or stainless steel, more movement being obtained with brass. The bellows is brazed to the cover (bonnet) and disk. Figure 11.11*d* shows a valve using a diaphragm instead of a stem gasket. Strictly speaking, this is not a true diaphragm valve since such valves use the diaphragm to make the seat seal. However, this terminology is used rather loosely. The diaphragm can be an elastomer or metal. An elastomer such as Buna N or neoprene has considerable flexibility, but also outgases considerably and is permeable to various gases. On the other hand, metals are not too flexible although they have better out-gassing and permeability characteristics. Consequently, elastomeric gaskets are commonly used where considerable movement of the disk is needed. The comments concerning construction materials for the valve of Fig. 11.11*b* apply to those of Figs. 11.11*c* and *d*. The dashed lines again indicate in-line arrangements. The valves described are shown as having a screw drive. This could be replaced by a pneumatic or magnetic drive (using a solenoid). For pneumatic operation it is only necessary to mount a pneumatic cylinder above the valve, connected to the stem. With a solenoid, it is necessary to have a nonmagnetic valve body and a spring return. The solenoid can be wired so as to have the valve in a normally open or closed position when the solenoid is not energized, depending on the application. Consider the case of a cut-off (nonreturn) valve in the foreline, which must close when there is a loss of power to the diffusion pump. This is to avoid mechanical pump oil getting into the system. This contamination could also happen if the mechanical pump, for some reason, ceased to function. If the solenoid were energized to keep the valve open, loss of power would cause the valve to close due to the spring. It would not open again until the power was restored. Other types of valves besides disk-type could be used for this purpose. However, the disk-type valves are most common.

Flap valves also make use of a disk (flap), which is turned in some manner to seal against a seat rather than being moved parallel to itself, as has been discussed (gate valves). The basic principle of a flap valve is shown in Fig. 11.12. Here the flap or disk is turned through 90° by an appropriate operating mechanism acting through the valve housing. The arrangement of Fig. 11.12 turns the flap to the side of the valve. Alternatively, the flap could be turned about an axis through its center. (This type of valve is often called a butterfly valve.) The problem with this kind of valve is to make an effective seal. Instead of causing a simple rotation of the flap or disk, it is pos-

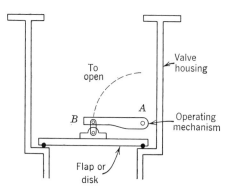

Fig. 11.12 Principle of flap valve.

sible to pull the flap away from the seat in opening and then rotate it. In closing, the procedure is reversed. Such a valve is made by Consolidated Vacuum Corporation, Rochester, New York, and is called a quarter-swing valve. With proper design a flap valve, in principle, offers such advantages as:

1. Simplicity.
2. Low gas flow resistance (depending on design).
3. Rapid operation.

In spite of these advantages, this type of valve suffers from certain inherent difficulties. Some of these are:

1. Difficulty in sealing against pressure in both directions.
2. Difficulty in designing adequate seals, particularly in the case of butterfly-type valves.

In addition, both the butterfly type and quarter-swing type have obstructions inside the valves so they do not offer an unobstructed opening as in the case of gate valves.

11.9   Globe, Needle, and Diaphragm Valves

Globe valves derive their name from their general appearance. The main features of such a valve are shown in Fig. 11.13a. It is the general appearance of the bonnet that gives rise to the name "globe." The arrangement shown in Fig. 11.13a indicates the use of a bellows. A gasket seal, such as O-ring or Wilson, could be used. Globe valves are also made with the stem at an angle to the general direction of gas flow. Actuation is usually by means of a handle but pneumatic operation can be used. These valves often have cast bronze bodies and brass internal parts although commercial valves making use of other materials are obtainable. The most common vacuum globe valves have openings ranging from ½ in. to about 3 in. or larger. Connections to the vacuum system are usually made by pipe threads with a suitable sealing material (glyptal, etc.). It should be noted that the basic principle of these valves is the same as that of the small in-line disk valves shown in Figs. 11.11b, c, and d. In fact, the distinction between these two kinds of valves is not clear and is usually made by appearance and construction. Globe valves do not have as low a resistance to gas flow as gate valves and, therefore, are not usually used between the final pumps and the vacuum chamber. They are commonly used

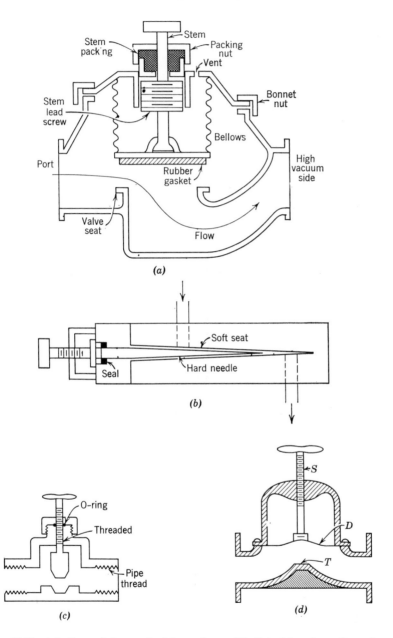

Fig. 11.13 (*a*) General form of globe valve. (*b*) Principle of needle valve. (*c*) Blunt-nosed "needle" valve. (*d*) Diaphragm valve.

in foreline and roughing lines and also as air admittance and isolation valves. For some applications, they are useful as throttling valves.

The primary function of a needle valve is to admit gases at a controlled rate into a vacuum system, i.e., to act as a leak valve. In certain circumstances, depending on the vacuum system and the particular design of valve, they can be used as air admittance and even throttling valves. The general principle of operation is shown in Fig. 11.13*b*. A needle is moved into a seat, the amount of gas flow being determined by the space between needle and seat. For fine control of gas flow, a long, hard needle made of some material such as stainless steel is used which seats in a soft material. The seating material is usually a soft metal such as soft solder, brass, or copper. The needle actually makes its own seat in the soft seat. Rubbers and other elastomeric materials could be used for the seat but fine leaks are not obtained by this method because of the permeability of the material. Various methods of moving the needle are used although the common method is a simple screw drive. Using a differential screw drive will give finer control of gas flow. Also, using two needle valves in series will give lower rates of gas flow. This is made possible by using the first valve to reduce the pressure differential across the second valve. Many small commercial needle valves use a rather blunt "needle," with the needle and seat polished to a fairly fine finish (see Fig. 11.13*c*). Such valves should be kept on hand since they can be used for a multiplicity of purposes.

A diaphragm valve uses a diaphragm (metal or elastomer) to seal against a seat. The actual sealant could be in the seat or could be the diaphragm itself, particularly if made of some elastomer such as a rubber. As has been noted, rubber and some other elastomers in general permit considerably more movement of the diaphragm than metals, although they are more permeable to gases. The principle of this type of valve is shown in Fig. 11.13*d*. The stem S is operated by means of a screw to move the diaphragm D away from or toward the seat T. With an elastomeric material it is possible to get a cross section of about the same area all through the valve, resulting in low resistance to gas flow. With a metal diaphragm the movement is more limited and an elastomeric seal has to be provided in the seat or in the diaphragm itself. Commercial vacuum diaphragm valves are available in sizes ranging from an opening of a fraction of an inch to one of several inches. Within the sizes available they are used in much the same way as globe or disk valves, i.e., in foreline and roughing lines, as air admittance and isolation valves, and to some extent as throttling valves.

11.10   Stopcock, Plug, and Ball Valves

Stopcocks have been used to control gas flow for many years and still find many applications in vacuum practice. With very small systems they can be used in the pumping line. In other cases they can be used in connection with vacuum accessories such as gauges. They can also be used to control the flow of gas into a system, although the control is not fine, and to act as admittance valves. The use of such devices is usually controlled by the amount of resistance to gas flow that can be tolerated. Both glass and metal stopcocks are available. The appearance of a typical two-way stopcock is shown in Fig. 11.14a. Usually the bore through the stopper is of smaller diameter than the tubes, although it is possible to obtain stopcocks with these diameters the same. Also, the bore through the stopper is usually diagonal, as is shown in Fig. 11.14b. For various purposes it is desirable to have two or more bores through the stopper as in Fig. 11.14c, which represents a three-way stopcock. Many other arrangements are available, such as tubes at right angles to each other with suitable bores in the stopper. Figure 11.14d shows a right-angle stopcock valve with a vacuum volume at the bottom of the valve. The volume helps in assuring a tight seal due to atmospheric pressure. The stopcocks shown are common glass types. Usually Pyrex is used. Stopcocks made of metal, usually brass, of the same general types are also available. The stopper and inside surface of the housing must have smooth surfaces. The seal between housing and stopper is accomplished with a vacuum grease. With glass stopcocks, a good seal is obtained when there is no apparent separation of housing and stopper, i.e., there are no air bubbles trapped and there are no flaws in housing or stopper. Such a seal is usually obtained by firmly pushing the stopper into the housing (with due care not to break the tubing connections) and turning it back and forth a few times. This same procedure can be used with metal stopcocks although there is no way to see into the valve. Instead of using vacuum grease for sealing, sometimes mercury is used when vapors from the grease are objectionable. Such a stopcock is shown in Fig. 11.14e. Here mercury in reservoir R seals off the top of the stopper. It will be noted that there are two bores in the stopper (C and D). C is used to connect tubes A and B while D connects B and a reservoir E when the stopcock is closed. By pumping on B it is possible to prevent leakage around the housing into a vacuum system connected to A.

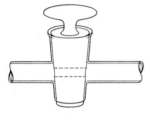

(a)

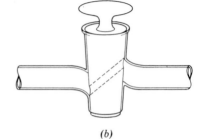

(b)

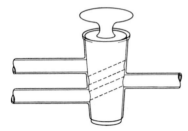

(c)

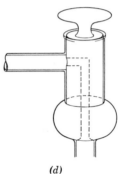

(d)

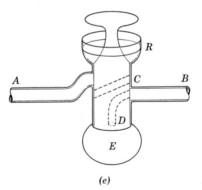

(e)

Fig. 11.14   (a) Two-way stopcock—straight bore. (b) Two-way stopcock—slanted bore. (c) Three-way stopcock. (d) Right angle stopcock. (e) Stopcock with mercury seal.

Stopcocks are relatively low conductance valves and are generally available only in fairly small sizes. Most laboratory supply houses will carry many sizes and varieties of glass stopcocks in stock, usually up to a few millimeters in bore diameter. A bore of around 10 mm is considered to characterize a large stopcock. Metal stopcocks are not carried widely in stock. The problem in going to large sizes of stop-cocks is partly in increased fabrication difficulties and partly in a rapidly increasing frictional force between housing and stopper. Also, disk valves, globe valves, and gate valves have many advantages over the large stopcocks.

Plug valves can be considered to be a form of stopcock. The main difference is in the use of gaskets. The general arrangement of a plug valve is shown in Fig. 11.15a. The O-rings A are in grooves cut into the plug (stopper), which is a cylinder. These O-rings make a seal with the inside of the valve. When they cover the inlet and outlet of the valve, gas can flow through the valve and there is no leakage because of O-rings A as well as O-rings B, C, and D. In operation O-rings B and C ride on the valve body and the cover plate respectively. The cover plate is attached to the body with screws. The shaft is part of the plug and can be turned by a wrench or a handle. One quarter of a turn is sufficient to change the valve from fully open to fully closed (or vice versa). Pneumatic operation is possible. The most difficult problem in fabrication is making the grooves in the plug. A method of doing this is described by Stark and Langsdorf (*Rev. Sci. Instr.*, **25**, 188, 1954). A wide variety of metals can be used for body, plug, and cover. Figure 11.15a shows pipe threads for connections. However, the body could be bored for tubing, or flanges could be used. In order to avoid the problem of cutting grooves in the plug the arrangement of Fig. 11.15b could be used. Here, inlet and outlet tubes project into the cylindrical cavity of the valve body so that O-rings can be slipped over the ends to seal against the plug. Of course, the ends of the tubes must be shaped to fit the plug. The tubing shown in Fig. 11.15b could be soldered or brazed into the system. Alternatively, the arrangement could be changed so that flange or pipe thread connections could be used. There is no upper limit to the size of such valves. However, the weight goes up rapidly as the size increases, as does the force necessary to operate it. The use of bearings will minimize the latter problem. Although these are straight-through valves, the length of path through the valve increases proportionally to the size of opening. Consequently, such valves are not too competitive with gate valves as the size increases.

Ball valves are also a form of stopcock. They differ from plug valves in that a ball is used instead of a cylindrical plug. This type is a straight-through valve like the plug and gate valves. Its main advantage over the plug valve is in greater ease of manufacture. It is much easier to machine a groove on the surface of a ball (spherical) than on the surface of a plug (cylindrical). However, this shape in-

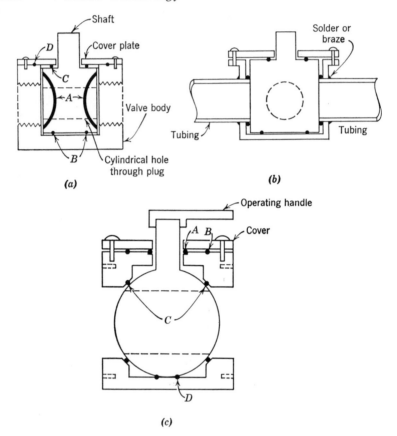

Fig. 11.15   (*a*) General arrangement of plug valve (open).   (*b*) Plug valve using tubing (closed).   (*c*) General arrangement of ball valve.

troduces other complications, such as having to use a housing with a spherical cavity to fit the ball.   Commercial ball valves follow the general arrangement shown in Fig. 11.15*c*.   The valve is shown in the open position.   When the operating handle is turned through a right angle, there is no opening through the valve, the sealing being done by gaskets *C*.   As shown, two gaskets are generally used so the valve can be operated with a pressure differential in either direction.   Gaskets *A*, *B*, and *D* will seal the valve from the atmosphere.   Usually these are O-rings.   Gaskets *C* can assume various shapes, depending on the specific design of valve.   The arrangement of Fig. 11.15*c* can be attached to the system by flanges but the specific designs can be such that soldering or brazing connections as well as pipe thread can be used. Various construction materials are used, including bronze, carbon steel, brass, and stainless steel. Several types of gasket material can be used, such as synthetic rubber, Viton A, Kel-F, and even Teflon if the design is suitable. The comments on size regarding plug valves are applicable here. Commercial valves ranging in size of bore from ½ in. to several inches are available. Because of the increase in impedance ball valves are not generally used on large systems between the diffusion pump and the chamber. However, in many cases the decision as to the type of valve to use must be determined by the particular system involved and by cost considerations. Ball valves are quite suitable for use in roughing lines and forelines. They can also be used as air admittance valves and to some extent as throttling valves although the control is not fine. Since they can be adapted to pneumatic operation, they can also be used as cut-off or isolation valves.

11.11   Bakable Valves

The valves discussed in preceding sections are types commonly used in vacuum systems operating down to 10^{-7} Torr or somewhat less. For systems operating at much lower pressures it is necessary to choose materials that have low outgassing rates and that can be baked to drive out gases and vapors from construction materials. Because of these requirements, valves for use at very low pressures generally do not make use of elastomers for gaskets, although there are several reports of operation in the very high vacuum range using elastomers that are first baked at temperatures that will not damage them and then cooled during operation. Naturally, using a minimum of elastomers will help. The general practice is to use all-metal valves. The problem here is obtaining reliable seals, particularly for moving parts. Various metal seals have been used, including aluminum, copper, tin, and indium. Although indium O-rings are extensively used, valves with such rings cannot be considered to be true ultra-high vacuum valves since indium melts at 156°C and this limits the temperature at which baking can be done. To effectively remove gases and vapors in a reasonable length of time it is necessary to bake at temperatures around 400 or 450°C. On the other hand, indium has a lower vapor pressure at room temperature than elastomers.

An early form of bakable valve for handling very pure gases is described by Alpert (*Rev. Sci. Instr.,* **22,** 536, 1951). The basic principle of operation is shown in Fig. 11.16a. The Kovar diaphragm B is brazed in a hydrogen atmosphere to a valve body A (a copper cup) and to the copper nose C. The driver assembly is used to move the diaphragm. By employing a differential screw arrangement, it is possible to exert pressures of several tons. The particular arrangement used by Alpert moved the nose about 0.010 in. per revolution and gave a total travel of 0.100 in. The valve openings and seat were ¼ in. in diameter. The nose had a 60° highly polished conical surface which formed its own seat in the copper when the valve was first closed. A valve of this sort must be mounted firmly because of the force needed to operate the driver assembly. Various modifications of this type of valve have been reported in the literature. Changes have been largely in the materials of construction and in the design of the driver assembly. A design reported by Thorness and Nier (*Rev. Sci. Instr.*, **32**, 807, 1961) is shown in Fig. 11.16*b*. This is a modification of refrigeration valves and the design aims to eliminate difficulties with solder

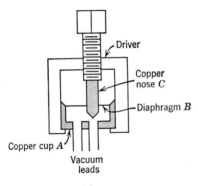

(a)

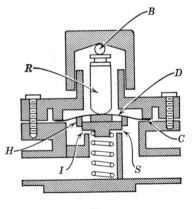

(b)

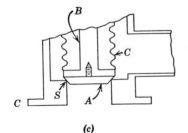

(c)

Fig. 11.16 (*a*) Principle of Alpert valve. (*b*) Thorness and Nier valve. (*c*) Lange valve.

joints. The diaphragm D is sealed by the copper gasket C rather than being soldered or brazed, the gasket being compressed by eight 10-24 cap screws. Various lubricants, including molybdenum disulfide and aquadag, were used on these screws. The diaphragm was 302 SS and 0.005 in. thick. A jig for making these diaphragms is described in the journal article. In closing the valve, the cap nut is screwed down against an ordinary steel ball B, which drives a push rod R against the diaphragm. The diaphragm forces the head H and head insert I (copper) against the seat S (stainless steel). To close the valve, 150 in.-lb of torque was initially required but after repeated baking this went as high as 250 in.-lb. It will be noted in Fig. 11.16b that a spring is provided for opening the valve.

The Alpert-type valve is limited in size by the diaphragm, which can only be moved so far without damage. Consequently, large conductances are difficult to obtain with this type of valve. An obvious approach is to use a bellows instead of a diaphragm. But there is still the problem of getting a reliable seal inside the valve. A method used by Lange (*Rev. Sci. Instr.*, **30**, 602, 1959) is shown in Fig. 11.16c. The copper nose A seals against the seat S which is part of the stainless steel valve body. The nose is driven by a stem B inside a stainless steel bellows C. The stem drive is arranged so the stem cannot twist. The upper part of the bellows is sealed by means of a gold gasket which is compressed to one-half its original diameter. In operating such a valve, higher torques are used on successive closings. This design of valve was made in port sizes ranging from $\frac{1}{2}$ in. to 4 in. The 2 in. size is claimed to have a closed conductance of 10^{-9} atmospheric l/sec or less. Another form of metal-to-metal seal used with a right-angle valve and sylphon is described by Wishart and Bancroft (*Vac. Symp. Trans.*, 1960). Here a conical spring steel washer seats against a part of the 304 stainless steel body. This type of valve was found to work satisfactorily as long as the seating surfaces had a fine finish. An interesting case of a metal-to-metal seal used with a sylphon is described by Parker and Mark (*Vac. Symp. Trans.*, 1960). The seat acts somewhat on the principle of a lathe. On closing the valve, the stainless steel cutter seat cuts a chip from the OFHC copper poppet. With successive closings the chip progresses up the poppet until the material is exhausted. The number of closures possible depends on the size of poppet. With the valve described (over 8 in. diameter opening) more than 300 closures were obtained. Advantages claimed for such a valve are: (1) long life, (2) relatively easy to rebuild when the seating material is exhausted, (3) can be used in strong magnetic fields, (4) tight closure (leakage less than 10^{-12} atm l/sec), and (5)

can be used at temperatures up to 450°C. On the other hand, the
particular valve described required a special hydraulic system to
achieve the necessary pressures.

The principal troubles with valves of the above type which use
solid metal-to-metal seals are the large forces needed to make the
seals and the difficulty in obtaining tight closures. To get around these

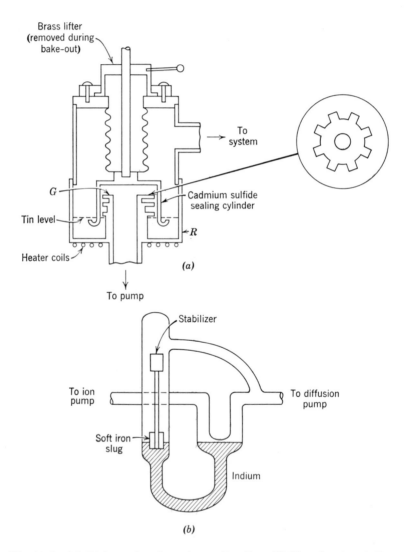

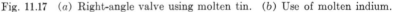

Fig. 11.17 (*a*) Right-angle valve using molten tin. (*b*) Use of molten indium.

difficulties, many attempts have been made to use liquid metal seals in valves. In such cases the seal can be made by melting the metal and then allowing it to solidify. The valve can be opened by remelting the metal. Clearly, the metal must be held in a container in a vertical position so it will not spill over. An example of a valve using molten tin is shown in Fig. 11.17a (Haaland, *Rev. Sci. Instr.*, **30**, 947, 1959). Tin has a vapor pressure of about 10^{-10} Torr at 500°C. As with all valves using a molten metal, some type of heater must be used. With tin, attention must be given to the corrosive action of this metal. With this particular valve, ordinary cold drawn steel was used. The tin was contained in a cadmium sulfide reservoir (*R*). It will be noted that splash guards (*G*), which are spaced vanes over notches, are included to prevent loss of tin. Gallium and indium were used instead of tin but these metals tend to alloy rapidly. A form of valve using molten indium is shown in Fig. 11.17b. This type makes use of a glass enclosed soft iron slug which floats in the indium. In this case the valve is open. When the slug is pulled into the indium by means of an external magnet the valve is closed due to a rise in the indium level. It can be left closed by letting the indium solidify. Such valves are claimed to be useful in ultra-high vacuum systems. However, the bake-out temperature is limited by the relatively low melting point of indium.

11.12   Some Special Valves

Many special purpose valves have been reported in the literature and it is not possible to cover them in any comprehensive manner. However, two special types of valves find fairly general use in vacuum practice. These are:

1. Greaseless valves.
2. Fast acting valves.

Ultra-high vacuum valves are examples of greaseless valves. However, there are various applications at somewhat higher pressures where special greaseless valves are essential, as in the case of working with very pure gases or vapors. Some examples of greaseless valves are shown in Figs. 11.18a, b, and c. Figure 11.18a shows a valve making use of a Pyrex ball containing a piece of soft iron (Metzler, *Rev. Sci. Instr.*, **33**, 130, 1962). This ball seals against a ground and lapped Pyrex seat in an enlarged part of the vacuum line. The methods of making the seat and the ball are described by Metzler. This type of

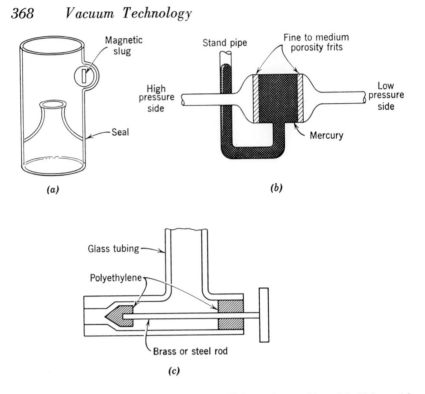

Fig. 11.18 (*a*) Valve using magnet. (*b*) Valve using a frit. (*c*) Valve with polyethylene seals.

valve can be baked to 450°C and is claimed to be suitable for operation at 10^{-9} Torr with the high pressure side at 10^{-5} Torr. It is also
suggested that Vycar, sapphire, aluminosilicates, etc., could be used
for higher temperature operation. Many designs of cut-off or check
valves using mercury have been reported, float valves being the most
common. An example of a check valve making use of glass frits and
mercury is shown in Fig. 11.18*b* (Smith, Posey, and Thomas, *Rev.
Sci. Instr.*, **30**, 202, 1959). This valve operates on the basis that mercury, because of its surface tension, will not pass through a glass frit.
In the valve shown in Fig. 11.18*b* the frit on the low pressure side prevents gas flow until the pressure of the gas on the high pressure side
exceeds the pressure exerted by the mercury in the stand pipe. The
mercury then flows from the space between the frits and rises in the
stand pipe, thus allowing gas to flow through the valve.

   Figure 11.18*c* shows a greaseless valve making use of glass tubing,
steel or brass rod, and polyethylene or polyvinylchloride tubing (Raats,

J. Sci. Instr., **34,** 510, 1957). Under pressure, the polyethylene "wets" the glass. A gas-tight fit between the threads of the metal rod and the plug of the valve is ensured by drilling and threading the hole in the plug slightly smaller than the rod. The rod is warmed to about 100°C and carefully worked into the polyethylene threads. Several variations of this design are described by Raats. Franks has described a form of automatic isolation valve (*J. Sci. Instr.,* **34,** 122, 1957). This valve admits air to a rotary backing pump and maintains vacuum in the remainder of the system in the event of failure of the rotary pump, say due to power failure. It operates by making use of gravity. An air admittance valve opens in the case of power failure. Alternatively, the air admittance valve may be actuated by a generator coupled to the pump. Many isolation valves make use of springs and electromagnets.

For various purposes, such as gas injection into high vacuums, for protecting vacuum systems in case of thin window breakage, etc., very high speed valves have been developed. Some of these valves will close in less than a millisecond. The usual design involves electromagnetic operation. Knudsen (*Rev. Sci. Instr.,* **27,** 148, 1956) describes a flap valve which is said to close in less than 10 msec. Another design is described by Gorovitz, Moser, and Gloerson (*Rev. Sci. Instr.,* **31,** 146, 1960). This valve is said to close in about 3 msec.

11.13   Care and Maintenance of Valves

The care of valves is pretty well determined by the particular design of valve and by the application involved. However, there are some points worth keeping in mind. When a valve is to be installed in a system, it should be inspected for any foreign material in the observable parts and checked for mechanical operation. Before installing the valve, the parts to be joined into the system should be cleaned thoroughly. Where soldering, brazing, or welding is involved, the recommended procedures for these operations should be followed (see Chapter 10). In the case of valves with flanges, these flanges should be inspected and cleaned. Particular attention should be given to gasket grooves, and gaskets should be inspected, cleaned, and given a light coating of vacuum grease (as appropriate). The position in which the valve can be mounted will be determined by the design of valve.

Any well-designed commercial valve should offer few maintenance problems. The usual maintenance problems are associated with movable seals. The common types of valve involve some means of operating the valve through the housing. Where O-ring, Wilson, chevron, etc., seals are used, the type of problem that might develop is wear of the gasket material and loss of vacuum grease. Trouble with such seals is determined by testing for leaks. Where trouble is indicated, the seal should be taken apart and all parts should be examined. If the gaskets are obviously damaged, they should be replaced. Particular attention should be given to seating surfaces. These parts should be made smooth and be cleaned thoroughly before assembly. Instead of gaskets, bellows or diaphragms are often used for the seal between the atmosphere and the vacuum. The failure of such devices is usually fairly definite since they will crack after so many operations. Failure is indicated by a sudden rise in pressure. Such valves are usually repaired by replacing the bellows or diaphragm with a type recommended by the manufacturer. Another possible source of trouble is the seal inside the valve that is used to shut off or control gas flow. As a rule it is necessary to completely disassemble the valve to make repairs. In assembling a valve the manufacturer's directions should be followed and proper inspection and cleaning of parts should be carried out. In many cases it may be necessary to return a valve to the manufacturer for repairs. Most commercial valves are sufficiently reliable that a regular maintenance schedule is not necessary.

TWELVE

Some Other Vacuum Components

12.1   Common Static Seals Using Elastomers

Static seals are fixed seals, i.e., seals that provide vacuum-tight joints but do not permit carrying out a motion inside the system. Such seals include the common gasketed joint using a rubber or special elastometer such as Viton A, Kel-F, or Teflon. The properties of various elastomers are discussed in Chapter 8, including comments regarding cements for joining them to make gaskets. Vulcanizing can also be used. Precautions in making gasket grooves as well as suitable dimensions for such grooves are discussed in Chapter 10. There is also some discussion in Chapter 10 of metal gasket seals, which are used primarily in ultra-high vacuum work. Metal gasket seals using wire are discussed in Section 12.3 while some special types of metal seals are considered in Section 12.5. The emphasis in Chapter 10 is on fabrication techniques that can be used in the vacuum laboratory or plant for making components. Here the emphasis is on some commonly used seals, many of which are available commercially in various versions.

Most commercial vacuum components come equipped with O-ring grooves with the appropriate O-ring gasket being specified or are designed to be sealed to a mating surface which carries the gasket. Synthetic rubber gaskets are most commonly used. However, various components can be obtained with other elastomeric or even metal gaskets. Special gasket designs, some of which are discussed below, are also available. Components such as pumps, baffles, traps, and gauges are often designed with O-ring grooves. In the case of piping or tubing, flanges are usually involved and it is usual practice to groove one flange for an O-ring. Sometimes two O-rings are used with a pump-out provided between the gaskets. The space between the gaskets can be pumped on in case of a leak in the inner gasket, thus avoiding a shut-down. This practice is not too prevalent because of the cost involved in making the second groove. Also, a properly made single O-ring arrangement is quite reliable. Instead of grooving flanges it is possible to use a grooveless gasket (sometimes called a flange gasket, although this term is misleading). The nature of this gasket is shown in Fig. 12.1*a*. Two rubber gaskets, *C* and *D*, are molded to both sides of a sheet of permanent-molded aluminum alloy *A*. The holes *B* in the ears take flange bolts, thus permitting positioning of the gasket. The raised ring *E* of the sheet (cast with sheet) serves to avoid overcompression of the gaskets. This is shown in section *E–E*, where the flanges have compressed the gaskets and are against the raised rings. Two gaskets with pump-outs are shown. However, it is possible to use just one gasket. The commonly used gasket material is Buna N. Other materials can be used, as long as they can be molded to the aluminum.

Bell jars (and other vessels) can be sealed to a base plate by means of an O-ring. However, this means cutting a groove in the base plate since most commercial bell jars are not provided with a groove. Of course, if the vessel is fabricated in the vacuum plant then a groove could be incorporated. In any case, it is expensive to cut grooves. An alternative is to use an L-gasket. This type of gasket is shown in Fig. 12.1*b* in connection with a bell jar. These gaskets can also be used on port covers. L-gaskets are available commercially in several gasket materials. Of course, reliance must be placed on atmospheric pressure for sealing since bolts cannot be used. An objection to L-gaskets is the exposure of a fair amount of elastomer to the vacuum. Another method of sealing vessels or port covers is shown in Fig. 12.1*c*. The O-ring *A* is kept from being pushed in by atmospheric pressure by means of the ring *B*, or plate *C*. The thickness of this ring is only about two-thirds of the diameter of the O-ring (for synthetic rubbers) so as to get a good seal. The outer edge of the ring could be cut in a V-shape in order to retain the O-ring and make assembly easier. This arrangement could be used with any type of flange that could be bolted down. In cases where it is desired to get into the very high vacuum region a double O-ring arrangement is used with a refrigerant circulated in the space between the O-rings. Of course, the temperature of the refrigerant cannot be so low as to impair the sealing properties of the O-rings.

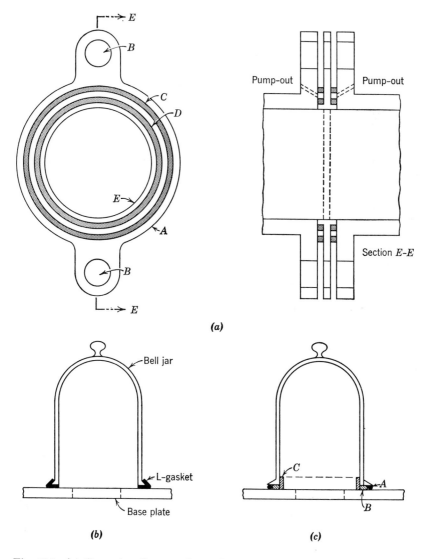

Fig. 12.1 (*a*) Grooveless flange gasket. (*b*) L-gasket. (*c*) O-ring with retaining ring or plate.

A common form of static seal using an elastomer is the *compression seal*. These are sometimes called *compression fittings* and certain forms are referred to as *stuffing boxes* or *packing glands*. They work on the principle of deforming the elastomer (usually rubber) so as to effect a tight seal on a wire, rod, or tubing and also on a retaining housing. A form of such seal commonly used for mounting vacuum gauge tubes is shown in Fig. 12.2a. The gauge tubulation is inserted into the opening in the housing A until it strikes the constriction B. The gasket C has a somewhat smaller opening than the gauge tabulation. Screwing down the compression nut E compresses the gasket against the tubulation by means of the compression ring D. Making the gasket of soft, gum rubber and with a fairly large area (say around $\frac{1}{4}$ in. thick) will lead to a tight seal as well as good mechanical support for the gauge tube (or other device being used). Of course, using a large gasket leads to more outgassing. Instead of a gasket of square or rectangular cross section, an O-ring could be used. This leads to less outgassing but mechanical support of the gauge tube is not as good. The dimensions shown in Fig. 12.2a are sufficient to take care of tubulations ranging from $\frac{7}{16}$ in. to $1\frac{3}{16}$ in. by increasing F from 1 in. to $1\frac{1}{4}$ in. Of course, the obstruction B could be eliminated and this seal could be used for tubing passing into the vacuum system. Such seals are usually soldered or brazed to the vacuum system. In some cases, pipe threads can be used with a vacuum cement (glyptal, etc.).

A simple form of compression fitting which is made commercially in various modified forms is shown in Fig. 12.2b. The compression nut A compresses the gasket C by means of the compression ring B. A seal is then effected on the tubing and on the body D. The body is usually soldered to the vacuum wall. Such seals are available to take tubes (or rods) ranging in size from $\frac{1}{16}$ in. OD to well over 1 in. OD. Overall lengths (or heights) of such seals range from $1\frac{1}{4}$ in. for the $\frac{1}{16}$ in. size to about $2\frac{1}{4}$ in. for the $1\frac{1}{8}$ in. size. Such seals can be used for vacuum gauges as well as for introducing tubes or rods into a system. Another example of a compression fitting is shown in Fig. 12.2c. This type of fitting can be used to introduce electrical leads, water-cooling tubes, etc., into the vacuum system. The principle is the same as for the seals of Figs. 12.2a and b. A medium hard rubber is used with such a fitting.

For many years rubber stoppers and pieces of rubber hose have been used to introduce leads or tubes into a vacuum system. These are essentially compression fittings and can be very useful in smaller systems, where the amount of rubber exposed is not critical. Figure 12.2d

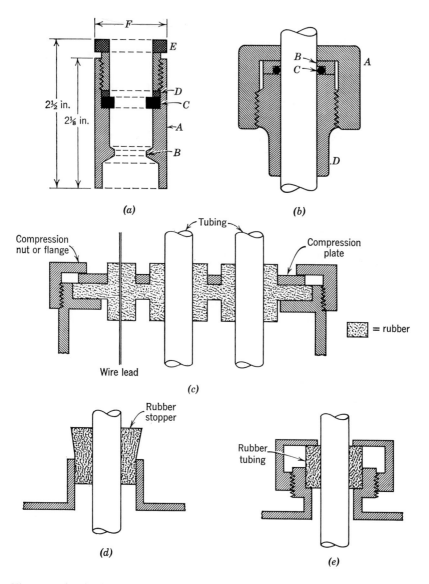

Fig. 12.2 (a) Stuffing box. (b) Form of compression seal. (c) Seals for tubes, rods, and wires. (d) Seal using rubber stopper. (e) Seal with rubber hose (or stopper).

shows a typical seal using a rubber stopper. A more permanent installation is shown in Fig. 12.2*e*, where a piece of rubber hose is used. This arrangement could be used to introduce electrical leads into a system by inserting the leads through a solid rubber stopper. These seals will permit some sliding, rotary, and wobble motions.

Joining of industrial glass pipe together cannot be done in the same manner as joining metal pipe, where grooved metal flanges can be used. However, industrial glass pipe can be obtained with a form of flange, the end of which is grooved smooth. The method of joining pipe with such flanges is shown in Fig. 12.3*a*. Metal rings *A* are bolted together to compress the flanges against a flat rubber gasket *B*. Usually some resilient material *C*, such as rubber, is used between the metal rings and the flanges to avoid damage to these flanges. Split metal rings are often used so they can be used when the glass pipes are in place. Doty (*Rev. Sci. Instr.*, **30**, 1053, 1959) has described a method of making grooves for standard O-rings. Glass working techniques were used with specially shaped (forming) pliers. Certain sizes of Corning Pyrex glass come supplied with depressions in the flanged ends. Branson and Sunderland (*Rev. Sci. Instr.*, **31**, 665, 1960) have used O-rings in these depressions to join pipe (using clamping rings) although Cobin (*Rev. Sci. Instr.*, **31**, 1165, 1960) has pointed out that these depressions are intended for use with special Teflon gaskets. Another method of joining glass pipe is shown in Fig. 12.3*b*, which is simply an adaptation of the arrangement shown in Fig. 12.1*c*.

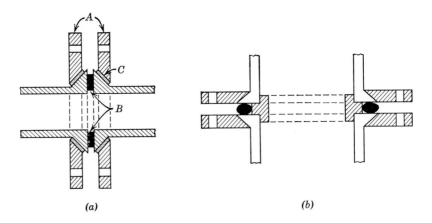

(a)   (b)

Fig. 12.3   (*a*) Use of flat gaskets with glass pipe. (*b*) Use of an O-ring and metal retaining ring.

Clearly flange gaskets (Fig. 12.1*a*) could also be used. It should be kept in mind that on some occasions it is possible to join tubing (glass or metal) by suitable cements or waxes. However, this is not a reliable method.

12.2 Sliding and Rotating Seals

These seals are designed to obtain sliding (translational) or rotary motions (or both) inside a vacuum system. Common methods for obtaining such motions are:

1. A cone joint.
2. A bellows sealed movement.
3. A Wilson seal.
4. A chevron seal.
5. Use of an external magnet.

A glass cone joint is simply a form of stopcock. The method of operation is shown in Fig. 12.4*a*. The rod *A*, of any convenient material, is used to produce the motion in the vacuum vessel. It can be attached at any angle. Clearly only a rotary motion is possible. Either glass or metal joints could be used. Figure 12.4*b* shows an example of a metal bellows used for sliding motion. Bellows are now available that have deep convolutions and can be compressed to as much as 40% of their normal length many times without cracking. The shaft shown in Fig. 12.4*b* might have to be furnished with a stop to prevent the atmosphere from collapsing the bellows when this is not desired. A method of using a sylphon bellows to obtain rotary motion in a vacuum system is shown in Fig. 12.4*c*. Shaft *A* turns shaft *B* which then turns shaft *C*. Shaft *B* (the wobble shaft) turns in a ball-and-socket bearing. The bellows *E* provides a vacuum seal. The housings *F* and *G* contain bearings for shafts *A* and *C*, respectively. The life of such a device is determined by the bellows since it is being flexed. The torque obtainable will depend on the angle of tilt of the wobble shaft, which is limited by the amount of distortion the bellows can take.

The most common devices used for obtaining motion in a vacuum system are Wilson and chevron seals. The elements of a Wilson seal are shown in Fig. 12.4*d*. The shaft *A*, which can be slid back and forth or rotated (or both), is sealed by gasket *C*. Gasket *B* is used as a seal only when gasket *C* leaks and the space between the gaskets is pumped on through the pump-out *D*. The gaskets are made of a

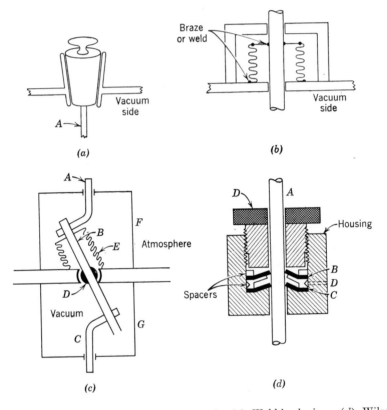

Fig. 12.4 (*a*) Cone joint. (*b*) Bellows seal. (*c*) Wobble device. (*d*) Wilson seal.

fairly hard rubber and a light coating of good quality vacuum grease is applied to them. For satisfactory operation, the following points should be observed:

1. The shaft must be smooth (a semipolished surface).
2. The gaskets must fit the inside of the housing.
3. The gaskets must have smooth edges. A sharp, circular cutter lubricated with soap solution can be used or the gaskets could be molded with all flash removed.
4. The holes in the gaskets should be about two-thirds the shaft diameter.

5. A good quality of vacuum grease should be used sparingly on the gaskets.

6. The compression nut should not be tightened too much—only enough to seal. A knurled nut is recommended to avoid someone using a wrench on it.

Wilson seals are generally made of brass although there is no reason why other metals, such as stainless steel, could not be used. The housing can be soldered or brazed to the vacuum system or pipe threaded and sealed with a suitable cement such as glyptal. They can be leak tested while under vacuum by using soap solution on the pump-out. If the inner gasket leaks (the one nearest the vacuum) air will be drawn through the pump-out and the soap film will be pulled in. This is not too sensitive a method. The usual methods of leak hunting, including the helium leak detector, can be used. Care should be taken not to clog the pump-out with soap or grease.

A chevron seal is a variation of a Wilson seal. Its main features are:

1. The use of V-shaped (chevron) gaskets.
2. The use of a spring washer to "load" the gaskets.

A compression nut spreads the gaskets and forces them tightly against both the shaft and the inside of the housing. Figure 12.5 shows the pertinent details of a type of chevron seal that has been widely used at the Lawrence Radiation Laboratory, University of California, Berkeley, California. The pertinent dimensions of this type of seal are also shown in Fig. 12.5. The compression nut and details of the housing are not shown in this figure. As in the case of the Wilson seal, this seal can be soldered or brazed to the vacuum wall or attached by a pipe thread, using a suitable cement. The comments concerning Wilson seals with regard to obtaining satisfactory operation are applicable here. Failure with Wilson or chevron seals occurs when the shafts become dry or when the temperature rises to excessive values. A higher temperature causes the greases to become fluid and their ability to seal poor fits between the gasket and the shaft is seriously diminished. The use of silicone grease is effective in controlling this effect. A second effect of elevated temperatures is the permanent set of the rubber, which makes it possible for the hole in the gasket to become too large for the shaft. This difficulty is minimized by proper choice of gasket material. Cooling the seals will, of course, prevent temperature rises. Without undue precautions, these seals can be operated at rotary speeds of 60 rpm or so for appreciable lengths of time. Using special methods of adding vacuum grease and cooling will lead to con-

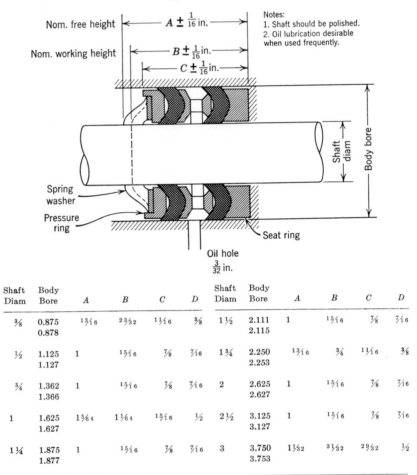

Notes:
1. Shaft should be polished.
2. Oil lubrication desirable when used frequently.

Shaft Diam	Body Bore	A	B	C	D	Shaft Diam	Body Bore	A	B	C	D
3/8	0.875 0.878	13/16	23/32	11/16	3/8	1 1/2	2.111 2.115	1	15/16	7/8	7/16
1/2	1.125 1.127	1	15/16	7/8	7/16	1 3/4	2.250 2.253	13/16	3/4	11/16	3/8
3/4	1.362 1.366	1	15/16	7/8	7/16	2	2.625 2.627	1	15/16	7/8	7/16
1	1.625 1.627	15/64	1 1/64	15/16	1/2	2 1/2	3.125 3.127	1	15/16	7/8	7/16
1 1/4	1.875 1.877	1	15/16	7/8	7/16	3	3.750 3.753	1 1/32	31/32	29/32	1/2

All dimensions are in inches.

Fig. 12.5 Chevron seal.

siderably higher rotary speeds. When used for sliding motion, the shaft should be greased periodically.

Many modifications of Wilson and chevron seals have been reported in the literature. However, regardless of the design, the following general comments apply to shaft seals:

1. There is little to choose between the various designs.

2. Spring-loaded washers are best. Such washers could be used in Wilson seals, instead of flat gaskets.

3. A thin vacuum oil is better than a vacuum grease (for very good finishes on the surfaces).

4. With a sliding shaft, most of the leak (80% or more) occurs on the inward stroke.

5. A slowly rotating shaft (60 rpm or less) has little more leakage than a stationary shaft.

6. Smooth surfaces are important.

Instead of flat gaskets of the type used in Wilson and chevron seals, O-rings have also been used, with varying success. The main trouble with such seals is that the surface contact of the O-ring is rather small and the mechanical support of the shaft is not too good. However, the mechanical problem can be solved to a large degree by using bearings along the shaft, outside the O-ring seal. In any case, O-ring seals have been used fairly widely for sliding and rotary motions. Figure 12.6a shows a device making it possible to swing a tube back and forth through fairly large angles (Brannon and Ferguson, *Rev. Sci. Instr.*, **25**, 836, 1954). It consisted of a brass sphere which was pinned to the vacuum wall. Seals were provided by means of an O-ring and a Wilson seal. The sphere and tube could be machined from a single piece of brass or the tube could be soldered to the brass sphere. Many attempts have been made to develop rotary seals that provide reasonably high speeds for extended periods. Wilson and chevron seals are commonly used to provide rotary motions but tend to leak at higher speeds. Also, the gaskets will wear in time and there will be a loss of lubricant (due to more heating in vacuum). An obvious solution to the latter problem is to continually supply lubricant. To prolong the life of a seal, materials other than rubber have been used. Toffer and Amrein (*Rev. Sci. Instr.*, **31**, 348, 1960) have described a form of rotary seal using moving and stationary parts separated by a thin film of vacuum oil. This design was based on an earlier design developed at Los Alamos National Laboratory, Los Alamos, New Mexico. The basic elements of this type of seal are shown in Fig. 12.6b. The stationary part consisted of meehanite cast iron and the moving part was a Tenite II ring (Tennessee Eastman Corporation, Clinton, Tennessee) which engaged an O-ring in a barrel, thus rotating the barrel. The oil supply and drain are shown in Fig. 12.6b. The mechanical driving system is not shown. This particular seal was developed for use in nuclear research. However, it should be applicable in a general way. This type of seal was operated at 137 rpm for 76 hr with no visible wear on the Tenite II. Cooling with a fan was found to be desirable. Many other types of rotary seals

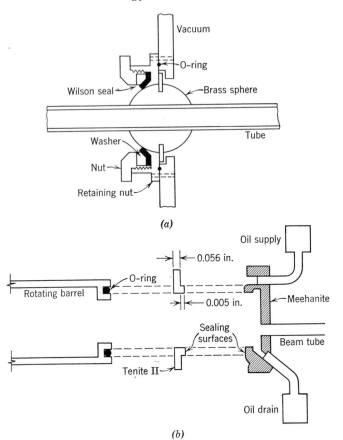

Fig. 12.6 (a) Seal for swinging motion. (b) Rotary seal with soft and hard matching surfaces.

have been reported. Some special types of seals for obtaining motion in a vacuum system are discussed in Section 12.5.

The use of magnets outside a vacuum system to control motion inside has been common for many years. Naturally the vacuum wall, or at least that part of the system where the magnet is to be used, must be nonmagnetic. Either a permanent magnet or an electromagnet is used to move a piece of magnetic material, usually soft iron, inside the system. As a rule only simple motions are possible by this method. However, it is possible to attain fairly complex motions by guiding the magnetic material in the vacuum system by mechanical arrangements. In any case, the forces possible with this method are usually quite limited. A simple example of a device using this method is the valve shown in Fig. 11.18.

12.3 High Temperature Seals

The increased interest in very high and ultra-high vacuums has given rise to greater interest in seals that can be baked. The baking process removes adsorbed and absorbed gases and vapors, making it possible to reach lower pressures. Seals to withstand high temperatures are also required in experiments not requiring the lowest possible pressures. Often, demountable mica windows must be used at high temperatures as, for example, in microwave work.

Although special types of seals are sometimes used to operate at high temperatures, the most common type involves the use of a metal gasket. Presently available elastomers cannot be baked to high enough temperatures for effective degassing. Also, the elastomer may deteriorate when operated for any length of time at elevated temperatures. Metal gaskets made of indium, lead, tin, copper, aluminum, and gold have been used. If a bake-out temperature of more than 400°C is required then indium, tin, and lead cannot be used since their melting points are 156°C, 232°C, and 327°C, respectively. However, they are still useful at elevated temperatures below their melting points since their vapor pressures are low compared to those of elastomers. The most commonly used metal gaskets are made of aluminum, copper, or gold, which have melting points of 659°C, 1083°C, and 1063°C, respectively.

Many special forms of metal gasket seals have been discussed in the literature. However, attention will be directed first to seals that have been studied and used extensively, viz.:

1. O-ring gaskets.
2. Flat gaskets.
      a. With knife edge.
      b. With step seal.

Metal O-rings can be used between plain surfaces or with a groove. The importance of carefully finishing grooves and surfaces has already been noted in Chapter 10. Careful attention must also be given to the surface finish of metal wires used for gaskets. No scratches or blemishes are permissible. In addition, the metal should be as pure as possible, e.g., OFHC copper should be used and 99.99% pure aluminum is much better than commercial aluminum, which is about 99% pure. The problems involved in obtaining tight metal gaskets are considerably more difficult than with elastomers. Some of these are (for metal surfaces):

1. Large pressures are needed to get a tight seal. This means that many more bolts are needed to effect the seal. No general rule can be given as to the increase in the number of bolts over elastomer seals. However, doubling the number of bolts is certainly not too conservative.

2. The gaskets and surfaces (including grooves) must be essentially polished.

3. Due to different expansions of the gasket and the mating surfaces, leaks may develop after cooling down from an elevated temperature.

4. It is usually not possible to use the gasket over again when the seal is taken apart.

Considerable work has been done with indium gaskets. Such gaskets are often made from wire of diameters $\frac{1}{32}$ to $\frac{1}{16}$ in. They can be sealed between two smooth flanges and are normally compressed to a thickness of about 5 mils for $\frac{1}{32}$ in. wire to 10 mils for $\frac{1}{16}$ in. wire. It is not necessary to seal the ends together. A cold weld is obtained when the ends are overlapped. In view of the fact that indium is quite soft, it is not necessary to use a large number of bolt holes as in the case of harder metals such as aluminum or copper. As an example, with a 7 in. diameter wire gasket and $\frac{3}{8}$ in. thick stainless steel flange it is possible to use 6 clamping bolts on an 8 in. diameter circle. The ends of an indium gasket could be melted together instead of simply being overlapped. Where an O-ring is to be mounted vertically, a retaining ring should be attached to one surface, of such a thickness as to allow for compression of the gasket. Indium and lead are sufficiently soft to allow their use in joining glass pipe. Green, Miles, and Richardson (*J. Sci. Instr.*, **36,** 324, 1959) have described a method for using a lead gasket to join industrial glass pipe. This is shown in Fig. 12.7*a*. The pipe was modified by grinding a taper to a uniform cone so the backing flanges pressed uniformly on the glass. Graphite impregnated asbestos inserts were cut to completely fill the space between backing flanges and the glass pipes. Also, the cast iron backing flanges were replaced by mild steel flanges. The ground sealing surfaces on the pipe were used without change. The lead wire was $\frac{1}{16}$ in. in diameter and the ends were butt joined using a suitable soldering paste. Pressures of about 2×10^{-8} Torr were obtained after baking at about 300°C. Careful polishing of the sealing surfaces would probably lead to considerably lower pressures. The authors also point out that for higher bake-out temperatures, 70/30 thallium/lead alloy (mp 380°C) could be used. Adam, Kaufman, and Liley (*J. Sci. Instr.*, **34**, 123, 1957) have described a method of making indium O-rings and the appropriate grooves for sealing brass and Pyrex surfaces together.

Aluminum O-ring gaskets are used between plain surfaces and in grooves. Gaskets made by twisting the ends together and clamping between stainless steel surfaces work satisfactorily when baked up to about 200°C and cooled. However, baking at much higher temperatures and then cooling usually results in leaks due to the different expansion rates of aluminum and stainless steel. By using butt-welded aluminum wire, which is cleaned with emery cloth and annealed before use, it is possible to bake at temperatures up to about 500°C and then cool without getting leaks. This is true even though the clamping bolts loosen. Apparently at these high temperatures the aluminum adheres to the stainless steel surfaces. Holden, Holland, and Laurenson (*J. Sci. Instr.*, **36**, 281, 1959) have described several experiments using butt-welded aluminum gaskets. They used 2 in. diameter gaskets bolted between stainless steel flanges of 3¾ in. OD and ½ in. thick. They point out the importance of adequate compression of the gasket. Butt-welded aluminum wire gaskets are commonly used with grooves. In this case it is important to smooth off the weld to the same diameter as the rest of the gasket. Annealing of aluminum gaskets is important to make them soft.

Copper O-rings are not commonly used. The problem usually arises from different expansion rates of the gasket and the mating surfaces. The importance of using OFHC copper, welding in an inert atmosphere, and annealing before use must be emphasized. Van Heerden (*Rev. Sci. Instr.*, **26**, 1130, 1955) describes a method of obtaining fairly reliable copper O-ring seals. He used 60 mil OFHC copper which was bent in a circle on a jig, with the ends sticking up. These ends were welded electrolytically in hydrogen and the resulting bead was filed and smoothed down with emery paper. The ring was then annealed in hydrogen at 950°C. Van Heerden emphasizes the need for using OFHC copper and welding and annealing in hydrogen, as well as having a good polish on the flanges. The O-ring could be fixed in position by using a flat ring of half the thickness of the ring screwed to one flange. The usual problem encountered was slackening of clamping bolts after baking (at 400°C) and cooling. This necessitated retightening of bolts. Instead of O-rings, specially shaped gaskets have been used with some degree of success. A common form is an annealed

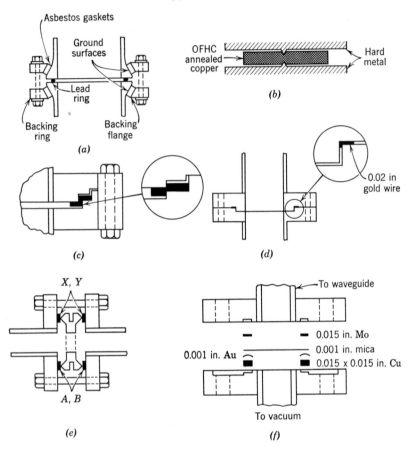

Fig. 12.7 (a) Joining glass pipe with a lead gasket. (b) Copper gasket and knife edges. (c) Step seal. (d) Corner gold ring seal. (e) Special form of copper gasket. (f) High temperature seal for mica window.

copper ring of diamond cross section. A good seal can be obtained by compressing the gasket between two stainless steel plates (polished) to 85% of the original apex-to-apex thickness. Temporary seals are obtained with such gaskets by compressing them between two flat plates of stainless steel. More permanent seals can be obtained by using such gaskets in a trough of rectangular cross section in one flange. The depth of this trough should be 85% of the apex-to-apex thickness.

Actually flat copper gaskets are more commonly used than O-rings. However, special arrangements are needed to use such gaskets. A common method is the use of a knife edge. Such a seal involves the use of sharp edges of hard metal (usually stainless steel) which are forced into the copper gasket. The principle of the method is shown in Fig. 12.7*b*. Van Heerden (*Rev. Sci. Instr.*, **26,** 1130, 1955) has described such a seal using a gasket 40 mils thick with knife edges having a 5 mil flat and 30° sides. Simple machining of the knife edges, with emery cloth used only to remove burrs, is sufficient. As with all copper seals, the use of OFHC copper and annealing in hydrogen (850 to 950°C) are important. Pattee (*Rev. Sci. Instr.*, **25,** 1132, 1954) has described a double knife edge seal for joining two stainless steel tubes. The knife edges were machined in the ends of the tubes and suitable flanges were used to press the edges into a flat copper gasket. In some cases a single knife edge can be used. An example of such a case is where the copper tubing of a Housekeeper seal is forced against a knife edge. The copper tube could be threaded into a flange which is bolted against the vacuum wall with the knife edge. The knife edge arrangement can also be used with metal foils. Ruthberg and Creedon (*Rev. Sci. Instr.*, **26,** 1208, 1955) have described such a seal using aluminum foil (0.001 in.) and Monel flanges.

Another method of obtaining a seal with copper is through the use of a step seal. The form of such a seal is shown in Fig. 12.7*c* (Lange and Alpert, *Rev. Sci. Instr.*, **28,** 726, 1957). The flanges were machined from stainless steel and heliarc welded to vacuum system components. The gasket was a washer shaped ring of OFHC copper about 0.040 in. thick, which was hydrogen annealed at 950°C after cutting. Twelve bolts were used on a 6 in. flange, ⅜ in. thick. These were tightened until the gasket was roughly one-half its original thickness. It is claimed that this type of seal could be carried through as many as 70 bake-out cycles at 450°C.

Knife edge seals suffer from several disadvantages. They involve more stringent machining and are much more easily damaged than step seals. The use of copper itself gives rise to difficulties since it must be annealed at 850 to 950°C. Consequently, attention has been given to other materials. Gold ring seals have been studied in some detail. At first glance it might appear that gold would be too expensive. However, a reduction in machining costs as against copper diamond gaskets and the high scrap value of gold make such seals attractive. A form of gold ring seal is shown in Fig. 12.7*d* (Grove, *Project Matterhorn*, Scientific Paper No. 15–NYO 8750). A 0.02 in. 24 carat gold gasket was used with a 16-μin. finish in the gasket area. A 0.010 in. axial clearance was used, i.e., the gasket was compressed to a thickness of 0.010 in. Also, there was a 0.002 in. radial clearance at the corner where the gasket was placed. It is claimed that this type of seal can be used for 40 to 50 bake-outs at 450°C without failure. This applies to seals with dimensions up to 8 in. diameter with loadings of 140 to 180 tons per inch of gasket. At the present writing, step seals using copper gaskets and gold ring seals can be considered to be competitive in ultra-high vacuum work.

Many forms of metal gaskets have been reported as a result of attempts to provide greater reliability. One such form is shown in Fig. 12.7*e*. (Robinson, *J. Sci. Instr.*, **34**, 121, 1957). The copper washers *X* and *Y* were 0.010 in. thick and were clamped between flanges by a steel ring on which two annular knife edges *A* and *B* were turned. The stiffness of the knife edge system must be sufficient to cause plastic deformation of the copper when the flanges are bolted together cold and still exert a positive force on the copper both during the annealing (on baking) and the subsequent cooling. Robinson also reports other forms of this type of seal. As has been noted, special seals to operate at high temperatures are often required. One such seal for use with a mica window is shown in Fig. 12.7*f* (Lindsay, *Rev. Sci. Instr.*, **32,** 748, 1961). As will be noted, a molybdenum (Mo) gasket 0.015 in. thick, a mica (synthamica) window 0.001 in. thick, a gold (Au) foil gasket 0.001 in. thick, and a copper (Cu) gasket (0.015 in. by 0.015 in.) are used. Molybdenum is used because it is nonmagnetic. Otherwise Kovar or Driver-Harris Alloy 146 could be used. The vacuum and backing flanges (SS 304) are suitably recessed to receive the copper and molybdenum gaskets, respectively. The copper was annealed at 500°C and water quenched, while the gold was annealed at 700°C. The flange-washer (molybdenum) interface and the screw threads were lubricated with colloidal carbon. It should be pointed out that the screws for clamping bolts are usually lubricated with a suitable material in the case of seals operated at high temperatures so as to avoid binding. Molybdenum disulfide is often used for this purpose.

12.4   Low Temperature Seals

The increased interest in low temperature work (cryogenics) as well as various special studies have led to the development of numerous types of seals. The problems associated with such seals are in some ways similar to those encountered with high temperatures, viz., deterioration of seal materials and leakage due to different expansion rates of construction materials. Many special seals have been reported. However, emphasis will be placed here on the following types:

1. Seals using elastomeric or plastic gaskets.
2. Seals using metal gaskets.
3. Seals using cements.
4. Glass-metal seals.

The most common type of low temperature seal is one involving some form of window, usually for optical purposes such as passing infrared or ultraviolet radiation. At first glance it might appear that elastomers and plastics would be unsuitable at low temperatures because of brittleness. However, by proper design it has been found that these materials can be used in certain applications. Weitzel, Robbins, Bopp, and Bjorklund (*Rev. Sci. Instr.*, **31**, 1350, 1960) have reported on the use of natural rubber, synthetic rubber (Buna N), and Viton A, as well as plastics such as Mylar and nylon. Flanges were designed to supplement the properties of the seal material. In all cases the final thickness was made small to minimize shrinkage. In the case of elastomer seals, 1 in. ID O-rings of $\frac{1}{8}$ in. diameter were used with a tongue and groove arrangement. The O-rings were compressed to a thickness of about 0.025 in. so as to obtain flat circular gaskets. The Mylar and nylon gaskets were 10 mils thick and were compressed to a thickness of 3 mils under a 1 in. diameter ring, with a semicircular projection on the ring of 0.084 in. diameter. The key to successful elastomer or plastic gaskets appears to be in having small final thicknesses.

An example of a low temperature window seal which can be used with copper diaphragm and plastic seals or with a simple copper seal is shown in Fig. 12.8a (Roberts, *J. Sci. Instr.*, **36**, 99, 1959). The copper diaphragm C is sealed to the window A by means of Araldite cement (Aero Research, Ltd., England). The retaining ring E makes seals with the copper diaphragm, a polythene gasket D, and the flange B on the tube F which is part of a cryostat. The gasket was 0.005 in. thick and 0.40 in. wide. This gasket was coated lightly with vacuum grease. This seal could be eliminated by soldering C to B with Wood's metal. It is claimed that this type of seal could be used at liquid hydrogen and liquid helium temperatures. An example of a seal using an indium gasket is shown in Fig. 12.8b (Willis, *Rev. Sci. Instr.*, **29**, 1053, 1958). This seal was developed for low temperature infrared work. The infrared cell was to be filled with liquid nitrogen and contained in an evacuated space. The elements of the seal (see Fig. 12.8b) include an indium gasket (0.075 in. wide by 0.055 in. thick),

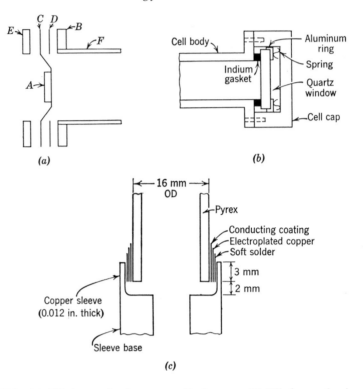

Fig. 12.8 (*a*) Window seal using copper diaphragm. (*b*) Window seal using indium gasket and springs. (*c*) Form of metal-to-glass seal.

a quartz window ($1\frac{3}{16}$ in. OD by 0.120 in. thick), a $\frac{1}{32}$ in. thick flanged aluminum ring used to center the window and to distribute the spring load, and a cell cap to compress the spring and load the indium gasket. A Belleville-type spring was used, which delivered a 130 lb force when compressed 0.010 in.

Considerable work has been done with epoxy resins to obtain tight low temperature seals. Hearst, Ahn, and Strait (*Rev. Sci. Instr.*, **30**, 200, 1959) have described the use of Resiweld Adhesive No. 4 (H. B. Fuller Company, St. Paul, Minnesota) for obtaining seals to such materials as aluminum, molybdenum, and tungsten. This is an epoxy resin with the usual two-component mixture. It was found unsatisfactory for sealing plastics. It is not known how effective other commercial epoxy resins might be in this regard. At least on the basis of experience with Resiweld Adhesive No. 4, it appears that, although epoxy resins make strong seals at room temperature, their strength decreases rapidly at low temperatures. Consequently, evacuation or letting down to air at low temperature should not be attempted. Hearst, et al., also describe the use of Kel-F O-rings at low temperatures. They used grooves such that the O-ring was compressed about 7%. They found that such rings tended to develop flats, and it was recommended that the rings be replaced when the seal was disassembled. Balain and Bergeron (*Rev. Sci. Instr.*, **30**, 1058, 1959) have described the use of several types of epoxy resins for use at low temperatures in connection with electrical lead-throughs.

Many forms of metal-glass vacuum seals for use at low temperatures have been reported. A form of such a seal is shown in Fig. 12.8c (deHaas, *Rev. Sci. Instr.*, **30**, 594, 1959). It involves the use of a deformable copper sleeve of the Housekeeper-seal type which is sealed to standard wall Pyrex tube. The cleaned end of a Pyrex tube (about 16 mm OD) was painted with duPont conductive coating material Type F No. 4666, allowed to set for about 5 min, and then heated at 500°C for 5 min. Electroplating of copper onto the conductive coating was carried out at a current density of approximately 3 mamp/cm² using a platinum anode in a saturated solution of cupric sulfate. The sleeve was machined from hard-drawn commercial copper to a thickness of 0.012 in., after which it was annealed. Annealing was done both by torching and water-quenching, and by heating in vacuum. Soft solder (50/50) was employed to seal the sleeve to the copper-plated Pyrex tube. Seals made in this manner were cycled successfully (sixteen times) between room temperature and 4.2°K. They withstood rapid immersion in liquid nitrogen. This seal is claimed to have an advantage over Kovar-Pyrex and Housekeeper seals since it can be constructed in the laboratory without the services of a glassblower. Szabo and Dimock (*Rev. Sci. Instr.*, **32**, 1256, 1961) have described a modification of the seal. Horwitz and Bohm (*Rev. Sci. Instr.*, **32**, 857, 1961) have described a metal-to-glass low temperature seal making use of an indium gasket. Many other types of metal-glass seals for low temperature operation are described in the literature.

12.5   Some Other Forms of Seals

Many seals intended for special applications have been reported. The types of seals that will be discussed briefly here are:

1. Window seals (room temperature).
2. Greaseless seals.

Viewing windows for vacuum chambers are usually installed with a suitable flanging arrangement and an O-ring. This method is commonly used to seal windows directly to the vacuum wall. However, the same general method can be used to seal a window to a tube. One way of doing this is shown in Fig. 12.9a (Moore, *Rev. Sci. Instr.*, **29,** 737, 1958). The end plate could be any vacuum-tight material which can be smoothed to effect a seal with the O-ring. Any elastomer could be used for the O-ring. The tube end was ground square and chamfered. The ID of the O-ring was equal to or slightly less than the OD of the tube. Window seals made with O-rings are quite reliable and are commonly used in applications where vapor from the elastomer is not objectionable. The simplest form of seal for attaching an end plate to a tube is obtained with a suitable cement or wax (see Chapter 8). In many cases, particularly for temporary setups, it is possible to use a vacuum putty such as Apiezon Q. A special form of seal for attaching calcium fluoride windows to tubes has been described by Greenblatt (*Rev. Sci. Instr.*, **29,** 738, 1958). This window seal is shown in Fig. 12.9b. The glass seat, which must be flat, is prepared either by ultrasonic drilling or by careful glass manipulation procedures. The shoulder serves to protect the finished seal from mechanical damage. The glass seat, shoulder, and adjacent areas on the window were painted with platinum paint, and then baked at 500°C. The calcium fluoride crystals were cooled slowly to avoid breakage. Silver chloride was then applied to the seat and inside shoulders of the glass. The glass was heated with a torch, and a strip of silver chloride was touched to the platinized glass. Finally, the window was placed in position in its glass seat and the entire assembly was put in an oven (room temperature) and heated to about 490°C. Although the temperature can be raised fairly rapidly, cooling must be done slowly. Many other forms of silver chloride seals for quartz and calcium chloride windows have been described.

Anderson and Stepp (*Rev. Sci. Instr.*, **33,** 119, 1962) have described a method for sealing a sapphire window to glass. Such seals are often made with Corning 1826 powdered glass. However, this type of seal suffers from obstruction of window area, low softening point, and damage from sodium vapor. The Anderson and Stepp seal used Corning 7280 glass, which has a coefficient of expansion close to that of sapphire. The 7280 glass was sealed to Pyrex 7740 glass in small diameter tubing by using 3320 glass as the intermediate. The 7280 glass was then fused to the sapphire. Martz (*Rev. Sci. Instr.*, **32,** 214, 1961) has described a method for sealing sapphire windows to glass by using epoxy resins. The resin used was EPON 828 (Shell Chemical

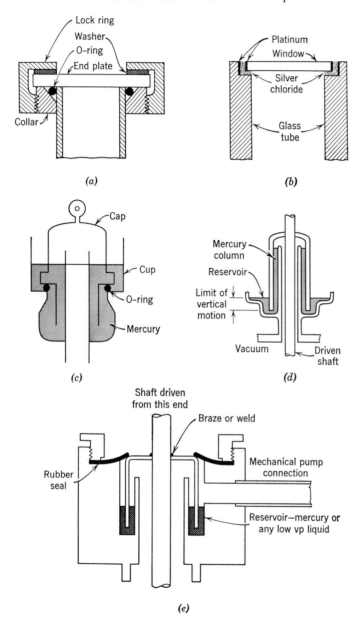

Fig. 12.9 (a) Form of end plate seal with elastomer O-ring. (b) Window seal using silver chloride. (c) Static seal with mercury and O-ring. (d) Seal using mercury exposed to atmosphere. (e) Mercury seal isolated from atmosphere.

Company) with a filler of lithium aluminum silicate powder (Carborundum Company, Latrobe, Pa., Lithafrox Grade 2121). The hardener
was diethylenetriamine (Union Carbide Chemical Company). The
formula used, in terms of parts by weight, was:

EPON 828 epoxy resin	10
Lithium aluminum silicate	4
Diethylenetriamine	1

In many vacuum processes hydrocarbon vapors cannot be tolerated.
Consequently, vacuum greases must be avoided in seals and other
components. Of course, various forms of metal gasket seals could be
used. Adams, Kaufman, and Liley (*J. Sci. Instr.*, **34,** 123, 1957)
have described a seal using indium wire to seal Pyrex to brass. This
variety of seal was designed to seal by atmospheric pressure. In
many cases mercury vapor is not harmful in systems where hydrocarbon vapors cannot be tolerated. An example of a seal using mercury
is shown in Fig. 12.9c (Gaunt and Redford, *J. Sci. Instr.*, **36,** 377,
1959). This particular seal was designed for remote control. Only
mercury and glass were exposed to the vacuum although the seal was
made with a rubber O-ring. With the apparatus at atmospheric pressure, the cap floats in mercury and it can simply be lifted off. The
O-ring is lifted with it because of its reasonably close fit on the cap.
When the apparatus is evacuated, the cup sinks and the O-ring makes
a seal. Of course, mercury seals have been commonly used in vacuum
practice for many years. However, in many mercury seal cases
vacuum greases are exposed to the vacuum. The advantage of mercury over metal gaskets for systems that must be free of hydrocarbon
vapors is particularly evident with rotary and sliding seals. Brueschke
(*Vacuum*, Vol. 11, No. 516, p. 255, 1961) has described methods for
transmitting motion into ultra-high vacuum systems. Figure 12.9d
shows the method used for obtaining rotational motion and limited
translational motion. Since the mercury is exposed to the atmosphere,
this seal is cumbersome, the height of the mercury column being 76 cm
(at sea level) when the system is evacuated. In order to avoid this
problem, Brueschke devised another type of seal, which is shown in
Fig. 12.9e. A Wilson-type seal is used to seal from atmosphere. Again,
the translational motion is limited. Where mercury vapor is objectionable, cold traps must be used. Of course, low vapor pressure
liquids other than mercury could be used.

Several attempts have been made to develop greaseless seals using
polytetrafluoroethylene (PTFE). The problem with this material is
its cold flow. Billett and Bishop (*J. Sci. Instr.*, **35,** 70, 1958) have

described a seal for rotating shafts. This seal involved several (four) PTFE washers (0.062 in. thick), separated by metal spacer washers (0.062 in. thick). Polishing of metal surfaces was emphasized. Also, to obtain tight seals, it was necessary to have the PTFE washers fit tightly against the shaft and the inside of the housing. Such seals were operated satisfactorily up to 200°C. Other types of seals involving the use of PTFE have been reported. In general, it is necessary to use thin sheets of PTFE in order to avoid difficulties with cold flow characteristics.

Seals to transmit motions into a vacuum system can be considered to be forms of bearings. Of course, bearings are often needed inside a vacuum system. In this sort of situation the problem of lubrication becomes progressively more important where continued, reliable performance is required. Although such lubricants as vacuum greases and oils, graphite, and molybdenum disulfide have been commonly used, in time there is a loss of lubricant. This can lead to binding and cold-welding of parts. This is a particularly important problem in the case of space vehicles. Some of the seals discussed above could be used in this connection in place of bearings. Also, seals which are continuously lubricated could be used in certain circumstances. It is to be expected that many new and novel bearings will be developed during the next few years.

12.6 Couplings, Hose Clamps, etc.

A supply of small couplings is highly desirable in any vacuum laboratory or plant. These couplings can be used for a multiplicity of purposes, such as: (1) temporary small hose connections, (2) vacuum gauge couplings, and (3) connections for temporary test gauges and other devices. It is best to standardize on such couplings to avoid having to stock many different sizes. A ¼ or ⅜ in. size is very useful. A form of coupling used for many years at the Lawrence Radiation Laboratory, University of California, Berkeley, California, is shown in Fig. 12.10a. The vacuum seal is simply a rubber washer. Nipples for rubber tubing and metal tubing are shown. The metal tubing is soldered or brazed into the union and the nipple. Both the union and the nut usually have hex shapes so that wrenches can be used for tightening. The union shown has a pipe thread on one end. In some cases this is eliminated (half-union) so the union can be soldered (or brazed) to a surface. Such couplings can be made in various sizes and in several modified forms. For example, the openings in union and nipple could be different sizes. These couplings are very useful for connecting water-cooling lines for use in a vacuum system as well as for a variety of other purposes. A solid nipple can be used to seal off a line.

Many special designs of couplings have been reported in the literature. Several forms of modified standard flare fittings have been reported. Figure 12.10*b* shows one form of such a fitting (Heinrich, *Rev. Sci. Instr.*, **29**, 1053, 1958). Such fittings were used on ¼ in., ⅛ in., and ½ in. tubing for closures, tube-to-tube seals, and valve-to-tube seals. In making tube-to-tube seals, two such fittings are soldered together. In Fig. 12.10*b*, the male part of the flare fitting *F* is machined down and an O-ring *O* is compressed by means of the normally used copper gasket *G*. Gilvey (*Rev. Sci. Instr.*, **24**, 984, 1953) has described the use of modified plumbing unions. Figure 12.10*c* shows both the modified and unmodified union. The union was modified by machining a flat in the end of part *B*. Such unions have been used with standard tubing, pipe, elbows, and tees. Cast brass, wrought iron (¼ in. to 2 in.), stainless steel (⅛ in. to 4 in.), Monel (⅛ in. to 4 in.), and nickel (⅛ in. to 4 in.) tubing and piping have been used.

Many forms of small couplings are available commercially and the type desired can usually be obtained through local supply houses. In many cases it is desirable to join pipes. Of course, welded flanges with O-rings or grooveless flange gaskets could be used. Standard pipe couplings also make satisfactory permanent vacuum seals if they are carefully threaded and the threads are coated with a vacuum cement, such as glyptal. In the case of temporary pipe connections it is more convenient and economical to use some type of coupling without welding or using a cement. Special types of couplings using gaskets are available commercially.

On many occasions it is necessary to use rubber or reinforced plastic hose, which must be clamped to some form of coupling. The coupling is soldered or brazed to some part of the apparatus (see Fig. 12.10*d*). A popular form of hose clamp is shown in in Fig. 12.10*e*. The threads of the screw *S* engage the perforations in the ring *R*. Turning the screw tightens or loosens the ring. Usually the screw is arranged so it can be pivoted (say at *P*) so as to quickly open the clamp. Of course, other types of clamps are also available commercially. Apart from small couplings, flanges, and hose clamps, a supply of small solder fittings including T's and elbows should be available. The ½ in. and ⅝ in. sizes find many uses. Of course, larger sizes can be carried in stock when they are used fairly routinely.

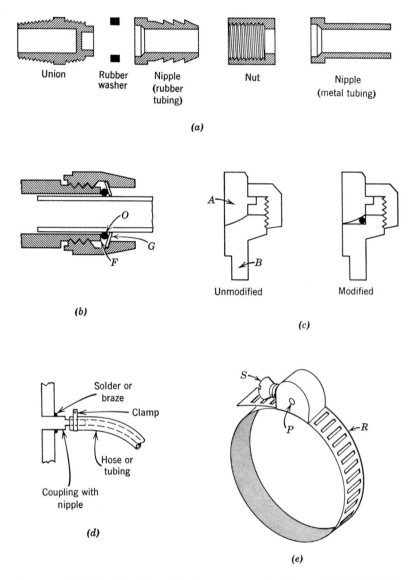

Fig. 12.10 (*a*) Couplings for rubber hose or metal tubing. (*b*) Modified flare fitting. (*c*) Use of plumbing unions. (*d*) Rubber (or plastic) tubing with nipple. (*e*) Hose clamp.

12.7   Electrical Lead-Throughs

It is usually necessary to bring electrical leads into a vacuum system for various purposes, such as heating filaments or furnace elements, supplying voltages to grids, etc. The seals used for this purpose are often called *insulated seals*. Some seals of this type have already been discussed in Section 12.5, although those seals were intended primarily for optical or thermal insulation purposes.

The types of seals for electrical lead-throughs can be broken down into two general classes of usage:

1. High current, low voltage.
2. Low current, high voltage.

The first type of usage is associated with the operation of heating elements, such as in vacuum furnaces, while the second type is usually associated with various electron devices, such as in speeding up electrons in vacuum. The distinction between the two types of applications is somewhat arbitrary. However, high current will be taken to mean tens of amperes while low voltage means a hundred volts or less. It must be kept in mind that although a seal can be used in the atmosphere at certain voltages, it may "break down" in vacuum. Actually, the electrical breakdown is due to a discharge in the residual gas in the system. The voltage at which breakdown occurs will depend on the pressure, the distance between electrodes, and the material and shape of the electrodes. However, the main concern here is with general types of insulated seals, which are normally rated in air. The insulating materials used in seals must have the proper vacuum, mechanical, and electrical properties. The principal vacuum property involved is the vapor pressure, which must be low. Mechanically the insulator must be rugged to be able to stand mechanical and thermal shock. The electrical property involved is the resistance.

To carry large currents at low voltages, a seal must have large leads, and the amount of insulation needed is not critical. Because of heating it may be necessary to water cool seals. Sometimes hollow tubing is used, with water circulated through the tubing. The water must be discharged through insulated tubing (Saran, etc.) to avoid grounding the electrical supply. In many cases, rubber insulation is used. Such seals can be designed not only for high currents but also for high voltage service. A form of general purpose seal with rubber has already been shown in Fig. 12.2c. This can be useful for a rather wide range of currents and voltages. However, all insulation here is provided by the rubber. Many commercial seals are available that use an insulating material other than rubber, such as glass or porcelain, which are then sealed to the system by rubber gaskets. A form of seal, using a plastic insulation and a single rubber O-ring for vacuum sealing, is shown in Fig. 12.11a (Edwards, *J. Sci. Instr.*, **35**, 111, 1958). Several advantages are claimed for this particular design. Although bakelite and plexiglass were used, presumably other materials could be used. Bakelite and plexiglass are not good where a high vacuum is needed. Edwards used this seal in connection with a vacuum furnace. The principal parts of this seal are: *A*, brass ring with O-ring seal; *B*, inside of vacuum wall; *C*, copper lead-in; *D* and *E*, bakelite or plexiglass; *F*, knurled locking ring; *G*, vacuum-tight soldered joint; *O*, O-ring. For more insulation, the O-ring groove could be eliminated. In this case, a loose-fitting Teflon ring could be used to hold the O-ring in place. It is claimed that this type of seal will handle 325 amp, with a minimum breakdown voltage of 360 v at a pressure of 0.1 Torr. This is for a ½ in. diameter copper lead-in and a 0.130 in. thick O-ring. Porcelain-metal seals that can be sealed with rubber gaskets to the system are available commercially. Alternatively, soldering or brazing could be used.

As the operating voltage is increased, the quality of insulator as well as the spacing of leads (for multiple seals) from each other and from the surroundings must be considered. Ceramics (particularly porcelain), glass, quartz, artificial sapphire, and various special insulators are satisfactory. Methods of bonding metal to ceramics and the use of "liquid" gold or platinum for glass or ceramics have been discussed in Chapter 10. In that same chapter, methods of making seals with tungsten, platinum, Dumet metal, Kovar, and Fernico, as well as Housekeeper seals, were discussed briefly. In general, it is more economical (considering materials and time) to use commercially available insulated seals, unless the requirement is very special.

Seals in a wide assortment of sizes and shapes of the Kovar, Housekeeper, iron-nickel alloy (Dumet metal, Fernico, etc.) types are available commercially. Small seals using iron-nickel alloys are particularly useful in a vacuum laboratory. Such seals are often marketed under the terms *hermetic seals* and *compression seals*. Many of these seals are made for soldering (or brazing) directly to standard vacuum (electronic) tube envelopes by providing a suitable metal body. A single terminal seal of this type is shown in Fig. 12.11b. This particular seal shows a straight lead (terminal) and a flanged body. Figure 12.11c shows a multiple terminal hermetic seal with a plain body and pierced terminals while Fig. 12.11*d* shows this type of seal with a skirted body and hooked terminals. Other body and terminal types are available. Seals with more than twenty terminals are also available. Many of these seals are rated at 2000 v (or more). The size of lead used is usually 60 mils or less and this limits the amount of current

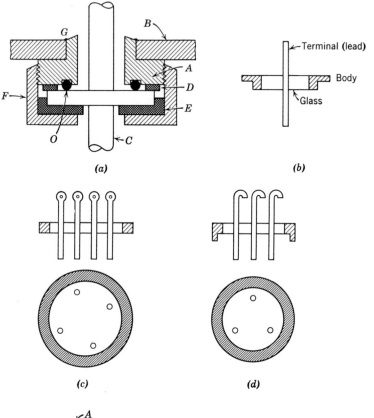

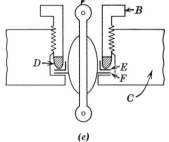

Fig. 12.11 (*a*) Form of insulated seal using one O-ring. (*b*) Hermetic seal with single lead. (*c*) Multiple terminal hermetic seal (plain body). (*d*) Multiple terminal hermetic seal (skirted body). (*e*) Demountable Kovar seal.

that can be handled. However, several amperes can be handled at lower voltages.

Although Kovar and nickel-iron alloy seals are usually soldered or brazed directly to the system, sometimes it is desirable to make a demountable seal. A method of making a demountable Kovar-Pyrex seal has been described by DeVilliers (*Rev. Sci. Instr.*, **29**, 527, 1958). This design was developed to avoid the use of rubber, which was replaced by aluminum foil. A Stupakoff Kovar seal, Type A (Stupakoff Ceramic and Manufacturing Company, Latrobe, Pa.) was used, although the method would work with other types of seals. The design of this type of demountable seal is shown in Fig. 12.11*e*. The Kovar seal *A* is provided with a metal skirt *F*. Nut *B* is screwed into the base plate *C* to compress a specially shaped ring *D* against an aluminum foil gasket *E* (0.01 in. thick) and the skirt so as to obtain a vacuum seal. The ring *D* was shaped in such a way as to keep the glass from cracking and to provide a small area for high pressures. It is claimed that such a seal can be used up to the softening point of glass and at 4000 v dc.

Insulated seals for most purposes can be obtained commercially. However, where special types of seals are needed, the fabrication methods described in Chapter 10 can be used. Also, many special types of seals have been described in the literature. Grover (*Rev. Sci. Instr.*, **31**, 349, 1960) has described a method of making Housekeeper seals with platinum rather than copper. The problem here is a matter of coefficients of expansion. In many cases, tungsten leads are continued into a vacuum system by means of liquid platinizers. Certain difficulties with this method led Wieder and Smith (*Rev. Sci. Instr.*, **29**, 794, 1958) to develop a special type of seal. Often, spark-plug-type fittings of laminated mica construction are useful for carrying electrical leads into a vacuum system. Such fittings are usually threaded into the system, a seal being effected by the use of solder or soft copper gaskets. Wade (*British report AERE R/R 2271*) has described several types of seals that involve joining suitable metals to quartz, ceramics, and artificial sapphire. Wade gives details of the methods involved.

12.8 Leak Devices and Cut-Offs

In many cases it is necessary to introduce a specific gas, such as oxygen, hydrogen, or helium, into a vacuum system. Of course, a needle valve can be used to introduce any gas. However, there are limitations to needle valves, particularly with respect to sensitivity. Consequently, many special devices have been developed, some of which are available commercially. Standard leaks are available that have a sensitivity of around 75 µl/hr (or less). Some commercial types make use of a carefully flattened copper (or German silver) tubing, the degree of flattening determining the sensitivity. The principle of such a leak is shown in Fig. 12.12a. The tubing is flattened at F. The clamp C can be used to adjust the size of leak. This is the simplest form of such a leak. A leak can also be made by clamping small rubber tubing, preferably with a wire inside the tubing.

A type of leak device found in vacuum plants and laboratories is a standard leak for calibrating helium leak detectors. Such devices admit helium to the system at a known rate. The most common form makes use of a small helium reservoir with a thin section of glass or quartz. When the leak is put in the system and the system is pumped down so that there is vacuum on one side of the glass or quartz, helium will pass into the system. The sensitivity will depend on the area and thickness of the thin glass or quartz section, as well as on the pressure in the helium reservoir. The elements of a typical helium leak of this type are shown in Fig. 12.12b. The housing H (usually brass) has a thin glass or quartz section W at one end (sealed appropriately) and a capped tube R at the other end. R can be used to refill the reservoir. V is a valve that can be closed to cut off the flow of helium into the system when this is desired. The section C can be made to fit a standard vacuum coupling. Commercial helium leaks of this type have flow rates as low as 10^{-10} std (atmospheric) cc/sec. The rate of leak of helium through the glass section will depend on the glass used (see Chapter 8). Ordinary borosilicate glass, e.g., Corning 7740, is much more permeable to helium than various other glasses.

It is sometimes necessary to introduce pure oxygen into a vacuum system at a controlled rate. The usual way of doing this is by means of a silver tube which is heated somewhere between 500 and 800°C. One form of such a device is shown in Fig. 12.12c (Whetten and Joung, *Rev. Sci. Instr.*, **30**, 472, 1959). The silver tube had a length of 6 in., an ID of 0.125 in., and a wall thickness of 0.010 in. It was attached to a glass vacuum system through the Fernico-glass seal F. The tube was heated by a spiral of 20 mil Nichrome wire, ½ in. in diameter and with six turns to the inch (H). The spiral was connected by 15 mil tungsten hooks C, connected through insulating glass beads to two 60 mil nickel rods R. These two rods made electrical connections to the Nichrome. The third lead L of a three-lead stem was attached to a

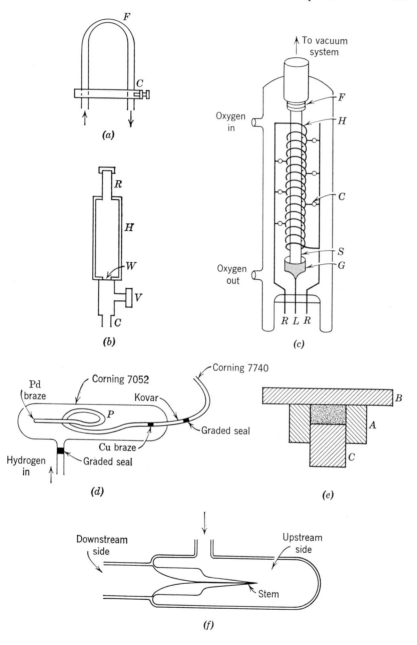

Fig. 12.12 (*a*) Leak with flattened metal tubing. (*b*) Helium standard leak. (*c*) Silver-oxygen leak. (*d*) Palladium-hydrogen leak. (*e*) Leak using powdered solder. (*f*) Special form of capillary leak.

Fernico cup G that fitted loosely over the end of the silver tube to provide mechanical support. The end of the tube was sealed off by melting the silver. The whole assembly could be baked with the rest of the apparatus. The tungsten hooks could be replaced by a less oxidizable material. The flow of oxygen through the silver tube could be controlled by changing the temperature of the tube, the flow being greater at higher temperatures. The flow is essentially zero at room temperature.

Many heated metals can be used to introduce hydrogen (or deuterium) into a vacuum system, including nickel, platinum, palladium, copper, iron, and molybdenum. However, palladium is most commonly used since, for the same flow rate, it is possible to use lower temperatures than for the other metals. Temperatures between 70 and 800°C or higher can be used. However, the palladium should not be exposed to hydrogen much below 150°C because of the formation of the hydride phase. Although the rates of flow of hydrogen through palladium quoted by various investigators differ considerably, a value of around 10 μl/cm^2/mm thickness at 300°C is fairly representative. The arrangement of Fig. 12.12c could also be used for hydrogen, when a palladium tube is used instead of a silver one. Many versions of this arrangement have been reported. The heater could be installed inside a palladium (or silver) thimble or the heating could be done by external heaters. An example of such a device is shown in Fig. 12.12d (Katz and Gulbransen, *Rev. Sci. Instr.*, **31**, 615, 1960). The palladium tube P was 11 in. long with a 0.125 in. OD and 0.061 in. ID and was 99.5+% pure, according to the supplier. In making the device, contamination was kept to a minimum—gloves were even used in bending the tubing. The copper (Cu) and palladium (Pd) brazes were made in hydrogen. The external heating was carried out by means of a Nichrome tape winding. Temperatures between 213 and 379°C were applied. The hydrogen was normally supplied at atmospheric pressure. Some workers in the field prefer nickel to palladium, in spite of the higher temperature required for a given flow rate. Besides admitting oxygen, hydrogen, or deuterium at given flow rates, these devices can be used to obtain the three gases in pure form.

Of the above devices, only the flattened metal tubing and the pinched rubber hose are applicable to admitting any gas. However, other devices to admit any gas to a vacuum system at a controlled rate have been reported. It has already been noted that needle valves can be used for this purpose, greater sensitivity being obtained by using two of them in series. Both needle valves and adjustable flattened metal capillary tubes require frequent calibration and may become plugged.

Many devices have been described that make use of gas diffusion through a solid medium. Of course, the oxygen, hydrogen, and helium leaks described above are of this type. The fact that such leaks are often selective to gases is sometimes cited as an objection. However, some types will pass several gases. An example is shown in Fig. 12.12e (Jenkins, *J. Sci. Instr.*, **35**, 428, 1958). This leak makes use of powdered solder. The solder is put in a pre-tinned hole in a mild steel block *A*, one end of the hole being covered by a flat steel plate *B*. A silver-steel rod *C* is inserted in the other end of the hole, with a sliding fit, and the whole assembly is compressed. The solder powder cold welds to itself and to the tinned bore of the hole. After compression the flat plate and rod are removed, leaving the end of the hole filled with the compacted solder powder. Suitable solder powder can be obtained by washing out from a proprietary soldering paste. The pressure necessary with a 4 mm hole can be achieved in an ordinary engineer's vise. The leak size will depend on the pressure, and reproducible leaks could be obtained by using a pressure indicator. These leaks could be attached to the vacuum system by low melting point vacuum wax or cold setting epoxy resin. An alternative method, using a glass-metal tubular seal, is described by Jenkins. This type of leak is claimed to give leak rates as low as 10^{-10} std cc/sec, for a solder compact 4 mm in diameter and 1 mm thick. Of course, porous materials other than powdered solder could be used. Almond (*J. Sci. Instr.*, **35**, 70, 1958) describes a leak using a sintered disk, 1 cm in diameter. Flow rates are not given.

Several capillary leaks have been described that depend on different expansion rates of a plug and the body in which the plug is inserted. An adjustable leak is obtained by varying the temperature of the plug. Smither, however, (*Rev. Sci. Instr.*, **27**, 964, 1956) describes a form of this type of leak which uses no plug. A glass capillary, with an external heater, was used in his work. As the capillary was heated, the viscosity of the gas increased while its density decreased. This resulted in a decreased flow rate. Devices depending on the differential expansion of a plug and the housing, or on the change in physical properties of the gas, do not give very low flow rates. A method of constructing small fixed leaks with predictable flow rates, starting with Pyrex capillary tubing, is described by Gordon (*Rev. Sci. Instr.*, **29**, 501, 1958). The form of such a leak is shown in Fig. 12.12f. It is claimed that such leaks can be made with flow rates as low as 10^{-10} std cc/sec.

Sometimes it is necessary to admit known quantities of gas into a system. This is often the case with studies in gas (or vapor phase) chromatography. Most devices used for this purpose trap a known volume of gas at a known pressure (usually atmospheric), which is then admitted to the system. Stopcocks (or other forms of valves) and mercury cut-offs are often used for control purposes. Standard techniques (see A. Weissberger, *Technique of Organic Chemistry*, Interscience Publishers, Inc., New York, 1957, Vol. III, 2nd ed. Part II, "Laboratory Engineering") will introduce volumes between 2 and 25 cc at atmospheric pressure. Various methods have been devised for introducing smaller volumes. Littman and Berkowitz-Mattuck (*Rev. Sci. Instr.*, **32**, 1154, 1961) have described a method using calibrated capillary tubing with stopcocks and mercury, which is claimed to handle volumes as small as 0.05 cc.

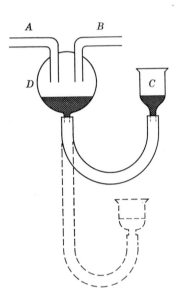

Fig. 12.13 Simple mercury cut-off.

Cut-offs are simply devices to cut off the flow of gas when so desired. In a sense, any valve that can be used to stop gas flow is a cut-off. Many other types of cut-offs have been developed for various purposes. Perhaps the simplest type is a mercury cut-off. Its principle of operation is shown in Fig. 12.13. The tubes *A* and *B* connect different parts of a vacuum system which are to be isolated from each other at times. The parts are isolated when mercury covers the ends of the tubes. In the method shown, the mercury level is controlled by movement of the open bulb *C* which is connected by flexible rubber (or plastic) hose to bulb *D*. When the cut-off is open and the system is evacuated, bulb *C* must be lowered so that the difference in levels of mercury in *C* and *D* is determined by the atmospheric pressure. Many variations of this design have been used. Often an auxiliary pump is employed to control the level of the mercury. In most cases either tubing or stopcocks are used, which contribute vapor to the system. In many studies it is necessary to eliminate all traces of vapors from greases. This has led to various forms of greaseless cut-offs. Miller, Pritchard, and Weston (*Rev. Sci. Instr.*, **31**, 466, 1960) have described a form of greaseless cut-off using floats and a sintered disk.

12.9   Protective Devices

Devices to protect pumps or gauges are fairly common in vacuum practice. Of course, special equipment may also be protected from damage, but little attention will be paid to this matter here because of the wide variety of such equipment. There are two causes of possible damage to diffusion pumps:

1. A large rise in pressure, which could damage the pump fluid.

2. A loss of cooling water or a drastic reduction in the rate of flow of the cooling water.

In case 1, oil or oil decomposition products could get into the system, necessitating a time-consuming and costly clean-up job. The pump would also have to be cleaned. The same general situation prevails with mercury, although mercury does not decompose. However, mercury does oxidize. This type of situation is usually avoided by using a vacuum gauge to close the high vacuum valve above the diffusion pump. Many commercial gauges can be obtained with a relay (often a meter relay) so that when the pressure reaches a preset value the relay closes (or opens). The relay is used to close the high vacuum valve. This is readily done with pneumatically controlled valves. A latching relay could be used so that the valve could only be opened after unlatching the relay. A thermocouple or Pirani gauge, equipped with a relay, is adequate to protect the pumps since a pressure rise to several hundred microns for a very short time will not hurt the pumps. Sometimes it is desirable to keep oil from the mechanical pumps from getting into the diffusion pumps in case of power failure. This can be done by using an isolation valve together with an air admittance valve. In case of power failure the isolation valve closes, thus isolating the mechanical pumps from the diffusion pumps. At the same time the air admittance valve opens to atmosphere. A time delay relay can be incorporated in the circuit so that when power comes on again, the mechanical pumps operate for a few seconds before the isolation valve is opened. Of course, the air admittance valve is closed as soon as power comes on. Sometimes it is also necessary to close the high vacuum valve above the diffusion pump.

The problem of protecting diffusion pumps against a sharp drop in cooling water flow or complete stoppage of the flow is more difficult. Many commercial devices operate on the basis of the pressure drop across an orifice. The pressure drop is adjusted to a value corresponding to the proper rate of flow. When the pressure drop becomes less than a particular range (to take care of minor pressure fluctuations), the device can be used to actuate a relay which cuts off power to the diffusion pumps. Microswitches that are actuated by a bellows, which in turn is controlled by the pressure differential, could be used to control the relay. The trouble with such an orifice is that it can become clogged by dirt or by corrosion. Consequently, the pressure drop would remain at an apparently safe value even when the flow rate was inadequate. A form of bellows-operated switch is shown in Fig. 12.14*a* (Preston, *J. Sci. Instr.*, **36**, 98, 1959). A T of metal tubing *A* was fastened to a base plate *B* and connected into the cooling line *C*. The side tube of the T was attached to a metal bellows that was closed at its other end. This closed end was free to expand under water pressure to actuate the microswitch *E*. The switch position could be adjusted through the screw drive *G*, which pivoted the switch around the single screw mounting *F*. Sensitivity of operation depended on the stiffness of the switch and bellows and the impedance of the water line outlet. The latter could be controlled by the clamp *H*, the former by adjusting screw *G*. For ¼ in. ID tubing the switch could be made to operate at flow rates between ¼ and 2 l/min, making and breaking with a sensitivity of 0.1 l/min.

Figure 12.14*b* shows a device which is claimed to eliminate difficulties with clogged orifices (Burford, *J. Sci. Instr.*, **37**, 490, 1960). This device basically works on the principle of an orifice with a variable area. The pressure drop across the orifice *A* causes the movable member *B* to move away from the needle *C*, thus enlarging the effective orifice. At constant flow, equilibrium is reached. An increase in pressure differential at the orifice causes the orifice area to increase until a new equilibrium state is reached. Solid matter (small particles) accumulating at the orifice causes an increase of orifice area and allows the obstructing material to pass through the orifice. The system then returns to the normal state. The moving member *B* actuates a microswitch *D* which controls the electric supply to the apparatus being protected (such as a diffusion pump). The amount of movement of the orifice varies with flow and also increases with any tendency to blockage. Additional microswitches can be used to indicate excessive movement by means of pilot lights. It is claimed that this device is undamaged by flow rates many times normal operating values, the opening of the orifice preventing any great increase of pressure differential.

A fairly reliable device for protecting apparatus from insufficient water cooling or failure of such cooling is sometimes neglected. This

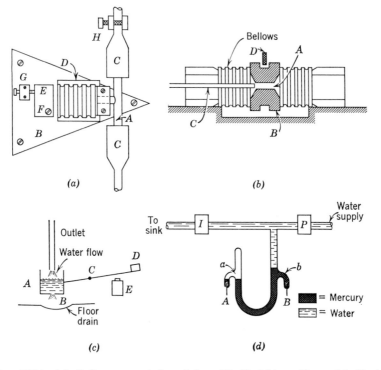

Fig. 12.14 (*a*) Bellows operated switch. (*b*) Variable orifice. (*c*) Bucket switch. (*d*) Device using mercury U-tube.

is the bucketswitch. The principle on which it operates is shown in Fig. 12.14*c*. Water from the outlet of the apparatus being cooled flows into a small bucket *A* with a hole *B* in it. The size of this hole is adjusted so that the bucket is kept about three-quarters full when the flow of water is right. The bucket is attached to a lever which is pivoted at *C*. There is also a weight *D* at the other end of the lever. When the flow of water is correct (or even too high) the bucket drops to the position shown in Fig. 12.14*c*. However, when the flow rate drops, the bucket is no longer kept nearly full, so it rises (due to weight *D*). The lever then closes (or opens) a microswitch *E* which can be used to cut off power to the diffusion pump or other apparatus being protected, either directly or through the use of another switch (relay) with a larger power capacity. Various forms of switches can be used in connection with the lever. In many cases a mercury switch is attached to the lever. When the bucket rises the switch opens (or closes) to control the electric supply. There is not too much chance that anything can go wrong with this type of protective device except clogging of the hole, which is unlikely if the design is right, and friction developing at the pivot. As with all cooling water protective devices, corrosion-resistant materials should be used in the construction. A stainless steel bucket is preferred to one made of brass or copper.

The bellows and bucket switches are most commonly used to protect apparatus against failure of the watercooling system. However, many special types of protective devices have been reported in the literature. One example is shown in Fig. 12.14*d*. The device basically consists of a U-tube with a fixed impedance I. The diffusion pump or other device being cooled is designated by P. Electrical contacts A and B (tungsten) are sealed into side arms on the U-tube. When the flow is adequate the pressure at the junction of P and I raises mercury in the closed limb of the U-tube so as to make an electrical connection between A and B. Total failure of the water supply or choking of the cooling system causes the pressure to fall. This breaks the electrical circuit at a, which disconnection can be used to cut off the diffusion pump or other apparatus. When the pressure increases due to blockage or choking at I, the electrical contact at b is broken. In the experimental form of this device the impedance I was rubber pressure tubing with a conventional screw clip. With a water head of 35 cm Hg, I was set to limit the flow to 600 cc/min. The electrical circuit was broken when the flow fell below 500 cc/min. A large capacitor across a relay coil delayed action for about 3 sec to avoid pressure transients. It should be noted that a thermoswitch on the barrel of the diffusion pump is sometimes used as a protective device. When the temperature of the pump rises above a preset value (too little cooling water) power is cut off by the thermoswitch.

In many cases it is necessary to protect vacuum gauges against a sudden rise in pressure. This is particularly true of thermionic ionization gauges which do not have non-burnout filaments. A common method of protection is to incorporate a sensitive relay in the output circuit of the gauge. Often a meter relay is used to indicate pressure readings, which can be used for information purposes and also for protection. However, there is an intrinsic time lag between a pressure change and its registering on the meter, so damage to the gauge tube can still occur in the case of a rapid pressure change. Special circuits for protecting thermionic ionization gauges have been reported. However, damage can be minimized by following good vacuum practice or by using non-burnout gauge tubes. Most cold cathode, Pirani and thermocouple gauges withstand atmospheric pressure and need not be protected.

THIRTEEN

Conventional Vacuum Systems

13.1 The Nature of Such Systems

These are the common systems found in industry and in the laboratory. Their design is such that the system as a whole (sometimes exclusive of pumps) cannot be baked, i.e., raised to high temperatures
to remove gases and vapors from the inside walls. Now the lowest
pressures cannot be obtained without removing these gases and vapors,
since they will gradually be released and will contribute to the pressure.
But, as is pointed out in Chapter 14, it is possible to keep such gases
and vapors in the walls by using low temperatures (cryogenic pumping). Conventional vacuum systems cover a pressure range from
near atmospheric to around 10^{-8} Torr, the particular operating pressure being determined by the types of pumps and techniques used. The
lowest pressures attainable without controlling the gases from the
walls are determined partly by the nature of these gases and partly by
the contribution of vapor pressure by the pump fluid. Pump design
and the use of cold traps are the common methods used to try to
minimize the effect of these gases and the pump fluid vapor. The
lowest pressures have been obtained using diffusion pumps (mercury
or oil) backed by rotary oil-sealed pumps.

Emphasis is placed here on conventional systems operating in the pressure range from a few microns down to 10^{-6} Torr or somewhat less. Mechanical pumps (usually rotary oil-sealed) are commonly used to obtain pressures in the micron range. However, in various applications Roots pumps or steam ejector pumps are used. The choice of pumps depends on the vacuum process involved and on the size of the system. As an example, where large quantities of water vapor are involved it might be possible to use a rotary ballast pump. Primary attention is given to rotary pumps and diffusion pumps, which are most commonly used. As will be recalled from Chapter 3, rotary pumps may be single-stage or compound. A single-stage pump will achieve pressures of a few microns while a compound pump will give pressures around 10^{-4} Torr or less. The actual pressures obtained will depend, of course, on the particular pump being used. It should be pointed out that the vapor pressure of the oil used in the pump is a limiting factor. Unless this vapor is trapped the pressure will not reach the lowest value often claimed by the manufacturer.

13.2 Basic Design Considerations

The principal factors involved in setting up a vacuum system are:

1. The operating pressure.
2. The pump-down time.

These factors determine the choice of pumps and various accessories. The operating pressure is generally determined by the particular vacuum process that is involved. Obtaining lower operating pressures means using greater care in fabrication and cleaning techniques. The pump-down time is determined by four factors: the volume being pumped, the impedance of connecting lines, the amount of gas and vapor being released, and the speed of the pumps. For a given pump-down time, the pumping speed will increase as the volume increases. As was shown in Chapter 3, the pumping speed used in calculating pump-down time is that at the vacuum vessel. The actual speed of the pumps must take into account losses in connecting lines. This was considered to some extent in Chapter 3 for round pipes or tubes with a single valve and elbows. Some rules in this connection are summarized in Section 13.3. The effect of baffles and traps on the design of systems is also discussed in this section. Actually, the impedances of baffles and traps have already been discussed in Chapter 11. The amount of gas and vapor being released from the walls is a difficult factor to estimate and reliance usually has to be placed on experience. The pump-down time is not particularly dependent on vapors until the pressure of the system approaches the pressures of these vapors. It should be pointed out here that the characteristics of the pumps must be such that various quantities of gas per unit time (throughput) at different pressures can be handled. Furthermore, the pumps must be capable of handling bursts of gases and vapors. Prime attention will be given here to systems that have been cleaned according to the techniques discussed in Chapter 9, which involve vacuum materials with relatively low vapor pressures and which are used in processes involving no large amounts of gas or vapor.

13.3 Conductances of Components

Some general rules for estimating the conductances of various components will be summarized here. These values can be used in designing general purpose vacuum systems. The basic components commonly used in vacuum systems are: piping or tubing, valves, water-cooled baffles, and liquid nitrogen traps. Actually it is the effect of these components on pumping speed that is of concern. It is assumed that the pumping speed at the vacuum vessel has been found by using the methods discussed in Chapter 3. This is the speed required of the forepumps to bring the pressure down to a value where the diffusion pumps can be turned on. It can be assumed that the flow is viscous in this pressure range. After the diffusion pumps are operating, the flow can be assumed to be molecular. With larger systems it may be necessary to consider the type of flow as related to pipe size.

Some general rules regarding the conductances and effects on pumping speed of various components are as follows:

1. *Piping or Tubing.* Use Fig. 3.7 to obtain the molecular conductance and eq. 3.1 to obtain the viscous conductance. The required pumping speed for the forepumps is obtained from Fig. 2.10 (using the viscous conductance) while the required speed for the diffusion pumps is obtained using the molecular conductance, plus all other conductances in the system.

2. *Elbows.* Use an equivalent length of straight pipe.

3. *Valves.*

a. *Globe or Disk Valves.* Consider as a length of straight pipe five times the dimension of the valve in the pipe. This is a safe figure for valves that have essentially the pipe area throughout their flow region, in spite of a curved flow pattern. Such valves are generally used in the viscous flow region.

b. Gate, Plug, and Ball Valves. Consider as a length of straight pipe equal to the dimension of the valve in the pipe. Generally the molecular conductance is of concern.

4. *Baffles.*

a. Baffle Plates. Use the values given in Table 11.1A. A pumping speed loss of 35% is usually a safe figure to use for such baffles.

b. Chevron and Multiple-Ring Baffles. Use a pumping speed loss of 40%.

5. *Cold Traps.* Table 11.2A can be used to obtain the conductances of thimble traps of the types shown in Figs. 11.3a, b, and c. In general, a pumping speed loss of 50% is safe for this type of trap.

The above rules concerning baffles and traps are quite general and cannot be applied to the design of other components. Here it is necessary to make measurements. Another source of loss of pumping speed is the entrance to piping at the vacuum vessel. However, this effect is taken into consideration in the curves of Fig. 3.7 for large vessels. The overall loss of pumping speed between diffusion pumps and vacuum vessel is due to all conductances involved. Often these include piping, a valve, a baffle, and a liquid nitrogen trap. The total conductance of this combination is given by

$$\frac{1}{U} = \frac{1}{U_P} + \frac{1}{U_B} + \frac{1}{U_T} + \frac{1}{U_V} \qquad (13.1)$$

where U = total conductance
U_P = pipe conductance
U_B = baffle conductance
U_T = trap conductance
U_V = valve conductance

An example of finding the total conductance is given in Section 13.4. The required pumping speed in terms of that needed at the vessel is then found from eq. 2.9 (or from Fig. 2.10).

13.4 Designing a Vacuum System

The specific design of a vacuum system will depend, of course, on its function. However, there are some rules that are generally applicable. These are:

1. Choose a size of vessel appropriate to the vacuum process. Too large a vessel will lead to unnecessarily large pumps.

2. Choose a convenient pump-down time, i.e., the time required to bring the system down to a pressure at which the process can be carried out or at which the diffusion pumps can be turned on. This time is not important where the system is pumped continually. However, it is quite important where the system is cycled.

3. From the volume of the vacuum vessel and the pump-down time calculate the pumping speed required at the vessel.

4. Keep the length of roughing line to a minimum. Knowing this length, choose a diameter that will result in less than a 20% loss of pumping speed. Valves and elbows in the roughing line will have to be considered.

5. Choose an appropriate roughing pump. If this pump is to serve as a backing pump (forepump) then it cannot be chosen until the characteristics of the diffusion pumps to be used are known.

6. Estimate the required pumping speed at the vessel for the pressure at which the vacuum process is to be operated. This is sometimes quite difficult since it depends on the quantity of gas and vapor being released. When this quantity (the throughput) is known, it is easy to obtain the required pumping speed since it is simply the throughput divided by the operating pressure. The required pumping speed can be estimated for clean, leak-free systems and this is discussed below.

7. Calculate or estimate the conductance of all components between vacuum vessel and diffusion pumps.

8. Using the above conductance and the pumping speed at the vessel, calculate the required speed of the diffusion pump. The diffusion pump can then be chosen, using a safety factor on speed to take care of gas bursts and with characteristics such that it can handle the gas throughput throughout its operating range.

9. The backing pump must be able to handle the gas throughput and to maintain a tolerable forepressure.

EXAMPLE 1

A general purpose vacuum system is to be constructed and it is determined that a 14 in. diameter (ID) bell jar will give about the right volume for the experiments being considered. A commercial bell jar is available which is 24 in. high (inside) and has a hemispherical top (see Fig. 13.1). The total volume is simply the volume of the cylinder up to the hemisphere plus the volume of this hemisphere, or

$$\pi \times 7^2 \times (24 - 7) + \frac{1}{2} \times \frac{4}{3} \pi \times 7^3 = \pi \times 49 \times 17 + 2\pi \times 114$$

$$= 3331 \text{ in.}^3$$

$$= \frac{3331}{1728} = \text{about } 1.9 \text{ ft}^3 \text{ or } 55 \text{ l.}$$

A reasonable pump-down time is taken to be 10 min. This is the time to pump from atmospheric to 20 μ pressure, which is assumed to be the pressure at which the diffusion pump can be used. It is assumed that this is a clean system so the method outlined in Section 3.6 can be used. The pumping speed at first is given by $(2.3 \times 1.9)/10 = 0.44$ cfm. To take care of the final pressure (20μ) this must be multiplied by some factor, which can be estimated to be about 5 (from Section 3.6). The required pumping speed is then

$$5 \times 0.44 = 2.2 \text{ cfm}$$

Allowance must still be made for the pumping loss in the roughing line. A maximum pumping speed loss of 20% is allowable in the roughing line. The pump speed, S_p, required is then given by

$$S_p - \frac{20}{100} S_p = 2.2$$

or

$$S_p = 2.75 \text{ cfm}$$

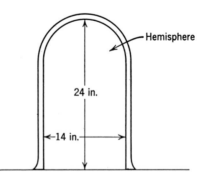

Fig. 13.1   Typical bell jar.

The conductance of the roughing line can be estimated from Fig. 2.10 as about 10 cfm. A more accurate value can be obtained by using eq. 2.9 and solving for U. Then

$$2.2 = \frac{2.75U}{2.75 + U}$$

or

$$6.05 + 2.2U = 2.75U$$

and

$$0.55U = 6.05$$

which gives

$$U = 11 \text{ cfm}$$

This is an average value of viscous conductance which increases as the pressure increases. Therefore, the pumping speed loss is considerably less than 20% until the micron region is approached.

Suppose the roughing line is 6 ft long and a globe valve is included. Take the length of this valve in the line as 5 in., which is fairly typical of such valves for a $1\frac{1}{2}$ or 2 in. line. This makes the effective length of the roughing line $6 + 5 \times \frac{5}{12}$ = about 8 ft. The diameter of pipe to use must now be determined. For viscous conductance the following formula can be used:

$$U = 2.98 \frac{D^4}{L} \bar{P} \tag{13.2}$$

where D is the diameter in inches, L is the length in inches, and $\bar{P}$ is the average pressure in the pipe in microns. The conductance, U, comes out in liters per second. To convert to cubic feet per minute, multiply by 2.12. First try a 1 in. pipe. The ID of such pipe is 1.049 in. Using eq. 13.2,

$$U = 2.98 \times \frac{(1.049)^4}{8 \times 12} \times 20 = 0.76 \text{ l/sec} = 1.6 \text{ cfm}$$

The average value of the pressure is taken as 20 μ. Actually it is somewhat higher than this, which would lead to a somewhat larger conductance. However, the above value is very much smaller than the required 11 cfm. The average pressure has to be about 140 μ to obtain this value. It will be safer to use a larger size of pipe, say $1\frac{1}{2}$ in. The ID of this pipe is 1.61 in. The conductance for this pipe comes out to be 4.1 l/sec = 8.7 cfm. This is still below the 11 cfm required.

However, as a safety factor, a pump with higher capacity than the 2.75 cfm required would be used, say twice this value or about 5 cfm. This will take care of the volume of air that must be pumped from the roughing line. Where the same pump is used to pump air out of the diffusion pump and other components this speed should still be adequate. Of course, this volume could have been estimated and added to the bell jar volume in the previous considerations. It must also be kept in mind that, with this type of system, liquid nitrogen is normally added to the trap when the pressure reaches around 100 μ. Doing this will bring the pressure down rapidly since there is always some water vapor present, even in a thoroughly cleaned system, unless it is baked.

Assume that the system includes a water-cooled baffle of the chevron type and liquid nitrogen thimble of the type shown in Fig. 11.3. A gate valve is also included between the vacuum chamber and the diffusion pump. The necessary piping is kept to a minimum. *For a clean, dry system the required pumping speed at the chamber, in liters per second, can be taken to be numerically equal to the volume of the chamber, in liters.* Consequently, the required pumping speed at the chamber is 55 l/sec. For a system of this type, involving a chevron baffle, a gate valve, a thimble-type liquid nitrogen trap, and minimum piping, the diffusion pump speed can be taken to be four times that at the chamber. Consequently, the required diffusion pump speed is $4 \times 55 = 220$ l/sec. Several commercial 4 in. oil diffusion pumps are available that have speeds above this value over a fairly wide range of pressure. If it is desired to operate at a pressure of 10^{-6} Torr or less, it is necessary to use a multi-stage pump, generally of the fractionating type. It should be kept in mind that the liquid nitrogen trap is very effective in pumping vapors and also that the outgassing rate decreases as pumping proceeds.

The factor of 4 used above can be checked by choosing a specific baffle, trap, and valve. Suppose the bell jar is connected to the gate valve, which is followed by the trap and baffle (see Fig. 13.2). Also, suppose the total length of 4 in. piping is 20 in. and the width of the gate valve is 4 in. The total equivalent length of piping is 24 in. or 2 ft. Take a pumping speed loss of 30% for the baffle. If S_p is the diffusion pump speed then the speed at the bottom of the trap is $[(100 - 30)/100]S_p = 0.7S_p$. Take a pumping speed loss of 40% for the trap (see Table 11.2B). Then the pumping speed at the top of the trap is $[(100 - 40)/100]0.7S_p = 0.6 \times 0.7S_p = 0.42S_p$. From Fig. 3.7 the conductance of the 2 ft of 4 in. piping is about 220 l/sec. Therefore, the pumping speed at the bell jar is given by

$$S = \frac{0.42 S_p \times 220}{0.42 S_p + 220} = \frac{0.42 \times 220 \times 220}{0.42 \times 220 + 220}$$

$$= 65 \text{ l/sec}$$

The ratio of diffusion pump speed to speed at the bell jar is $220/65 = 3.4$, which is less than the factor 4.

Suppose the type MCF-300 Consolidated Vacuum Corporation pump is chosen as the diffusion pump. According to the manufacturer's data the throughput at 20μ is around 700 μl/sec and this is the maximum quantity to be handled. The limiting forepressure for this pump under full load is 230 μ. Therefore, the required speed of the forepump is $700/230 = 3$ l/sec $= 3 \times 60 = 180$ l/min $= 180/28.3 = 6.36$ cfm. Allowing a maximum pumping speed loss of 20% in the backing line results in a required pump speed of about 8 cfm ($S - \frac{1}{5}S = \frac{4}{5}S = 6.36$ and $S = 8$). Suppose the backing line is 4 ft long and includes a globe valve. From the considerations of the roughing line, $1\frac{1}{2}$ in. pipe should be adequate. This is still true even when an air admittance valve of reasonable cross section is included in the line. In order to take care of gas bursts the pump might be chosen with a speed somewhat greater than that indicated. This prevents the limiting forepressure from being exceeded. A value of 9 or 10 cfm might be used. A Kinney KS-13 or a Welch 1397B are examples of pumps that could be used. Such a pump will reduce the pump-down time considerably. Actually, a 5 cfm pump could be used for both roughing and backing if the pressure were reduced to a micron or so by the use of liquid nitrogen before using the diffusion pump. This would reduce the throughput that has to be handled by this pump (and the mechanical pump). It should be possible to reach pressures well below 10^{-7} Torr with this type of system.

EXAMPLE 2

Consider a vacuum system which has a source of gas in it, say as an ion source such as might be used in a mass spectrograph. The requirements of the system are:

1. The ion source is to operate up to 3 μ.
2. The opening between the ion source and the vacuum chamber is 0.8 cm².
3. The pressure in the vacuum chamber must be less than 10^{-4} Torr.
4. The volume of the chamber is 30 l.
5. The distance between the chamber and the diffusion pump is 3 ft.

6. The distance from diffusion pump to backing pump is 12 ft and includes a globe valve.

Assume that the diffusion pump is connected to the chamber by piping of suitable diameter (no baffle or cold trap). A straight-through valve could be included as part of the piping. Assume, further, that the chamber is roughed down by the backing pump through the diffusion pump. A conductance or pumping speed can be associated with the opening in the ion source. This is given by $S = U = 11.6A$ l/sec, where A is the area of the opening in square centimeters. The "speed" of the opening is $11.6 \times 0.8 = 9.28$ l/sec. The flow of gas (throughput) is given by the speed times the pressure: $9.28 \times 3 = 27.8$ μl/sec. This throughput must be handled by the pumps. It is assumed that outgassing of the chamber contributes a negligible amount of gas compared to the above value. The pumping speed required at the chamber is given by the throughput divided by the chamber pressure. Assume a chamber pressure of 0.1 μ (10^{-4} Torr). The speed required is then $27.8/0.1 = 278$ l/sec. Many commercial 6 in. pumps have speeds of around 500 l/sec at 10^{-4} Torr. If such a pump were used then the diameter of piping would have to be chosen so that the speed at the chamber was 278 l/sec. The conductance of the piping can be determined from eq. 2.9 by solving for U. When this is done, the expression for U is

$$U = \frac{S_m S_n}{S_m - S_n}$$

where

$$S_m = 500$$

and

$$S_n = 278$$

Then

$$U = \frac{500 \times 278}{500 - 278} = 626 \text{ l/sec}$$

This value could also be found from Fig. 2.10. Using Fig. 3.7, it will be seen that the 3 ft length of pipe must have an ID of about 7 in. Brass tubing of $7\frac{1}{4}$ in. OD (7 in. ID) would probably just be adequate. However, to take care of the entrance to the 6 in. pump (if used), which offers resistance to gas flow, it will be best to use 8 in. pipe. For more accurate values the formulas in Appendix C could be used.

The limiting forepressure for a 6 in. diffusion pump can be taken as about 100 μ. The speed required at the outlet of the diffusion pump must be the throughput divided by 100 microns, or 27.8/100 = 0.278 l/sec. The actual pressure at the outlet of the diffusion pump is reduced as the pumping speed is increased. The diameter of pipe required can be determined as in the case of Example 1. However, sometimes it is easier to calculate the conductance for a given diameter and pressure and then compare this with the pump speed. As long as the pipe conductance is at least four times the pump speed, the pipe size is adequate. Consider the use of a small backing pump with a speed of 0.5 l/sec, which is still larger than the requirement of 0.278 l/sec, allowing for 20% loss. The pressure at the pump is 27.8/0.5 = 55.6 μ. Taking the globe valve into consideration (length of 5 in.) makes the pipe length about 14 ft. First consider 1 in. pipe (ID = 1.049 in.). Then, by eq. 13.1, the conductance of the pipe is 2.98 × [$1.049^4/(14 \times 12)$] × 55.6 = 1.21 l/sec. This is only about two and a half times the pump speed so the pipe size is not adequate. Next try 1½ in. pipe (1.61 in. ID). In this case the conductance is 2.98 × [$1.61^4/(14 \times 12)$] × 55.6 = 6.7 l/sec. This is over thirteen times the pump speed so the pipe size is quite adequate. Usually a somewhat larger size of pump is used to take care of gas bursts. Suppose a pump with a speed of 1 l/sec were used, e.g., a Welch 1405. This would reduce the pressure to about 28 μ, which would drop the conductance to about 3.4 l/ sec. This figure is somewhat less than four times the pump speed but can still be considered to be adequate. It is unlikely that a gas burst will raise the pressure at the outlet of the diffusion pump above the limiting forepressure.

The only question remaining is in connection with pump-down time. The method used in Example 1 can also be used here. Suppose the vacuum chamber is pumped down through the diffusion pump and the volume of this pump and connecting lines is estimated to be 40 l. The total volume is then 40 + 30 = 70 l. The final pressure is 28 μ so a factor of about 4.8 can be used. The pump-down time is then (2.3 × 70)/60 × 4.8 = 12.9 min. This is a reasonable value.

The above examples cover only two selected cases of vacuum system design. There is so much variation in the requirements of vacuum systems that it is difficult to set down any detailed set of rules for design. However, there are some guide lines which are often useful. These are:

1. Use accepted vacuum techniques in fabrication and cleaning (see Chapters 9 and 10).

2. When the throughput is known, the pumping speed at the vacuum

chamber is simply the throughput divided by the pressure in the chamber.

3. When the throughput is not known, take the pumping speed at the vacuum chamber (in liters per second) as numerically equal to the volume of the chamber (in liters).

4. When the diffusion pump is connected to the chamber by piping alone, use a pump speed twice the speed at the chamber. This assumes the shortest possible length of piping of sufficiently large diameter.

5. When a baffle and trap are included between the diffusion pump and the chamber, use a pump speed four times the speed at the chamber. The baffles and traps used should not have such low conductances that a factor of more than 5 is needed. In many cases, actual speed measurements may have to be made.

6. Choose pumping lines of such length and diameter that their conductance is about four times the pump speed.

7. Use roughing and backing pumps with about two or three times the minimum requirement in order to allow for gas bursts.

13.5   Setting Up a New Vacuum System

After a system has been designed and assembled, certain steps are necessary to put it into operation. The type of system to be discussed here is of the general nature of the type described in Example 1, Section 13.4. It was indicated there that a multi-stage, fractionating oil diffusion pump would be used, thus making it possible to reach pressures well below 10^{-7} Torr. However, even with nonfractionating pumps and using a cold trap (liquid nitrogen) it should be possible to attain pressures of 10^{-6} Torr or somewhat better. Of course, as the desired pressure is lowered more attention must be given to the techniques of cleaning and fabrication and to the choice of materials. Using a minimum of elastomeric materials and employing special materials such as Viton A instead of synthetic rubber are always helpful. The material for the base plate is not critical in the pressure range under consideration. Brass, aluminum, mild steel, or stainless steel could be used. If a bell jar were used, it would be sealed to the base plate with an L-gasket. Alternatively the base plate could be grooved for an O-ring, although this is more expensive. If some other form of chamber were used, it could be grooved for an O-ring or the base plate could be grooved. It is usually cheaper to groove the chamber. Special forms of gaskets are available that make it possible to avoid grooving (see Chapter 12). The backing pump can be connected to the diffusion pump by tubing, piping, rubber hose, or plastic hose reinforced by a spring. The roughing line should be tubing or piping since the chamber "sees" part of this line. Sometimes an oil reservoir is included above the mechanical pump to prevent cracked oil from the diffusion pump from getting into the mechanical pump oil. If a mercury pump were used, a cold trap would be included between this pump and the mechanical pump. This is often done even with oil diffusion pumps.

In setting up the system, attention should be given to convenience of operation. It is best to mount components in a chassis of some form that can be provided with casters for mobility. The top of the chassis includes the base plate and provides some work area. Switches to start and stop the pumps and the controls for the vacuum gauges should be mounted on the front or side of the chassis for ready accessibility. Furthermore, it should be possible to control the valves from outside the chassis, either by mechanical means for hand-operated valves or by suitable switches for pneumatic or solenoid-operated valves. The liquid nitrogen trap should be mounted so that it can be cleaned and filled readily. Also, the pumps (mechanical and diffusion) should be mounted so that the oil can be changed without undue difficulty. In the case of mechanical pumps it is a good idea to have them mounted in such a way that the oil level can be inspected. Shock mounting these pumps is advisable to minimize vibration. When the chassis is pretty well enclosed, a fan should be provided to cool components. Of course, this cooling should not be such as to cool the diffusion pump too much. It is also advisable to circulate the cooling air through a filter to cut down on dust and dirt. With any general purpose vacuum system many optional features can be incorporated, including an automatic liquid nitrogen filling system, protective devices for the diffusion pumps and ionization gauges, valves to protect the mechanical pumps in case of power failure, etc. How many of these extra devices should be employed depends on the functions of the system and on budgetary considerations.

13.6   Starting Up a Vacuum System

Some typical steps that are involved in starting up a general purpose vacuum system can be summarized as follows (see Fig. 13.2):

1. Be sure the diffusion pump and mechanical pump are filled with recommended types and quantities of oil.

2. Close valves V2, V3, V4, and V5. V3 and V4 are generally types that are normally closed. V5 may be a pneumatic type.

3. Start the mechanical pump.

a. Continue pumping until the pump stops gurgling (sounds hard).

b. If the pump does not sound hard after a few minutes, look for large leaks. These may be indicated by the hissing of air. Gaskets should be checked for misalignment. At this stage it may be necessary to stop the mechanical pump, let the system down to air (open valve

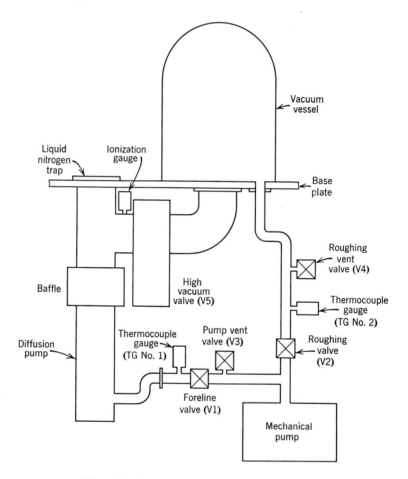

Fig. 13.2   Typical conventional vacuum system.

V3), and disassemble parts of the system between V5 and the mechanical pump to repair leaks.

4. After step 3*b* (if necessary) turn on TG No. 1 when the mechanical pump sounds hard. Within a few minutes it should read 100 μ or less. If it does not continue to drop in pressure reading, it may be necessary to do some additional leak hunting using some of the methods discussed in Chapter 15. TG No. 1 could be used for this purpose with an appropriate detecting material (acetone, etc.). If a helium leak detector is used, an appropriate flange must be provided in the foreline to which the detector is connected prior to step 3.

5. Leaks should be located and repaired so that the reading of TG No. 1 keeps going down until it reaches a few microns.

6. Close V1 and open V2 (mechanical pump running). Repeat steps 3*a* and 3*b* (as necessary). Repeat steps 4 and 5, using TG No. 2.

7. Close V2 and open V1.

8. When TG No. 1 reads a few microns, turn on the diffusion pump cooling water, if this part of the process is not automatic. For most pumps, a value between 50 and 100 μ is used. Sometimes the water is turned on when TG No. 1 reaches a preset value. At this stage water usually flows through the water-cooled baffle.

9. Turn on the diffusion pump heater.

10. If the system has protective devices in the water lines, check to be sure they operate correctly.

11. Most 4 in. oil diffusion pumps will take around an hour to get up to temperature and start pumping, the actual time depending on the design.

12. Fill the liquid nitrogen trap about *one-quarter* full. This can be done while the diffusion pump is heating up.

13. Turn on the ionization gauge when TG No. 1 reads 1 μ and stays there for 10 min or so. Many ionization gauges are designed so that the emission current can be adjusted. Start with a low emission current and increase it *very slowly* so as to avoid possible discharges. If the gauge goes off scale, turn it off and wait a few minutes, then try again. The above steps are not necessary with a cold cathode gauge.

14. Wait at least 10 min after step 9 before filling the liquid nitrogen trap completely.

15. Close V1 and open V2. Pump until TG No. 2 reads near 1 μ.

16. Close V2 and open V1.

17. Open V5.

18. Check for small leaks by appropriate leak-hunting procedures (Chapter 15).

19. After several hours of pumping, the pressure should be around 10^{-6} Torr or less, the exact time and pressure value depending on the design of the system.

Several variations of the above procedure could be used. The whole system could be pumped on at first. Then, if the pressure did not drop to a low enough value in a reasonable time, various parts of the system could be valved off for leak-hunting purposes. Also, the part of the system including the liquid nitrogen trap and diffusion pump (from V5 to V2) could be checked out for leaks first and then the diffusion pump could be heating up while the remainder of the system was checked. Systems involving the use of mercury diffusion pumps differ from the type of system shown in Fig. 13.2 largely in the trapping arrangement. A cold trap (dry ice or liquid nitrogen) is inserted in the foreline to keep oil vapor from the mechanical pump from getting into the diffusion pump. This trap also acts as a pump for other vapors, including mercury vapor, which is thereby stopped from getting into the atmosphere and the mechanical pump. Such traps are sometimes used with oil diffusion pumps. The foreline trap is normally filled (or the refrigeration system turned on) when the mechanical pump sounds hard. This trap then helps to pump vapors. The trap above the diffusion pump is filled while the pump is heating up. Although this trap can be of either the dry ice or Freon refrigeration type, the type chosen sets a lower limit on the pressures that can be obtained due to the vapor pressure of water. The vapor pressure of water at the temperature of dry ice ($-78°C$) is about 0.5 μ. To get down to pressures around 10^{-6} Torr or better, a liquid nitrogen trap must be used.

13.7   Starting and Operating Tight Vacuum Systems

Once a system has been checked out as vacuum tight the start-up procedure is relatively simple. Several of the steps listed in Section 13.6 for the system shown in Fig. 13.2 can be omitted. A recommended procedure for this type of system, starting with the system at atmospheric pressure, is:

1. Close valves V2 and V5. (It is assumed V3 and V4 are normally closed types. If not, they must be closed.)
2. Start mechanical pump and operate until it sounds hard.

3. Turn on TG No. 1. It should be on scale (for the usual range of about 1 μ to 1000 μ).

4. The pressure should drop within a matter of minutes to 100 μ or less.

5. Turn on diffusion pump cooling water, if its coming on is not automatic, when the pressure is between 50 and 100 μ, depending on the design of pump. The manufacturer's literature will generally indicate when this should be done.

6. Turn on the diffusion pump heater.

7. If the system has protective devices in the water lines, check to be sure they are operating correctly.

8. At this stage one can simply wait until the diffusion pump reaches operating temperature, which may take an hour or so for most pumps. Alternatively, if the vacuum chamber has been modified during shutdown and there is a possibility of leaks, then V1 can be closed and V2 opened. TG No. 2 indicates the state of the vacuum and if the pressure doesn't drop rapidly, an appropriate leak-hunting procedure can be followed. Once the leaks have been located and repaired, V2 is closed and V1 is opened so that the diffusion pump is again being backed.

9. When the diffusion pump reaches operating temperature and starts to "take hold" the pressure indicated on TG No. 1 will start to drop rapidly.

10. Fill the liquid nitrogen trap about *one-quarter* full and follow steps 13, 14, 15, 16, and 17 of Section 13.6. This could be done prior to step 9. Step 18 of that section may also have to be carried out for the vacuum chamber.

If the system has been shut down with essentially no changes made in the system, except for perhaps removing the vacuum chamber, then the start-up procedure is much the same as above except that the steps involved in checking for leaks can be omitted. Rather, after step 7, wait for the diffusion pump to heat up and then perform steps 9 and 10 above, followed by steps 13, 14, 15, 16, and 17 of Section 13.6 (step 18 may also be necessary). Most vacuum chambers, which may be among the items being tested, are sealed by means of a rubber gasket (O-ring, L-type gasket, etc.). Clearly, if the chamber doesn't pump down rapidly the place to look for leaks is around this gasket. In many cases the force providing the seal is due to the atmosphere; therefore, pressing the sealing surfaces together while pumping may give a tight seal. If this does not work (after checking for other leaks which may have opened up), the chamber should be removed from the system and the gasket inspected. If it is damaged (cuts, scratches, etc.) it should be discarded and a new one used, after careful inspection of both this gasket and the gasket groove or sealing surfaces. Often a light coating of good quality vacuum grease is applied to the gasket. Leaks through gasket seals are often caused by foreign particles, such as metal filings.

13.8   Shutting Down a Vacuum System

The procedure to be followed in shutting down a vacuum system will be determined to some extent by the exact nature of the system. However, there are certain points common to all systems involving the use of diffusion pumps (oil or mercury). These are:

1. Do not let the diffusion pump down to air while it is hot. Even silicone oils will deteriorate after continued exposure to the atmosphere while hot.

2. When not running, the mechanical pumps should be at atmospheric pressure.

3. Clean and dry all cold traps and re-install in the system.

4. Wherever possible let the system down to atmospheric pressure with a dry gas, e.g., air or nitrogen. This will save start-up time later.

Specific steps which might be taken in shutting down a system can be illustrated by reference to the system of Fig. 13.2. Typical steps are:

1. Close valves V2, V3, and V5.

2. Open valve V4. If possible, this valve should be connected to a source of dry air.

3. Turn off diffusion pump heaters, ionization gauge, and thermocouple gauges. *Leave the cooling water on.*

4. The pump may take as much time as an hour or so to cool, depending on the design. If one's hand can be held to the bottom of the pump without discomfort, the pump has cooled sufficiently for the next step.

5. Turn off the mechanical pump.

6. Open valve V3 (to dry gas, if possible).

7. Remove the liquid nitrogen trap, saving whatever liquid nitrogen is left, and clean and dry this trap. Put the trap back in the system.

8. Turn off the cooling water.

If a system does not have a high vacuum valve (like V5) then the diffusion pump must be allowed to cool down before the system as a whole can be let down to air. It is important that the mechanical pump be raised to atmospheric pressure to prevent oil from getting into parts of the system. It is good practice to close vent valves after they have been used. When starting up again, an open vent valve might well give the indication of a large leak. This is particularly true of needle valves. The above steps apply in general to systems using both oil and mercury diffusion pumps except that in the use of mercury pumps a foreline trap is usually involved. This trap can be treated in the same manner as the main trap (step 7 above). It is possible to shut down a system without cleaning the traps. However, as they warm up, the ice evaporates into the system, which results in a considerably longer start-up time. When the cold traps hold enough nitrogen to last overnight, or an automatic filling system is used, it is possible to just leave the system pumping. It is good practice to turn off all vacuum gauges except in cases where the main valve (between chamber and diffusion pump) is controlled by a gauge. Shutting down a system involving only mechanical pumps is a simple matter since there are no diffusion pumps to worry about.

13.9   Maintenance Problems with Vacuum Systems

Maintenance problems associated with mechanical and diffusion pumps have been discussed in Chapters 3 and 4, respectively. The actual maintenance schedule for pumps depends on the particular system in which they are used. A maintenance problem associated with any system making use of cold traps is the cleaning of these traps. The problem is—how often should traps be cleaned? No simple answer can be given since there are too many factors involved. The biggest factor is the amount of vapor being pumped by the traps, and this depends on the nature of the vacuum process, and to a lesser extent on the design of the system. Some processes involving the handling of large quantities of vapor may require that cold traps be cleaned every few hours. On the other hand, with clean systems involving little vapor, it may be necessary to clean the traps only every few days. After some experience, a regular schedule for cleaning traps can be set up. An indication that a trap is getting "iced" up is given by a rising pressure. This is due to the fact that the outer layer of ice is warmer than the refrigerant being used, because of the insulating properties of this layer. The cleaning schedule should be set up so that the ice layer doesn't become too thick. A thickness of $\frac{1}{16}$ to $\frac{1}{8}$ in. is tolerable for systems operating in the 10^{-6} Torr region. After cleaning, the trap should be thoroughly dried before being put back in the system. An electrical heating unit with a fan behind it is quite suitable for this purpose although other methods can be used.

Apart from pumps and traps, the various gaskets used in a system require some maintenance. This is particularly true of movable seals. Such seals should be inspected periodically for damage and to see if they are dry. If undamaged but dry, they should be cleaned and given a light coating of a good quality vacuum grease. Sealing surfaces and grooves should be examined at the same time. Damaged gaskets should be replaced. Parts moving through movable seals should be greased periodically. Properly prepared and installed static gaskets hardly ever give any trouble. When parts held together by gaskets are disassembled, the gaskets, seating surfaces, and grooves should be cleaned and the gaskets greased before reassembly. Any parts subject to aging, such as rubber tubing, should be periodically inspected for deterioration. A properly designed, constructed, and operated vacuum system should give rise to few maintenance problems apart from the type noted above. Of course, special features incorporated in a vacuum system can give rise to peculiar difficulties. To avoid unnecessary maintenance problems, particular attention must be given to the proper operational procedures. Accidentally letting a system down to air with the diffusion pumps hot can damage the pump fluid and can also result in getting pump fluid into the system. Cleaning up the system is a tedious and time-consuming operation. Too high a rise in pressure can burn out an ionization gauge (thermionic type), unless it is of the non-burnout variety. A protective circuit can be used to avoid this.

13.10   The Small Vacuum Laboratory

In setting up a small vacuum system (or systems) there are certain requirements that apply to the room which is to be used. This room should be of adequate size for the type of work to be performed. It should also be free of dust and dirt and should be well ventilated. In fact, the degree of ventilation must be considered carefully. Ventilation is important to minimize the hazard due to the various vapors arising from liquids commonly used in vacuum practice such as solvents and mercury. On the other hand, too much circulation of air around the vacuum system can lead to cooling of pumps with resulting deterioration of performance. At the very least, the power requirement will go up. Also, drafts around the system can lead to considerable difficulty in leak hunting.

The room should be equipped with benches and tables appropriate to the work being performed. Also adequate small tools should be available for repair and assembly. However, machine tools (lathes, drill presses, mills, etc.) should not be used in the room. There is too much danger of oil or grease getting on parts, and also, metal dust or filings can cause considerable trouble. The room should be equipped with compressed air, gas, and water supplies. Drains in the floor are very convenient. Also, adequate floor or wall electrical outlets of the proper voltage and power capacity should be provided.

The actual vacuum components that should be kept on hand often raises a question. Certainly the nature of the vacuum work will determine the exact numbers and types of components that should be stocked. However, there are certain items that are common to many systems. These include:

1. One refill of oil for each pump being used (mechanical and diffusion). The containers should be sealed to avoid water getting into the oil.

2. An assortment of O-rings. In general these will be of synthetic rubber. Special elastomers like Viton A, Teflon, etc., are expensive to carry in stock. Sometimes it is advisable to have synthetic rubber O-ring cord available so O-rings can be made up as needed.

3. A supply of cleaning solvents including acetone, MEK, carbon tetrachloride, benzene, alcohol, and ethyl ether.

4. An assortment of vacuum greases, the quality depending on the nature of the vacuum work being conducted. These greases should be kept covered when not in use.

5. A supply of waxes and cements for temporarily joining components into a system. Useful waxes are Picein and Dennison's. Common cements to stock are DeKhotinsky, bakelite cement, beeswax, glyptal, and silver chloride. Apiezon Q is useful for making temporary seals. Where O-rings are to be made up, the cements appropriate to the elastomers being used should be available. Newer cements and waxes are constantly being developed and, as they come into use in the laboratory, they should be carried in stock (after adequate testing).

6. Where mercury devices, such as McLeod gauges, are used, a supply of cp mercury should be available.

7. A supply of rubber hose and/or reinforced plastic hose for connecting pumps and other components.

8. An assortment of small valves, fittings, and electrical lead-throughs. Valves of the ⅛ in. size or needle-type valves are very useful and inexpensive. The fittings should be suitable for use on water tubing and vacuum hose as well as for other purposes. It may be necessary to carry in stock a small supply of both high current–low voltage and low current–high voltage lead-throughs, depending on the work being carried out. An assortment of hermetic seals with various numbers of terminals is useful.

9. A small supply of seals. Stuffing boxes, Wilson seals, and chevron seals are useful.

10. A supply of soft and hard solders, with appropriate fluxes, should be available. With hard soldering, a suitable source of heat, such as a gas-oxygen flame, is needed. Soldering irons and gas-air flames are needed for soft soldering.

11. When a considerable amount of glassblowing is to be carried out, a special setup should be arranged, including adequate working space, special tools, and adequate heat sources. For electronic work, an assortment of parts such as wire, resistors, capacitors, meters, tubes, etc., should be available.

12. In many cases it is advisable to have spare parts available for gauge circuits and leak detectors. These spare parts can be specified by the manufacturer.

FOURTEEN

Very High and Ultra-High Vacuum Systems

14.1 The Nature of Such Systems

In the previous chapter, conventional vacuum systems were discussed. As has been noted, such systems involve the use of a vapor pump (oil or mercury) backed by a mechanical pump (usually rotary oil-sealed). With reasonable care in choice of construction materials and in methods of cleaning parts, pressures between 10^{-6} and 10^{-7} Torr are achieved fairly routinely. By exercising extraordinary care and using suitable cold traps and low vapor pressure pump oils, pressures as low as 5×10^{-8} Torr have been obtained. But to achieve pressures much below 10^{-6} Torr different techniques are generally chosen. The principal modification involves driving out gases and vapors from internal surfaces by heating (bake-out procedure). The bake-out temperature is determined by the construction materials. Since the higher the temperature, the faster gases and vapors are driven off, rubber gaskets, mercury cut-offs, and standard stopcocks are not generally used. Sometimes, however, metal and glass parts are baked while elastomeric gaskets are cooled (usually with tap water) to protect them. The types of pumps and components needed to achieve very low pressures are covered in previous chapters. A temperature around 450°C is commonly used to bake hard glass and certain metal

433

parts. Stainless steel is commonly used. Brass should not be used because of the high vapor pressure of zinc at elevated temperatures.

A somewhat arbitrary division of vacuum systems between the very high vacuum region (10^{-6} to 10^{-9} Torr) and the ultra-high vacuum region (10^{-9} to 10^{-11} Torr or less) is made here. Actually, several of the techniques and components are common to both regions. The bake-out procedure in some form is used with most systems. In some ways a division between small (less than a few liters) and large (up to a thousand liters or more) might be more appropriate. Of course, it is not possible to arbitrarily divide techniques between small and large systems. The lowest pressures (less than 10^{-11} Torr) are obtained more readily with small systems. This is primarily due to the fact that such systems can be baked out quite effectively. One general comment can be made regarding pumps used to obtain very low pressures. Vapor pumps (oil or mercury), with mechanical backing pumps and suitable cold traps, combined with a bake-out procedure, can be used to obtain very high vacuum conditions, but ultra-high vacuum conditions are only obtained with great difficulty. The limiting pressure of such a system is determined by backstreaming of pump fluid and vapors from the pump. Although this can be minimized through proper design of cold traps and vapor pump it can never be entirely eliminated. The general procedure for attaining ultra-high vacuum in small systems is to use conventional pumping systems during the bake-out procedure, and then achieve the final low pressure by the use of another form of pump, usually a getter-ion pump of the sputter-ion type.

14.2   Small Very High Vacuum Systems

The vapor pump–mechanical pump combination will first be discussed since vapor pumps are capable of handling large quantities of gas (throughputs) over a wide pressure range. The basic arrangement can be used with various sizes of systems although primary attention is given here to small systems. Constructional materials are either a hard glass (such as Pyrex) for small systems or a stainless steel for large systems. Other metals, such as mild steel, can be used as long as they will stand the baking temperature (around 450°C) without damage and they do not produce too high a vapor pressure. Appropriate valves, seals, etc., which are bakable, are used. The pumps are made of hard glass or a steel (preferably stainless). Multi-stage pumps are commonly used. For a vacuum chamber around 10 l in size, a pumping speed of 60 to 120 l/sec is adequate. A good three-stage fractionating pump, with adequate trapping and baking, can produce a pressure of 10^{-9} Torr against a backing pressure of 10^{-1} Torr. The higher the pressure drop across the vapor pump, the greater will be the backstreaming of pump fluid. To minimize this backstreaming the backing pressure can be reduced by using a vapor pump between the mechanical backing pump and the main vapor pump. A single-stage mercury pump is sometimes employed for this purpose.

A two-stage rotary oil-sealed pump with a nominal speed between 50 and 100 l/min and with an ultimate pressure less than 10^{-3} Torr is adequate for small systems. To avoid oil from the rotary pump getting into the system, a liquid nitrogen trap is used between this pump and the second vapor pump. Sometimes a reservoir (about 3 l or so for a 10 l chamber) is included between the rotary pump and the second vapor pump. When the backing pressure is below 10^{-3} Torr and there is little gas being pumped, the rotary pump can be shut off (being let down to air by an admittance valve) and valved off from the reservoir. The reservoir then acts as the backing pump and can serve this purpose for many hours. A liquid nitrogen trap is used between the first vapor pump and the vacuum chamber. This trap must be bakable. A Pirani or thermocouple gauge is useful between the trap and second vapor pump in order to get an idea of the initial rate of pump-down and also to indicate when the vapor pumps can be turned on. An ionization gauge is generally used between the trap and chamber. Since conventional gauges will be operating near their limit, it is best to use a Bayard-Alpert type of gauge. Valves on the high vacuum side of the system must be bakable. A typical system of the above type is shown in Fig. 14.1. The reservoir is shown in dotted outline since such a reservoir is often omitted with such systems. Also shown are the two diffusion pumps (DP), two liquid nitrogen (LN) traps, a water-cooled baffle, and an oven for baking the vacuum system. Re-entrant-type liquid nitrogen traps are often used. Spherical traps offer less resistance to gas flow but are less effective in trapping. Sometimes two (or three) traps are used so that one can be used to trap pump fluid vapor while the other is being baked. As shown in Fig. 14.1 the Bayard-Alpert (BA) gauge is mounted outside the portion of the system to be baked. For a volume of around 10 l, the main diffusion pump could be a 2 in. metal fractionating pump or an appropriate glass pump. The second diffusion pump is chosen to maintain a low pressure at the outside of the main diffusion pump. A pressure of about 10^{-6} Torr is often used. Valves between the second

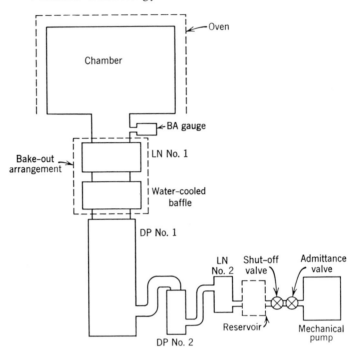

Fig. 14.1   Very high vacuum system.

diffusion pump and the rotary pump could be glass stopcocks or metal valves. On the high vacuum side any valves used must be bakable. Connecting lines should be as short and as of large diameter as possible, with a minimum of bends. For a 10 l volume the lines should be at least 1 in. in diameter. The basic system of Fig. 14.1 can be extended to much larger systems, although new problems are then introduced (see Section 14.6).

14.3   Small Ultra-High Vacuum Systems

The limitation in lower pressure obtainable with a system of the type shown in Fig. 14.1 is due to pump fluid vapor and gases released from the top part of the diffusion pump. The designation "diffusion pump" for vapor pumps (mercury or oil) will be used from this point on. Because of its high vapor pressure compared to oils, mercury is more difficult to trap effectively. Consequently the bulk of the work in the very high and ultra-high vacuum regions has been carried out with oil diffusion pumps. The early work on small ultra-high vacuum systems, which was carried out by Alpert and co-workers at Westinghouse Research Laboratories (D. Alpert, *J. Appl. Phys.*, **24**, 860, 1953) involved the use of systems with volumes of less than about 1 l, an oil diffusion pump without a liquid nitrogen trap, and hard glass construction, except for the presence of bakable valves and glass-metal seals. This system reduces the pressure (after baking to around 450°C) to a pressure of about 10^{-8} Torr. Final pumping is carried out by means of a getter-ion pump. Alpert et al. used a Bayard-Alpert gauge for this purpose. The general features of an Alpert-type system are shown in Fig. 14.2. The main features to be noted are the use of an oven to bake the part of the system which is to be maintained in the ultra-high vacuum region (dotted lines). The various parts of the system to be baked are mounted on a frame and an oven can be

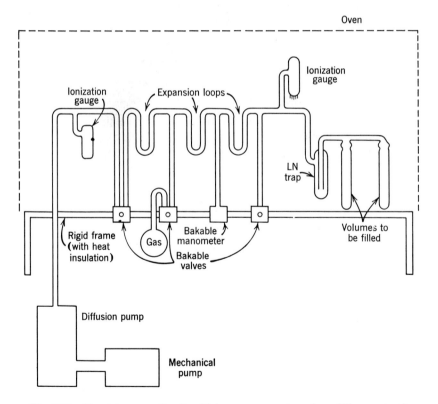

Fig. 14.2 Alpert-type small ultra-high vacuum system (gas filling system).

lowered over them for baking. The loops shown in the tubing are used to avoid damage due to the heating and cooling cycle. The main features of the Alpert design are the arrangement of the system so that ovens can be used for bake-out and the use of glass and metal parts suitable for heating to high temperatures. For the low pressure obtained, the conventional ionization gauge is, of course, useless. Many modifications of the type of system shown in Fig. 14.2 have been made by workers in the field. Both oil and mercury diffusion pumps have been used with liquid nitrogen traps. Also, several ovens are often used so that one or more of these can be removed from certain parts while other parts are still being baked. This is often done when it is desired to bake a liquid nitrogen trap and then fill it while other parts are still being baked.

The quite small systems used by Alpert and co-workers could be pumped down to very low pressures quite adequately with a Bayard-Alpert gauge. However, in making larger systems of this type, say a few liters in volume, this gauge becomes a limiting factor. This type of gauge saturates fairly rapidly with volumes of a few liters when operated at around 10^{-8} Torr. One solution to the problem is to pump the system down to around 10^{-9} Torr before turning on the gauge. This can be done with a very high vacuum system such as that shown in Fig. 14.1. An alternative method is to use a small commercial getter-ion pump of the sputter type. In any case the system and gauge (or sputter-ion pump) must be thoroughly baked before turning on the pump. Also, the pump is only turned on after the system has been sealed off from the conventional pumps. For small systems using rather small glass tubing it is relatively easy to collapse the tubing uniformly and get a good seal. However, with systems of a few liters capacity the size of tubing involved (1 in. diameter or more) introduces problems. The easiest way is to fuse a glass constriction in the connecting line. This connecting line should be installed during initial assembly and should be as short as possible. The tubing should be previously heated so as to increase its wall thickness and decrease its internal diameter (see Fig. 14.3). It is fairly easy to seal at this constriction when required. However, trying to fuse ordinary tubing 1 in. in diameter or more is very difficult. The chances are that the tubing will collapse prematurely at one spot and cause a leak. Glass gives off a lot of gas when being heated. Therefore, the constriction should be heated to near the softening point as many times as necessary to eliminate any pressure rise on fusing. Doing this three or four times is usually sufficient. A bakable metal valve is often used in both glass and metal systems. This valve has

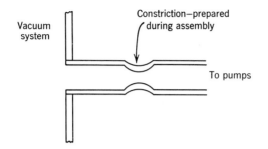

Fig. 14.3   Constriction to seal off vacuum system.

to be baked before final closing to isolate the vacuum chamber from the diffusion pump. After the pressure has reached a value of around 3×10^{-9} Torr or so (10^{-7} to 10^{-8} Torr for systems of less than 1 l in volume), the vacuum chamber should be sealed off or valved from the diffusion pump as rapidly as possible. The Bayard-Alpert gauge is then turned on at a fairly high emission current, say around 10 mamp. Pressures of 10^{-11} Torr or less are then achieved, although the final value depends on the size of vessel and the particular gauge being used. The pumping speed of such a gauge is quite low and it may take as much as a day to get down to 10^{-11} Torr with a system of several liters volume.

To decrease the time required to get down to 10^{-11} Torr, a gettering material is often used in conjunction with the Bayard-Alpert gauge. This getter is evaporated at the same time as the gauge is turned on and is very effective in pumping gases such as oxygen, carbon monoxide, and water vapor, which are commonly residual in degassed systems. The gauge then takes care of the rare gases. Titanium and molybdenum are usually preferred as getters since they have higher vapor pressures near the melting point. They are commonly used in the form of filaments or coils mounted in an auxiliary tube. An electrical resistance unit is usually used to evaporate them on the walls of the tube. The getter should be evaporated immediately after the vessel has been sealed off from the diffusion pump and until such a time that a copious deposit of material is formed on the inside of the tube. The Bayard-Alpert gauge is operated at the same time. After gettering, the pressure should reach a value of about 10^{-10} Torr. With the gauge operating, a pressure of 10^{-11} Torr is reached after a few hours of operation. It should be noted that in much of the early work on ultra-high vacuum systems, copper foil traps were used (see Chapter 11).

14.4 Bake-Out Arrangement for Small Systems

With small systems, it is possible to bake with small standardized ovens. If such ovens cannot be obtained commercially, they can be made quite readily in the laboratory. A typical oven might be around 16 in. by 16 in. by 12 in. in height, although the height will be determined by the parts to be baked. The oven might be made of thin (say $\frac{1}{16}$ in.) aluminum sheet shells, separated by about 2 in. The shells are supported at the edges by asbestos board and the space between them is filled with some insulating material such as glass wool. Heating is most simply accomplished by means of strip heaters attached to the inside walls. To distribute the heat uniformly, the heaters are clamped to metal plates. The heaters should be so arranged as to produce a uniform temperature in the oven. Typically, with eight heaters, the arrangement of them could be four on each side. For the above type of oven, the power required is about 1 kw to maintain the temperature at 450°C. The general form of such an oven is shown in Fig. 14.4. Whatever design of oven is used, it should be lightweight for easy handling. A temperature regulator should be used to maintain the temperature within a few degrees of that required. The ovens described by Prescott (*Rev. Sci. Instr.*, **33**, 485, 1962) are inexpensive, easy to construct, and adaptable to various arrangements.

The vacuum parts should be mounted on a base plate, on which the oven rests when in use. This plate must be rigid and act as a good heat insulator. Asbestos board of appropriate thickness could be used. Connections through such a board should fit snugly to avoid

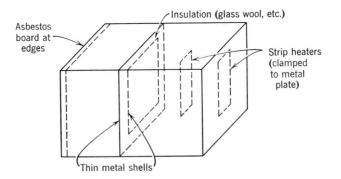

Fig. 14.4 Typical oven for baking parts.

movement of vacuum parts with respect to each other. Alternatively two base plates forming a shell filled with insulation could be used. Bakable metal valves must be mounted very rigidly because of the forces required to operate them. For the same reason the base plate must be mounted firmly to a frame. In mounting parts, due allowance must be made for thermal expansion during the bake-out process. Expansion loops should be provided between rigidly mounted parts. Sections of tubing more than about 9 in. in length should not be rigidly clamped at both ends. One end should be allowed to "float," i.e., it should be supported but not clamped. The above remarks are applicable to systems used to produce very high and ultra-high vacuums.

14.5   Bake-Out Procedure for Small Systems

The procedures discussed here are aimed primarily at systems of a few liters volume but are generally applicable to quite small systems. The procedures for very high and ultra-high vacuum systems are essentially the same. The degassing rate is strictly a function of the temperature and the material. To remove gases from inside materials at any reasonable rate a temperature between 400 and 500°C is required. At lower temperatures the complete removal of gas cannot be accomplished in any reasonable time and the ultimate pressure will probably not reach the ultra-high vacuum region. Certain vacuum parts, such as filaments, evaporating and bombarding coils, etc., may be operated at temperatures well above the bake-out temperature. At these high temperatures the parts will evolve gas, so they should be given additional bake-out treatment. Of course, it does no good to bake any parts unless the gases evolved are pumped away. The degassing procedure depends considerably on the exact nature of the system. A procedure for a system of the type shown in Fig. 14.1, together with certain modifications when used on ultra-high vacuum systems, is described below.

The general steps to be followed are:

1. Start rotary pump.
2. Check for large leaks with Pirani (or thermocouple gauge).
3. Immerse LN No. 2 in liquid nitrogen, or fill, depending on type.
4. When the pressure reaches 10^{-3} Torr or less, turn on diffusion pumps (and cooling water).

5. Put oven around parts to be baked and switch on when the diffusion pumps are operating and the pressure is around 10^{-5} Torr.

6. When the oven is up to temperature, degas parts not under the oven. These parts *must* include LN No. 1. Sometimes two or more liquid nitrogen traps are used. In this case at least one of them must be outside the oven so that it can be baked separately and then cooled to liquid nitrogen temperature. The liquid nitrogen trap, and any other parts outside the oven, can be baked by:

a. Hand torch. Heat part to a temperature compatible with its strength. For glass, this is its softening point. The heating should be continued for at least 1 hr.

b. Heating tape. This tape is wound around the part to be baked and a suitable electrical current is passed through it. Several hours of heating are required. The temperature is usually about the same as in the main oven.

c. An auxiliary oven. This oven is maintained at the same temperature as the main oven and heating is for several hours.

7. Immerse liquid nitrogen trap (LN No. 1) in liquid nitrogen (or fill, depending on type). For small systems, immersion-type liquid nitrogen traps are usually used.

8. Degas ionization gauge further. This is necessary since the gauge is to be operated at high temperatures. One method is:

a. Apply about 750 v ac or dc between filament and grid.

b. Adjust electron bombarding current by controlling the temperature of the filament (through the filament heating current).

c. About 70 w is required to raise the grid and ion collector to around 1200°C.

d. Avoid using too much electrical power, which can result in an electrical discharge or overheating of the electrodes.

e. Bake for an hour or more.

NOTE: One of the newer types of ultra-high vacuum gauges discussed in Chapter 6 could be used.

9. Turn on ionization gauge.

NOTE: This is done only after filling the liquid nitrogen trap (LN No. 1). Otherwise, pump oil getting to the gauge will be decomposed. Also, the gauge should not be used with the envelope above 150°C because of increased electrical conductivity along the glass walls. At this stage the pressure should be about 5×10^{-6} Torr (main oven still on).

10. Continue baking with the main oven for 10 hr or so (overnight if possible).

11. Degas vacuum parts which are to be operated at high temperature—filaments, evaporating coils, etc. Degas at temperatures near or, if practicable, in excess of those they will experience. Heat other metal parts at temperatures compatible with their strength and vapor pressure characteristics.

12. If all metal parts can be heated by resistance heating this can be done while the oven is still on. The advantage lies in the fact that gases evolved by the metal parts are not immediately readsorbed by the walls of the system, but are pumped away.

13. Sometimes induction or electron bombardment heating techniques are used. In this case the oven must be removed while the metal parts are being degassed. Due to readsorption of gas, it may be necessary to alternate baking with the oven and degassing metal parts by induction or electron bombardment until there is no further evolution of gas.

After the above steps have been carried out, with the oven cool and the metal parts hot, the pressure should be less than 10^{-8} Torr. When everything has cooled off, the pressure should approach 10^{-9} Torr. The degree of degassing will be less if pressures somewhat higher than this are adequate. It should be noted that in step 2 above, a Tesla coil is often used to look for leaks in a glass system (see Chapter 15). If a reservoir is included between the second diffusion pump and the rotary pump, then after the pressure is down to its ultimate value, the cut-off valve is closed, the rotary pump is turned off, and the admittance valve is opened.

If a small ultra-high vacuum system is pumped initially with a system of the general type shown in Fig. 14.1, then all of the steps listed above are applicable. The ionization gauge shown is used to follow the pressure down to the ultimate value obtained with this pumping arrangement. However, some type of getter-ion pump (sometimes a Bayard-Alpert gauge) is attached to the system and must be baked out with the rest of the system. If a gauge is used, it is further degassed in the manner described in step 8 above, after the oven has been removed. The system is then sealed off from the very high vacuum system by fusing a constriction in a glass connecting line (short) or by closing a bakable metal valve in this line (which could also be metal). The connecting line, including constriction or valve, must be baked with the other parts. A check for small leaks is then made (see Chapter 15) and the getter-ion pump turned on. The pressure should drop fairly rapidly to around 10^{-11} Torr or less. The lowest pressure attainable will be determined largely by the thoroughness of the degassing. In some cases a more or less conventional pumping system without a liquid nitrogen trap is used. Several of the steps listed above can be omitted. These omissions occur primarily because of the fact that there is no liquid nitrogen trap. Consequently there is no need to have the ionization gauge outside the oven baked separately, since it doesn't need to be used until the system has been isolated from the diffusion pumps. All parts down to the top of the diffusion pump are baked at the same time. Steps 6, 7, 8, and 9 can be omitted. The oven is removed and cooled after several hours of baking (overnight, if possible). The system is then sealed off from the diffusion pump and the getter-ion pump is turned on.

14.6   Large Low Pressure Vacuum Systems

The systems considered in previous sections had volumes of not more than about 10 l (approximately $\frac{1}{3}$ ft³). Here systems up to 1000 l (about 40 ft³) or more are considered. The basic problems involved in attaining pressures in the very high or ultra-high regions are essentially the same as for small systems, viz.:

1. Contaminating vapors from the pumping system.
2. Gases and vapors evolved from the internal surfaces of the system.

These problems become much more difficult to solve as the system increases in size. Probably the most difficult problem to solve is 2. This is due to the difficulty of adequately baking out walls and components. Not only does the power needed for baking go up drastically as the size of the system increases but, also, instead of using a simple oven, more elaborate methods of applying heat must be used. An alternative method is to cool the walls and components (where possible) so as to drastically reduce the evolution of gases (cryogenic pumping). This method is usually combined with some other form of pumping, as is considered later in this section.

The demand for large systems has stemmed, to a great extent, from simulation of outer space conditions and designs of new nuclear particle accelerators. The effort to date has covered the very high region but little success has been achieved in obtaining large ultra-high vacuum systems. Undoubtedly the demands will rise for systems larger than the sizes being considered here. This may be done by the methods discussed here or by entirely new methods. In order to get into the ultra-high vacuum region it is necessary to adequately bake not only the vacuum chamber and accessories but also the pumps used to obtain the final low pressure, or at least certain critical parts, such as the top of a diffusion pump. Basic methods used to achieve pressures in the very high vacuum region are by means of:

1. Getter-ion pumping system with baking.
2. Diffusion pumping system with baking.
3. Differential pumping system with baking.
4. Turbomolecular pumping system (molecular drag pumps) with baking.
5. Cryogenic pumping system with auxiliary pumps.

Many modifications of these methods have been tried and no attempt is made here to cover them in detail. Only some examples will be discussed.

GETTER-ION PUMPING SYSTEM WITH BAKING

The characteristics of getter-ion pumps have been discussed in Chapter 5. With large pumps of this type the limitation in lowest pressure attainable is probably associated with the bake-out procedure. The problem as far as vacuum chamber and accessories are concerned is common to all large systems regardless of pumping system. However, various forms of getter-ion pumps present their own peculiar difficulties as far as bake-out is concerned. Pumps based on evaporation of getters and ion pumping, e.g., the Evapor-ion pump, are difficult to bake thoroughly because of their basic design and constructional materials. Consequently, to date most of these have been operated around 10^{-7} Torr. However, it has been possible to get down to around 10^{-9} Torr by baking the pump and prepumping to 10^{-7} Torr using a mercury diffusion pump and liquid nitrogen trap. Getter-ion pumps based on the cold cathode gauge principle (sputter-ion type) can be baked somewhat more readily. However, although many commercial designs of pump can be baked to 450°C, the magnets are usually the limiting factor. The allowable temperature of bake-out for the pump-magnet system is more likely to be around 250°C, which does not give very effective bake-out in reasonable lengths of time. The problem of separating the pump from its magnet is often a time-consuming one and in some cases the design of the whole system may prohibit it. Newer design techniques may eliminate this particular problem. But even with the above limitation, it is not too difficult to obtain pressures of 10^{-8} Torr with such pumps. With a very thorough bake-out and degassing procedure, as was discussed in connection with small very high vacuum systems, it is possible to get pressures in the ultra-high vacuum region. However, the release of gases from pump surfaces as the pumping proceeds may become a matter of some importance. One of the big advantages claimed for getter-ion pumps is the "clean" vacuum obtainable without the use of cold traps, i.e., the absence of contaminating vapors such as the oil vapors in oil diffusion pumps and turbomolecular pumps. It may well be that to obtain effective operation in the ultra-high vacuum region it will be necessary to combine getter-ion and cryogenic pumping. For ease of starting a system and operating without interruption, it has been suggested that two sputter-ion pumps be used, one being baked while the other is pumping. Clearly the initial investment would thus be increased.

DIFFUSION PUMPING SYSTEM WITH BAKING

Many modifications of conventional diffusion pumping systems have been made in an attempt to reach low pressures. The principal design changes have been made to prevent pump fluid vapors from getting into the vacuum system. Both mercury and oil diffusion pumps have been used, with more attention given to oil diffusion pumps, particularly for the larger systems. Not only must attention be given to vapor from the diffusion pump but also oil vapor from the rotary oil-sealed pump which is usually used must be considered. Common methods employed to minimize backstreaming of pump fluid vapor are:

1. Keep the pressure rise across the diffusion pump at a minimum. This is usually done by backing this pump with a small booster diffusion pump. However, proper design of the main pump may have the same effect.

2. Use bakable cold traps and baffles above the diffusion pump. Oil from the rotary pump is trapped by a liquid nitrogen trap between this pump and the diffusion pump in the foreline. It will be noted that the essential components are the same as in the system shown in Fig. 14.1. Systems have been used with only a trap and no baffle.

It is important that the traps and baffles used present as little resistance to gas flow as possible and still be effective in preventing vapors from getting into the system. A water-cooled chevron-type baffle has been used quite commonly with oil diffusion pumps. The cold traps are usually liquid nitrogen traps of the no-creep or spherical types (see Chapter 11). Sometimes as many as three spherical-type traps have been used. Copper foil traps have not been found to be effective in large systems since they saturate too quickly.

A system using an oil diffusion pump with liquid nitrogen trap and a booster diffusion pump for backing is described by J. C. Simon (*Vac. Symp. Trans.*, 1960). Oil diffusion pumps were used on the basis that mercury is more difficult to trap and constitutes a hazard. The chamber was 40 ft³ in volume and was made of austenitic stainless steel plate $\frac{1}{4}$ in. thick which was arc welded and fully stressed before machining the flanges. The actual chamber was $40\frac{1}{2}$ in. in diameter and 66 in. long, with dished doors (ASME standard) at the ends. The diameter of the pumping line was 14 in., and a 10 in. Leybold DO-2001 oil diffusion pump was used with DC-704 pump oil. Two liquid nitrogen traps were used, the one nearest the pump being cooled first after baking and then filled so as to trap oil vapors. An elbow water-cooled baffle was used between the first trap and the pump. The main diffusion pump was backed by a 2 in. diffusion pump, which, in turn, was backed by a 30 cfm rotary gas ballast pump. Bake-out was carried out with strip heaters at a temperature of 250°C. Some seals were water-cooled during the bake-out process. With this system it was possible to reach 6×10^{-9} Torr in about 16 hr. The use of water-cooled seals, which are not damaged by baking, has been fairly widespread. This makes it possible to use various elastomer materials such as Viton A. However, when these seals are used, pressures much below 10^{-8} Torr cannot be obtained, although the final value will depend to a large extent on the specific design of the system.

Instead of employing liquid nitrogen traps, various sorption materials such as zeolite and activated alumina have been used as traps. These materials must be refrigerated to trap mercury vapor but not to trap pump oil vapors, which offers a considerable advantage. Biondi (*Vac. Symp. Trans.*, 1960) used an 8 in. trap consisting of staggered trays of zeolite and a water-cooled chevron baffle on a 4 to 5 in. oil diffusion pump. The trap was baked with the rest of the system. A pressure of around 2×10^{-10} Torr was obtained at this point, which rose to about 7×10^{-10} Torr in 70 days. It took 100 days for the pressure to reach 5×10^{-9} Torr. The effectiveness of the trap apparently depends on the backstreaming rate, which depends in turn on the pump and pump oil being used. Biondi's trap appeared to be effective after exposure to nitrogen, carbon dioxide, hydrogen, helium, and argon. It should be noted again that neither liquid nitrogen nor these sorption materials will trap the rare gases.

DIFFERENTIAL PUMPING SYSTEM WITH BAKING

This type of system makes use of two vacuum chambers, one inside the other. The inside chamber is the "working" chamber and it is pumped down to the desired low pressure, often with a diffusion pumping system of the type discussed above (trap, baffle, diffusion pump, booster diffusion pump, and rotary pump). The space between the two chambers is pumped down, usually with a conventional diffusion pump system. This means that there is very little force acting on the walls of the inner chamber. The atmospheric pressure is supported by the outer chamber. Some features of a typical differential pumping system are shown in Fig. 14.5. This particular example shows two bell jars with two pumping systems. The inner system can be of the type discussed above. Fairly conventional systems have been used with attention given to adequate trapping. Figure 14.5 shows a chevron-type liquid nitrogen trap. Other types can be used as long as they provide good trapping efficiency without introducing much resistance

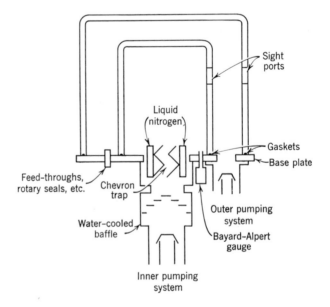

Fig. 14.5 Differential pumping system.

to gas flow and are of the no-creep type. Gaskets for sealing the vacuum chamber to the base plate can be elastomers or metal gaskets. However, using an elastomer for the inner chamber gasket will limit the bake-out temperature that can be used for the inner chamber. To overcome this limitation the gaskets could be water cooled. Also, the inner vessel could have a metal gasket while a water-cooled elastomeric gasket was used with the outer vessel so as to avoid damage from heat conducted through the base plate. With a differential pumping system (baked), using an inner chamber volume of a few cubic feet, it is not too difficult to obtain pressures around 10^{-9} Torr. The main advantages of this type of system are:

1. The inner chamber can be of lightweight construction, which results in drastically reduced power requirements for bake-out.
2. Small leaks in the inner chamber are not as critical as with single chambers because of a small pressure differential.
3. The outer chamber can be made of an economical material, such as mild steel, since its inner surface is not exposed to the working pressure of the system.

Perhaps the principal disadvantage of this kind of system stems from the additional constructional difficulties inherent in a double chamber design. There may be a considerable problem in arranging for feed-throughs, rotating seals, etc. Also, if the design of the inner chamber is such that it will not stand full atmospheric pressure then special leak-testing procedures may be necessary. However, even with these disadvantages, differential pumping systems may well find increasing favor with very large systems (tens or hundreds of cubic feet), largely on grounds of economy. As the size of a system increases, the cost of a single chamber made of stainless steel goes up sharply due to the increased wall thickness needed to withstand atmospheric pressure. The use of a thin inner chamber with good outgassing properties and a thick outer chamber of a lower priced material such as mild steel could result in substantial savings. The possibility of combining a differential pumping system with cryogenic pumping has also been considered.

TURBOMOLECULAR PUMPS

The "molecular drag" type of pump was discussed in Chapter 5. There a design of pump was mentioned that is based on the principle of molecular drag, which makes use of stationary and rotating disks of particular design. This type of pump is fairly new (W. Becker, *Vak-uum-Tech.*, Oct. 1958, pp. 149–152) and is often called a turbomolec-ular pump. Various advantages are claimed for this design of pump, including low ultimate pressure, high pumping speed, and very little backstreaming of oils from the bearings. Pressures as low as 10^{-9} Torr without the use of cold traps or bake-out procedure (but using metal gaskets) have been claimed, with a pumping speed loss of about 50% over somewhat higher pressures. It is questionable that most commercial pumps of this type will reach pressures this low. There is also some question as to the amount of oil vapor which gets into the main vacuum system. This may be the limiting factor with this type of pump just as in the case of oil diffusion pumps (together with the limitation of gas and vapor being released from the pump itself). The turbomolecular type of pump may well find considerable use in the very high vacuum region.

CRYOGENIC PUMPING

Cryogenic pumping can be achieved by the use of traps or by im-mersing the vacuum system (or parts of it) in the refrigerant. As was noted in Chapter 5, liquid nitrogen will not "pump" many gases, i.e., freeze them out so that their vapor pressures are negligible. Liquid helium provides the lowest temperature and, therefore, is the most ef-fective cryopump. Liquid hydrogen is not quite as effective but can still be useful. To obtain pressures in the very high and ultra-high vacuum regions using either of these materials it is not possible to start at any substantial pressure. Suppose a system is pumped down to 1 Torr by some other means, such as mechanical pumps. Then the pressures obtained are about 10^{-4} Torr with hydrogen and 4×10^{-6} Torr with helium. When liquid helium is used, its residual gases, helium and hydrogen, must be dealt with. These gases could be re-moved by some other means, such as getter-ion pumping. However, the best procedure is to pump the system down to much lower pres-sures before doing any cryopumping. A conventional diffusion pump–mechanical pump system would be used to obtain pressures around 10^{-6} Torr, at which point further reduction in pressure could be achieved by cryopumping. Of course, other pumps such as getter-ion or turbomolecular pumps could be used initially. By this general pro-cedure, pressures in the ultra-high vacuum region are achievable. In principle, cryopumps together with some other type of pumping system offer a method of obtaining ultra-high vacuum conditions in very large systems such as are required for space simulation.

14.7 Bake-Out of Large Systems

Many of the comments made in Sections 14.4 and 14.5 are applicable here. However, simply going to a larger size introduces a number of problems. Generally small systems are made of a hard glass since it is relatively easy to construct such systems. The temperature limit for bake-out is then between 400 and 500°C, depending on the exact glass that is used. As the size of the system increases, the constructional difficulties with glass increase rapidly. Also, problems are introduced in connection with annealing and baking without introducing strains. Consequently, metals are used for larger systems. As far as the bake-out procedure is concerned, the choice of metal is dictated by the following types of considerations:

1. Ability to stand the bake-out temperature without changes in mechanical properties.

2. Low vapor pressure at the bake-out temperature.

3. Adequate mechanical strength to withstand atmospheric pressure, at both the operating temperature and the bake-out temperature (a double chamber system is a special problem).

4. Mechanical properties making it possible to achieve a smooth finish.

Requirement 4 serves two purposes. A smooth surface means that the total area presented to the vacuum is minimized. A rough surface can have a surface area several times that of a smooth one with a resultant increase in the amount of adsorbed gas and, consequently, an increased bake-out time. Also, it is much easier to thoroughly clean a smooth surface. Because of these considerations, stainless steel is commonly used. It is not generally realized how much gas is released by an unbaked material. This only shows up as the pressure is reduced. For example, 1 cm² of unbaked stainless steel is equivalent to over 10^7 (10 million) cm² of thoroughly baked stainless steel. The practical effect of this can be shown as follows. Consider an unbaked stainless steel cylinder 30 cm in diameter (about 1 ft) and 60 cm in height (about 2 ft). To maintain a pressure of 10^{-6} Torr in this cylinder, assuming no leaks in the walls and negligible losses in the pumping line, a pumping speed of about 560 l/sec is required. However, to maintain a pressure of 10^{-8} Torr, a pumping speed of about 56,000 l/sec is required. That this latter pressure can be obtained in practice is due to the fact that it is only achieved after pumping for a considerable length of time and the rate of evolution of gas gradually decreases with time. Also, liquid nitrogen traps are used to assist in pumping. Stainless steel can be raised to quite high temperatures without changes in mechanical properties. However, most vacuum systems contain glass-metal seals and this limits the bake-out temperature to around the value used with glass systems, say 450°C. Of course, other constructional materials may lower the permissible temperature considerably.

The actual bake-out procedure used with large systems is determined by the nature of the system. As the size increases, power requirements go up rapidly. Also, in general it is difficult to use ovens (as with small systems) surrounding the system entirely. Some heating methods which have been used are:

1. Heating by passing an electric current through the chamber walls.
2. Use of induction heating.
3. Use of internally mounted radiation sources.
4. Use of resistance heaters (strip or tape).

Most vacuum chambers do not lend themselves to the use of method 1. Method 2 is awkward and expensive. Method 3 increases the likelihood of heating the apparatus excessively. With large systems, method 4 has found increasing use, mainly because of its flexibility. Various complex parts can be effectively heated by this method, particularly by the use of resistance tape. Such tape is obtainable from various manufacturers. Tape is preferred to wire since the heat contact with the surface to be heated is larger and it is easier to wind crosswise. Many parts of the system can be heated quite effectively in this manner. Insulation should be used to cover the tape and parts of the system which cannot be heated directly. Asbestos is commonly used for this purpose. The insulation serves to cut down on power consumption and help heat inaccessible parts. Good contact between the tape and the part being heated is important. Heating elements can be made up with resistance tape for certain common vacuum elements, such as flanges. Considerable ingenuity may have to be exercised in arranging a suitable bake-out system. For example, it may not be possible to heat certain parts (because of damage) and they will then have to be insulated. Also, it may be necessary to degas certain internal parts at higher temperatures than the system as a whole. Induction and resistance heating are often used for this purpose. Electronic tubes, such as ionization gauges, can be degassed by electron bombardment as in the case of small systems. The time required to effectively bake out a large system is determined by the nature of the system. The difficulty in obtaining ultra-high vacuum with such systems is largely associated with the bake-out procedure (including the pumps). Attempts have been made to bake the system by circulating hot fluids through tubing attached to the system. It is difficult to get high enough temperatures by this method although it can be useful in the very high vacuum region.

14.8 Construction Materials and Techniques

In achieving very low pressures, particular attention must be given to the choice of materials and to various fabrication techniques. As was noted in the previous section, materials which are to be raised to the bake-out temperature must be able to stand this temperature without damage. Another important consideration is the vapor pressure contributed by materials. Elastomeric materials are not acceptable in the ultra-high region for this reason. Also, they cannot be raised to bake-out temperatures. Brass is not acceptable because of the vapor pressure of the zinc in it. Also, loss of zinc at bake-out temperatures can change the properties of the brass. Soft glass will not stand the necessary bake-out temperatures. Another property of a material that must be considered is its permeability to gases, i.e., the ease with which gases pass through it. This property becomes particularly important as the pressure becomes lower. With the most thorough bake-out procedure and no contribution of vapors or gases by the pumps, the ultimate pressure is going to be determined by the permeability of the construction materials. The light gases, hydrogen and helium, are of particular concern in this regard. The permeabilities of various materials for certain gases are given in Chapter 8. Other vacuum properties of materials are also included in that chapter.

Not only are the vacuum properties of materials important at very low pressures, but also the mechanical properties become of increasing importance. Because of the use of bake-out with most ultra-high vacuum systems, the matter of thermal expansion becomes important. When heated, parts may separate, causing leaks. Large temperature changes across metal gasket seals should be avoided. Also, valves and gaskets must stand bake-out temperatures for many hours without leaking and without deterioration. Some aspects of this problem have been considered in Chapter 11. The matter of thermal expansion also leads to the necessity for free but supported movement of ends of parts, as necessary. Repeated heating and cooling of parts can lead to deterioration of mechanical properties. Also, repeated heating and cooling as well as deformation of parts such as valve seat gaskets can give rise to difficulties. At this moment this is no complete answer to these problems.

The internal surface finish of parts is also important since a smooth finish minimizes the amount of adsorbed gases and facilitates cleaning. The cleaning techniques discussed in Chapter 9 are generally adequate. A polished finish on materials such as stainless steel is recommended. Where this is not possible, electroplating is recommended. This removes inclusions in the surface and exposes holes just beneath the surface. The recommended welding and brazing techniques of Chapter 10 should be followed with more than usual care to avoid trapped pockets of gas and any porosity. If possible, components should be radiographed. After manufacture and cleaning, each section and component should be heated repeatedly to the maximum bake-out temperature and allowed to cool. The leak rate should be checked after each heat cycle to ensure that the part is capable of withstanding service conditions. It is not uncommon for welds to leak after the first heat cycle. Also, it is not always possible to detect small leaks until the outgassing rate is a minimum. Leak detection facilities are quite important. Various construction materials and techniques have been discussed in previous chapters.

14.9   Is There a Low Pressure Limit?

The development of ultra-high vacuum equipment and techniques was spurred initially by experimenters in certain areas of research. In particular, the study of surface phenomena required a "clean" vacuum with no adsorbed layers of gas. Of course, relatively small systems can be used for such research. Interest in large, very high and ultra-high vacuum systems has stemmed largely from space programs. There is no reason to believe that this interest will diminish in the future. The question might well be asked—Is there a lower limit to the pressure that can be obtained? In a strictly theoretical sense there is no lower pressure limit. However, practically, there are certain factors which combine to limit the lowest pressure obtainable. New techniques may make it possible to go to considerably lower pressures than can now be obtained. At the present time various experimenters have claimed pressures in the 10^{-13} to 10^{-14} Torr region. This is getting near the limit of presently available vacuum gauges (see Chapter 6). At the 1962 meeting of the American Vacuum Society, it was reported that gauges using electron multiplier detectors were linear to 10^{-17} Torr. The types of factors which must be considered in going to lower and lower pressures are:

1. Leaks in the vacuum walls.
2. Release of gases and vapors from surfaces inside the system.
3. The capability of various types of pumps to remove gases at very low pressures.
4. The ability to measure very low pressures.
5. The permeability of the vacuum wall materials.

The presently available leak detection methods probably can be used to detect small enough leaks to permit reaching pressures below 10^{-14} Torr (depending on the size of the system). The basic methods now used to eliminate factor 2 are baking and cryogenic pumping (or both), accompanied by some form of pumping system. It should be possible to remove essentially all absorbed and adsorbed gases by application of such methods with the proper choice of constructional materials. The matter of adequate pumps may well become the limiting factor. All presently known pumps suffer from certain limitations. In the case of molecular drag and vapor pumps, it is extremely difficult to eliminate all backstreaming of vapors. Getter-ion pumps depend on ion pumping to remove the rare gases. At very low pressures the chance of ionizing gas molecules becomes smaller. Also, there is some release of gas from the walls of these pumps, which becomes more important as the pressure is reduced. Cryogenic pumping will not remove helium or hydrogen. Vacuum gauges have been developed to read pressures down to about 10^{-14} Torr. New techniques may make it possible to read considerably lower pressures. As the pressure is reduced, the permeability of the walls of systems becomes an important factor. Although many materials are considered to be impermeable to gases at pressures around 10^{-9} Torr or so, this is no longer true at very much lower pressures. The solution is to carefully choose constructional materials for low permeability and to use differential pumping. Admittedly, there are many difficulties involved in obtaining very low pressures, particularly in large systems. However, there is every reason to believe that pressures well below 10^{-14} Torr will be achieved.

FIFTEEN

Finding and Repairing Leaks

15.1 When Is a System Tight?

The problem of finding leaks in a system can be very exasperating. Consequently, one should be reasonably sure that a leak exists before going into an elaborate leak-hunting procedure. As long as a system reaches the operating pressure, there is no need to be concerned with leaks in the system. If an unduly long time is involved in reaching operating pressure, this is an indication that gases or vapors are being released in the system. Actually, there are three possible sources of gas and vapor in a system:

1. Actual leaks from the atmosphere through the vacuum walls.
2. Gases and vapor released by materials inside the system (outgassing).
3. Release of trapped gases and vapors ("virtual" leaks).

Leak hunting is concerned only with the first source. Source 2 can be minimized by the choice of suitable construction materials (Chapter 8) and the use of proper cleaning techniques (Chapter 9). Source 3 can be minimized by the use of proper fabrication techniques (Chapter 10). Common causes of virtual leaks are double welds and double gasket systems. Readings of a vacuum gauge, made both when the system is untrapped and when it is trapped, will give some indication of how much condensable vapor is present. Traps are very effective in removing such vapors. Once it has been reasonably well established that actual leaks do exist, then one of the leak detection methods discussed in subsequent sections can be used. The particular detection system to be used will be determined by the type of vacuum system that is involved. It should be kept in mind that for many systems it is not necessary to use the most sensitive leak detection method to ensure that the system is sufficiently tight for the process involved.

15.2   Sizes of Leaks

The size of leak that can be tolerated is strictly a function of the system that is involved. The factors determining tolerable leakage are: (1) the operating pressure, and (2) the pumping speed. Consider the case of a small volume which must be maintained at a pressure of 10^{-8} Torr. Suppose a diffusion pump with a speed of 5 l/sec is used with this volume. The amount of gas per second that can be handled is then given by the pumping speed times the pressure. This amounts to 5×10^{-8} Torr l/sec or about 5×10^{-5} μl/sec. This value is the tolerable leak for such a system and can be referred to as the *leak*, the *rate of flow of gas*, or the *throughput*. Now consider the case of a much larger system, to be maintained at a pressure of 10^{-4} Torr. Suppose the diffusion pumps required have a speed of 10,000 l/sec. The tolerable leak is then $10^{-4} \times 10,000 = 1$ Torr l/sec or 1000 μl/sec. These are upper limits on the permissible leaks since the pumps may have to handle some gases and vapors being released in the system.

The leak unit, micron liters per second (μl/sec) is often called a *lusec*. This usage is quite common in Europe but not so common in the United States. Other units of leak rate are often used also. However, the most commonly used units are the above and atmospheric (or standard) cubic centimeters per second (std cc/sec). The principle governing leak hunting is the rapid detection and repair of leaks over and above the tolerable leak. In most cases there is no interest in the actual size of the leak. In those special cases where there is such an interest, calibrated leaks can be used. Rough calculations of the physical size of a leak can be made by assuming the leak to be cylindrical in shape. Of course, this is not generally true. However, for most purposes the calculations are sufficiently accurate. This matter has been treated by Guthrie and Wakerling (*Vacuum Equipment and Techniques*, McGraw-Hill Book Company, New York, 1949). It should be noted that the leakage rate of a gas through a particular leak depends on the nature of the gas. The pressure on the outside of a leak is atmospheric while inside it is essentially the pressure in the system. By consideration of the nature of the flow, it can be determined that for large leaks a gas with low viscosity flows through most readily. On the other hand, for very small leaks a high viscosity, light (low molecular weight) gas is best. Since the time for any gas to flow through a large leak is small anyhow, the choice of probe gas for such leaks is not too important. However, when dealing with very small leaks it is important to use a probe gas that flows readily through the leak. It is for this reason that the light gases (hydrogen and helium) are widely used in leak detection.

15.3   Methods of Leak Detection

Numerous leak detection methods have been reported in the literature. However, most of them are simply variations of the following methods:

1. Partial vacuum inside system, soap solution applied inside system.
2. Overvacuum or evacuated hood.
3. Sealing substance on outside of system, change of pressure inside.
4. Rate-of-rise measurements.
5. High pressure inside vacuum system, indicator on outside.
6. Spark coil used outside system.
7. Discharge tube attached to system.
8. Probe material on outside, change of apparent pressure or of nature of gas inside system.

Some of these methods are only useful in certain pressure ranges and in particular circumstances. The most sensitive leak-detecting systems are included in method 8. The type of performance desired of an "ideal" leak detector consists of the following qualities:

1. Capability for measurement of total leak and isolation of individual leaks.
2. Rapid response.
3. Temporary sealing of leak.
4. High sensitivity.

5. Applicability to any vacuum system without loss of vacuum, if necessary.

6. Simplicity.

7. Low cost.

8. Ease of maintenance and operation.

No presently available leak detector has all these desirable features. In general, the requirements of rapid response and high sensitivity lead to more complex and expensive equipment. The helium leak detector satisfies most of the above requirements, although other types of mass spectrometers may ultimately become competitive. Where high sensitivity is not required, relatively simple leak detection systems can be used.

Method 1 (see first list of this section) finds very limited application and little attention will be given to it here. The method can only be used with very large systems. The procedure is for a man, equipped with oxygen tank, to get inside the vacuum system, which is then partially evacuated, say to ¾ atm or so. The man then applies soap solution to the inside of the system and looks for the formation of soap bubbles. The term "soap solution" includes solutions of actual soap and also of detergents. It is evident that strict precautions must be taken in this method, including: (1) the man in the system must be under observation at all times, (2) a continuous and uninterrupted source of oxygen must always be available, and (3) extreme care must be taken to avoid having the valve between the vacuum system and the pumps opened wide by mistake. This method is only used to find large leaks in parts of the surface of a large vacuum system that are not accessible from the outside. The method should be resorted to only when all others fail.

In method 2 the pressure is reduced on portions of the outside surface of the vacuum system where it is suspected there may be leaks. The system is continually pumped while the procedure is carried out. When the pressure is reduced over a leak, a vacuum gauge will show a drop in pressure. The pressure can be reduced over a portion of the outside surface by appropriately shaped hoods which seal to the surface and are pumped out with a roughing pump. Naturally, such a hood must be capable of withstanding atmospheric pressure. The pressure in the hood can be several Torr, except for the smallest leaks. The difficulty with the method lies in making hoods that will seal to irregular or curved surfaces. This hood method has been modified for use with probe materials and the modification is described in detail in connection with the helium leak detector.

15.4   Sealing Substance Outside, Pressure Change Inside

This method involves covering outside parts of a system progressively with some material that will seal the leak. Once the leak has been covered, the pressure will drop. In this way leaks can be located and permanent repairs made. The procedure is to paint, brush, or spray the sealing substance over various parts of the system until a change in pressure is noted. Any of the gauges discussed in Chapter 6 can be used, the choice being dictated by the pressure. The sealing substance may temporarily or permanently seal the leaks. Some permanent sealants are: glyptal lacquers, shellac in alcohol, vacuum cements that are liquid at room temperature, such as Eastman Kodak Resin 910, cellulose acetate, etc. Some temporary sealants are: water, acetone, alcohol, and plasticene-like substances. Two effects result from the use of a liquid, however. First, after the initial closing of the leak, the pressure will drop. Second, as vapor enters the system (instead of air), the gauge will show a change in pressure. The change in pressure will depend on the nature of the vapor, on the type of gauge, and on the manner in which the gauge is used. The vapors from materials that are liquid at room temperature are readily condensable. Consequently, all gauges used with a cold trap will show a pressure drop when a leak is covered by a liquid. The particular liquid used (no cold trap) will determine whether the gauge shows an increase or decrease in pressure. Alcohol, acetone, and ether, which are commonly used probe liquids, all show an increased pressure reading with an ionization gauge.

The temporary sealing substances are quite effective for all sizes of leaks except the very smallest. For very small leaks, a permanent sealing material works satisfactorily. However, in repairing large leaks the material is drawn into the vacuum system and a seal cannot be obtained. Although the "permanent" sealing substances give fairly satisfactory results with leaks in metal plates, in soldered, brazed, and welded joints, and in glass systems, they are not as satisfactory as a final repair obtained by reworking the material of the vacuum system (soldering, welding, etc.).

15.5   Rate-of-Rise Measurements

This method is not satisfactory for routine leak hunting since it is too slow. Also, it can only be used to find leaks in parts of the system that can be isolated. However, the method is very useful for determining when leaks are present in a system and, in many cases, for finding the sizes of leaks. The general steps involved are:

1. Isolate the vacuum system (or a portion of the system) from the pumps by appropriate valves.
2. Measure the rate of rise of pressure with a suitable pressure measuring device.
3. Use a vacuum gauge suitable to the pressure range involved, e.g., if the pressure cannot be reduced below 100 μ, use a Pirani or thermocouple gauge.

The most common types of gauges used in vacuum systems are: Pirani, thermocouple, and ionization (thermionic and cold cathode). These are continuously reading gauges and are, therefore, most convenient. A discontinuously reading gauge, like the McLeod, can be used, but with more difficulty. The general procedure is to first find the rate of rise of pressure for the system as a whole. If there has been previous experience with the system when it was "tight," then the rate of rise will give an immediate indication of the condition of the system. When the rate of rise agrees with that for a tight system, then nothing more need be done. Suppose leaks are indicated and the particular system has a volume of 1000 l. Suppose, further, that the rate of rise is roughly 1 μ Hg/10 sec, starting at a pressure of 100 μ. The leak rate is given by: change in pressure per unit time times volume. In this case the measurement of the leak is given by

$$\frac{1}{10} \times 1000 = 100 \ \mu l/sec$$

This is considerably higher than the value for a tight system. Knowing the overall value of the leaks in a system serves as a guide in determining when the larger leaks have been found.

Once it has been determined that there are leaks in the system, information regarding the location of the leaks can be obtained by isolating various portions of the system, using available valving, and taking more rate-of-rise measurements. The results obtained may quickly indicate portions of the system where major leaks exist. The actual location of the leaks can then be determined by using one of the leak detection methods summarized in Section 15.3. Where a vacuum gauge exists in an isolated part of the system, the rate of rise is determined directly. However, even when there is no gauge, the rate of rise can still be determined. Take the volume of the system to be V and the volume of the isolated portion where the leakage is to be measured to be V_1. In addition, take the base pressure of the entire system as P_1. Close off V_1 at pressure P_1. At a later time, t, isolate the rest of the system from the pumps and open the valve between V_1 and the rest of the system. If at this time the pressure in the entire system is P_2, the rate of rise of pressure in V_1 is given by

$$\frac{1}{t}\left(P_2 \frac{V}{V_1} - P_1\right)$$

Take the case of $V = 1000$ l, $V_1 = 100$ l, and $P_1 = 50$ μ. First, V_1 is isolated for 20 sec, and then the system is isolated from the pumps and V_1 is opened to the system. The pressure P_2 in the system is then 60 μ. Under these conditions the rate of rise in V_1 is given by

$$\frac{1}{20}\left[60\left(\frac{1000}{100}\right) - 50\right] = \frac{1}{20}(600 - 50) = \text{about } 28 \text{ } \mu\text{l/sec}$$

Suppose that the rate of rise for the whole system is 100 μl/sec and that the leaks are distributed uniformly over the system. Then V_1 would have a leak of $100/1000 \times 100 = 10$ μl/sec. The value of 28 μl/sec indicates the presence of a large leak in V_1 which must be located.

This method can be quite sensitive since t can be made as large as desired. Vacuum components can be isolated for days to test for leaks. The vacuum gauge used for measuring the rate of rise could be trapped so that condensable vapors from outgassing would not be measured. The pressure rise then would be due only to actual or virtual leaks, but most virtual leaks consist of condensable vapors which *are* trapped. Even without previous experience with a system and with either trapped or untrapped gauges, considerable information about the tightness of the system can be obtained through rate-of-rise measurements. Figure 15.1 shows the general shape of a pump-down and rate-of-rise curve for a system involving a liquid nitrogen trap. The section AB represents pump-down by the pumps. At B liquid nitrogen is added to the trap and the pressure drops suddenly. After further pumping the pressure reaches a minimum and doesn't change (CD). The system is then valved off from the pumps and trap (at D). Now, the pressure starts to rise, the rate of rise gradually dropping off (DE). During this period, there is outgassing of materials in the system and vapors from virtual leaks. Finally, these sources become negligible and the pressure-time curve becomes straight (EF). The rise in pressure is now due to leakage into the system. The slope of the curve EF, i.e., the rate of rise of pressure, gives an indication of the amount of leakage.

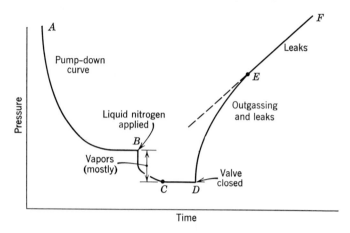

Fig. 15.1 Pressure-time curve.

15.6 High Pressure Inside, Indicator Outside

This method consists of pressurizing the vacuum component or system, usually with air or nitrogen, and finding leaks on the outside by an appropriate indicator. The method is not too generally applicable to complete systems, because of possible leakage through gasketed joints or damage to components. However, it can be very valuable in checking out components prior to installation in a system. Often, considerable economy can be achieved by checking components for leaks prior to carrying out final machining and other fabrication procedures. Of course, the fabrication procedure may expose other leaks. However, the method is valuable in that it finds bad "leakers." This is particularly true in the following categories: cast parts and welded, brazed, or soldered joints.

An indicator commonly used with this method is soap solution, which also includes solutions of detergents. The method is often called the "soap bubble" method. The procedure is to pressurize the component (or the whole system, if feasible). The house air supply is commonly used, although tanks of compressed gas can be used. The pressure that can be used will depend on the nature of the component. For small, brazed, welded, or soldered parts it is often possible to use pressures up to 80 psig or more. However, with large parts or components (and systems) involving gasketed parts or fragile parts it may be necessary to limit the pressure to 1 or 2 psig. Clearly, smaller leaks can be found by using higher pressures. Also, light gases will diffuse through small leaks more readily than air (or nitrogen). Helium is commonly used, hydrogen being too flammable.

After pressuring the pertinent part or system, solution is applied over the outside surface and this surface is examined for the formation of bubbles, which indicate the presence of a leak. To achieve maximum sensitivity, certain points should be observed:

1. Use a solution without a lot of bubbles in it.
2. Apply the solution gently so as to avoid forming bubbles in this way, which can be confused with those formed at a leak (i.e., flow the solution).
3. Use good lighting.
4. Observe each part of the surface to which the solution has been applied at least 5 min.
5. Mark each leak found with a waterproof crayon for possible later repair.

By proper attention to the above points, it is possible to detect leaks as small as 0.04 μl/sec at a pressure of about 60 psig. It is advisable to first apply solution to the parts that are most likely to leak, such as gaskets, welds, brazes, and solder joints. Considerable ingenuity may have to be exercised in making up special clamps and fittings for sealing the component and attaching the air (or gas) hose. The use of rubber sheets and C-clamps is not to be ignored. A variation of this method is to immerse the component in water. Agitation of the water is to be avoided.

In the case of large leaks, bubbles will not form. In this case leaks can be found by passing a flame slowly over the surface of the system (or component) and looking for a wavering of the flame. Alternatively, one can listen for the hissing of escaping air. Types of indicators other than soap solution have been used, although the soap method is probably most common. One method involves filling the system or component with certain chemicals in the gaseous or vapor state under pressure. Chemicals that are commonly used are: acidic materials, ammonia, and organic halides. Another chemical, which produces some recognizable effect when it interacts with the pressurized chemical escaping through a leak, is used as the detector. For example, if ammonia gas is used in the system, hydrochloric acid is used as an indicator. When the hydrochloric acid covers a leak, it reacts with the escaping ammonia gas to produce a white fog (ammonium chloride). With carbon dioxide or sulfur dioxide (acidic materials), ammonia can be used as the detector, the indication of a leak being fumes. Halides, particularly Freon, can also be used as pressuring gases, the detection being carried out by means of a special torch (halide torch method). This method has been described in detail by Guthrie and Wakerling (*Vacuum Equipment and Techniques*, McGraw-Hill Book Company, New York, 1949). The torch consists of a burner operating on acetylene or some other halide-free gas which is used to heat a copper plate red hot. Traces of the halide from a leak that are drawn to the burner through the air intake hose react with the hot copper plate, giving a bright-green flame that is characteristic of copper. Some gases used are: Freon, methyl chloride, methylene chloride, carbon tetrachloride, and chloroform. Freon is most commonly used. This method has been largely superseded by electronic halogen leak detectors.

The maximum sensitivities of the above chemical methods are about the same as that of the soap bubble method, i.e., 0.04 μl/sec. Some newer methods involving pressurizing for the purpose of detection have been developed during the past few years. Fluorescent penetrating dyes have been used, usually in liquid form and under pressure. However, such penetrating dyes will pass through leaks without a pressure differential. With this type of method, detection can be carried out by means of an ultraviolet light, which makes the dye fluoresce. Suitable developers can also be used to show up the dye.

A method that probably does not fall in any of the categories listed in Section 15.3 involves the leak testing of hermetically sealed components by use of a radioactive gas (Cassen and Burnham, *Intern. J. Appl. Radiation and Isotopes,* **9,** 54, 1960). Equipment for this leak detection system is manufactured by Consolidated Electrodynamics Corporation, Pasadena, California, under the name Radiflo. A maximum sensitivity of about 10^{-12} std cc/sec is claimed. The basic principle involves the detection of radioactive krypton-85 which has been allowed to diffuse into the leaky components. Krypton-85 emits both beta rays (electrons) and gamma rays. The gamma rays are the ones used in detecting leaks since they pass readily through most materials. In making use of this method, the components to be tested are placed in a tank, which is sealed and then evacuated to about 2 Torr. Diluted krypton-85 is then pumped into the tank under a pressure up to 7 atm. The radioactive gas diffuses into existing leaks in the components. After a prescribed "soaking" period (a few minutes to a few hundred hours), the krypton is pumped out of the tank and stored for reuse. A suitable combination of krypton pressure and "soaking" time may be selected to give the desired sensitivity. After this, an air wash is circulated over the components to remove any residual krypton from the external surfaces. The components are then removed from the tank. Those with leaks will retain some radioactive atoms, which emit gamma radiation. This radiation is detected by a suitable radiation counter (scintillation, etc.). The measurements taken make it possible to determine the leak rate. The presence of surface contamination absorbed by various surfaces (organic coatings, gaskets, etc.) can be determined by measuring beta ray activity with an appropriate instrument.

15.7   Spark Coils and Discharge Tubes

One of the oldest methods for finding leaks is by the use of a spark coil, usually a Tesla coil. The general appearance of a Tesla coil is shown in Fig. 15.2. The coil produces a high-frequency high voltage at the tip (metal). The body is insulated so that it can be held in the hand. The control is used to properly adjust the spark. This type of spark coil is usually about 1 ft long altogether, with a diameter of around 2 in., and produces several thousand volts of potential. Any spark coil that will produce high voltages can be used. Dowling (*J. Sci. Instr.*, **37**, 147, 1960) has pointed out that the high frequency transformer used in connection with television tubes can be used as a spark coil. He describes a suitable circuit.

The spark from such a coil will ground through metal parts and, therefore, spark coils cannot be used on all-metal systems. However, they can be quite useful on all-glass systems or even on metal systems containing glass parts. The spark tip can only be brought several centimeters from metal parts, since the spark will jump to the metal. Spark coils are used in two ways for leak detection:

1. By observing a change in color of the glow discharge produced by probe gases or vapors entering the leak.

2. By observing the spark jumping to the leak when the probe tip is 1 cm or so from it.

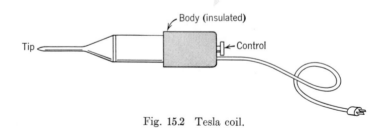

Fig. 15.2   Tesla coil.

In the first method, the spark coil tip is kept on one glass section of the system. Preferably this section should be between the diffusion pump and the backing pump in order to have a high enough pressure to maintain a glow discharge. The nature of this discharge will depend on the pressure and on the gases in the system. A discharge can usually be obtained between a few microns and a few Torr. The color is characteristic of the gases present. For air, this color is reddish or purplish. The exact color (as for other gases) depends to some extent on the glass used in the system. Pyrex or soda glass will show a yellow-green fluorescence while lead glass shows a blue fluorescence. In using this method, although the spark coil should be used on only one section of glass, the tip should not be held too long at one spot or it might puncture the glass. The probe material used can be a gas or a liquid. Some materials that are commonly used are: illuminating gas, ether, and carbon dioxide. With the first two materials the discharge takes on a whitish appearance. This is also the characteristic color of carbon dioxide (see Appendix D) but, possibly due to fluorescence of the glass, the color is often reported as bluish-green. The glass may be punctured either by holding the spark coil tip too long at one spot or by using too high a voltage. A spark gap ($\frac{1}{4}$ to $\frac{1}{2}$ in.) in parallel with the high voltage tip and ground will eliminate the latter difficulty. In the second method noted above, the spark coil tip is moved slowly over the outside glass surfaces until the spark is observed to jump to a particular point (the leak). The leak will show up as a whitish or yellowish spot, which is very clear against the glow discharge background.

Instead of a spark coil, a discharge tube (like a Geissler tube) can be used. The first method of leak detection discussed above then applies. Normally the discharge tube will be incorporated in the foreline for the purpose of obtaining adequate pressure. The pressure range required for a discharge is about the same as for a spark coil, viz., a few microns to a few Torr. The design of the tube is not critical—a simple glass tube with two sealed-in electrodes and a source of high voltage (d-c or high frequency). Once voltage has been applied to the tube and a glow discharge is evident, leak hunting can proceed as in the first method when employing a spark coil. The probe materials indicated for a spark coil can be used. Methane and alcohol are also commonly used. Carbon dioxide appears to be one of the most useful probe gases. It must be kept in mind that molecules that are heavy and have a high viscosity are not very suitable for leak hunting.

The spark coil and discharge tube methods of leak detecting are essentially only qualitative in nature. However, for glass systems (or glass parts) they can be quite useful, particularly where the operating pressure is not too low. Attempts have been made to use spectroscopic methods to improve the sensitivity of these methods. This improvement involves observing spectral lines characteristic of the probe material. With the advent of reliable electronic devices, such as d-c amplifiers and photomultiplier tubes, this method may ultimately be of importance in the field of leak detection. However, reliance on visual observation introduces the usual problems of personal error.

15.8   Probe Materials with Standard Vacuum Gauges

This is an example of the use of a probe material with an evacuated system, but employs some other detecting method than the gas discharge one. In a sense this method resembles the latter method and also the method using a sealing substance on the outside of the system. However, with a sealing substance, leak detection is normally carried out by observing a drop in pressure, either by sealing of the leak directly or by sealing with a liquid whose vapors are removed by a cold trap. In the method being discussed, permanent gases or vapors can be used but cold trapping is not used with the vacuum gauge. The particular behavior of the gauge with respect to the gas or vapor being employed is the basis of the method. Light, permanent gases are commonly used in order to get a rapid rate of diffusion through the leak. The particular gauge to be used will be determined by the pressure range involved. If diffusion pumps are not used (or are not pumping properly), Pirani or thermocouple gauges are commonly used. As is the usual case, light gases of low viscosity are the best for detecting the smallest leaks. It will be noted from Fig. 6.9 that such gases as helium and hydrogen show higher readings than air. This means that when these gases are used as probe gases, a leak will show up as a sudden increase in pressure reading. On the other hand, a gas like argon would show a leak by a decrease in pressure reading. Thermocouple gauges behave much like Pirani gauges with respect to their response to various gases. When the diffusion pumps are operating properly and the pressure is in the range of 1 μ to 10^{-7} Torr or so, then an ionization type of gauge can be used (thermionic, cold cathode, and Alphatron). The responses of these gauges to various gases and vapors are much the same. Table 6.2 shows that helium and hydrogen have a considerably lower response than air. Therefore, these gases are quite useful in leak hunting with this type of gauge. The low molecular weights of these gases are particularly advantageous. Argon shows a greater response than air, although it suffers from a high molecular weight, which results in a time lag in looking for leaks. A cold cathode gauge responds to various gases and vapors in much the same way as the thermionic ionization gauge. The performance of an Alphatron gauge is shown in Fig. 6.17. It should be kept in mind that certain gases and vapors can affect the performance of a gauge. This is particularly true of the thermionic ionization gauge, where "poisoning" of the hot tungsten filament can occur. One of the problems associated with the use of a vacuum gauge for leak detection concerns random pressure fluctuations. One method of minimizing this problem is to use a differential arrangement, in which two gauges are connected in a bridge arrangement in such a way as to cancel out pressure fluctuations (Blears and Lech, *Brit. J. Appl. Phys.*, **2**, 227, 1951). The probe gas is allowed to reach only one of the gauges. Pirani gauges are commonly used in such an arrangement. In general, the gauge deflection of a given probe gas depends on: (1) partial pressure of probe gas in the atmosphere over the leak, (2) change in flow rate through the leak when the probe gas is substituted for air, (3) change in flow rate through the pumps under these circumstances, (4) relative sensitivity of the gauge for air and probe gas and, (5) the effectiveness of absorption or condensation (for differential arrangements). Butane and carbon dioxide are often used as probe gases in the differential arrangement since these gases can be prevented from entering one of the gauges by means of a suitable cold trap.

15.9   Unique Response of Vacuum Gauges

When a probe gas is played over a leak, the gas entering the system consists of air as well as probe gas. A gain in sensitivity will be realized if only the probe gas is allowed to reach the gauge. The methods relied on to do this have generally involved the use of hydrogen with a selective absorber that permits only hydrogen to get to the gauge. Ionization gauges (thermionic or cold cathode) are usually used. Palladium has the property of passing only hydrogen when heated to about 700 or 800°C. Gauges using palladium are often called *palladium barrier ionization gauges*. Figure 15.3a shows one form of such a gauge that is used for leak detection. In this particular case, the palladium P is part of the ionization gauge and also acts as the anode. The palladium is heated by electrons from the filament F, which are speeded up and also focused by the grid G. The gauge tube is evacuated to a low pressure (say 10^{-7} Torr) or less and sealed off, usually with a getter. Bake-out of the gauge tube (prior to sealing and gettering) is advisable to eliminate readings due to subsequently released gases. However, it must be kept in mind that the gauge tube acts as a pump (getter-ion). As long as this pumping action takes care of released gases, good results can be achieved. The pressure is read by means of the ion current collected by the grid (negative). It is claimed that a hydrogen pressure change as low as 2×10^{-8} Torr can be detected. One of the main problems with such a gauge is caused by the presence of hydrocarbon vapors, which dissociate on the palladium surface and give rise to hydrogen, thus leading to spurious readings. Also, these hydrocarbon vapors produce carbide layers on the palladium, leading to a reduction in permeability for hydrogen. A cold trap between the gauge and the system is of help, although this

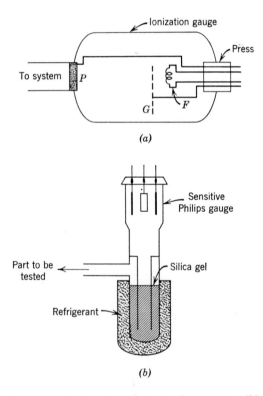

Fig. 15.3 (a) Form of palladium barrier ionization gauge. (b) Leak detector using silica gel.

will not eliminate hydrogen from decomposed oil in the diffusion pump. The smallest leak detectable with this system depends on the lowest pressure attainable in the gauge, on the pumping speed of the gauge, and on the permeability of the palladium to hydrogen. The detection of leaks as small as 3×10^{-7} Torr cc/sec (4×10^{-10} std cc/sec or 3×10^{-7} μl/sec) have been claimed.

Instead of palladium, it is possible to use another type of absorbent to pass hydrogen and block air. Silica gel, outgassed at 300°C and then cooled to liquid nitrogen temperatures, is commonly used for this purpose. Under these circumstances, silica gel readily passes hydrogen and the noble gases (helium, neon, argon, etc.) but not air. A system using silica gel with a cold cathode (Philips) gauge and hydrogen has been described by van Leeuwen and Oskam (*Rev. Sci. Instr.*, **27**, 328, 1956). The arrangement they used is shown in Fig. 15.3b. Van Leeuwen and Oskam claimed this system to be about a hundred times more sensitive than the palladium-hydrogen system. However, several hours were required to measure leaks of the order of 10^{-9} Torr cc/sec (about 1.3×10^{-12} std cc/sec or 10^{-9} μl/sec) and careful degassing of the leak detector and the tube to be tested had to be done. Another advantage claimed for silica gel is a long usage time (30 cc Torr of air) before it has to be degassed. The increased sensitivity of silica gel is claimed to be due to less gas evolution from the gel than from heated palladium, which results in lower pressures. Activated charcoal has also been used with a gauge for leak detection. Kent (*J. Sci. Instr.*, **32**, 132, 1955) has described the use of this material with a Pirani gauge. The smallest detectable leak claimed was about 4×10^{-7} Torr cc/sec (5.3×10^{-10} std cc/sec or 4×10^{-7} μl/sec). The above sensitivities (obtained by use of palladium or silica gel with ionization gauge and silica gel with Pirani) are much greater than those normally obtained (see Table 15.1).

The "poisoning" action of oxygen on a heated tungsten filament has been made use of in leak detectors. Thoriated tungsten is particularly sensitive to the effect of oxygen. Both diode (two-element) and triode (three-element) tubes have been used. Sensitivities of about 10^{-7} μl/ sec are claimed for this system. This makes it comparable to a hydrogen-Pirani gauge system using a selective absorber. Bloomer and Brooks (*J. Sci. Instr.*, **37**, 306, 1960) have described a simple leak detector using a thoriated tungsten filament. It should be noted that the above oxygen and hydrogen detection schemes can be adversely affected by release of gases from the walls of the apparatus. This can be minimized by a suitable bake-out procedure for the leak detector and other parts.

Table 15.1   Sensitivities of Some Leak Detection Methods

Leak Detector	Operating Range (μ)	Leak Size (std cc/sec)	Probe	Remarks
1. Spark coil (Tesla, etc.)	50–1000	—	Acetone, methanol, hydrogen, carbon dioxide	For glass.
2. Discharge tube	50–1000	—	Acetone, methanol, hydrogen, carbon dioxide	Residual gases confusing.
3. Pirani and thermocouple gauges	<100	10^{-4}–10^{-6}	Acetone, methanol, hydrogen	Pressure changes. Residual vapors.
4. Thermionic ionization gauge	<0.5	10^{-5}–10^{-7}	Gaseous hydrocarbons, hydrogen, oxygen, helium	Pressure changes. Residual vapors.
5. Halogen leak detector	<200	10^{-6}	Freon 12 or 22	Avoid halogen contamination.
6. Palladium barrier ionization gauge	<10^{-4}	10^{-6}–10^{-7}	Hydrogen	Avoid hydrogen contamination.
7. Tungsten diode or triode gauge	<0.5	10^{-6}–10^{-7}	Oxygen	Avoid oxygen contamination.
8. Helium leak detector	<0.1	10^{-10}	Helium	Most sensitive "standard" leak detector.
9. Pirani or thermocouple gauge with backing space	<100	10^{-5}–10^{-6}	Acetone, methanol, hydrogen	Time-consuming.
10. Thermionic gauge with backing space	<0.5	10^{-6}–10^{-7}	Gaseous hydrocarbons, hydrogen, oxygen, helium	Time-consuming.
11. Soap bubbles	Pressure—about 60 psig	5×10^{-5}	Air, helium	Observe for 5–15 min.

15.10 Halogen Leak Detectors

The halogens are chlorine, iodine, bromine, and fluorine. Materials containing these elements are usually called halides. The most common halide materials used in leak detection are those containing chlorine, and by far the most common specific substance used is Freon 12, although Freon 22 is used also. Electronic halogen leak detectors are often called simply *halogen leak detectors* or *Freon leak detectors*.

This type of detector makes use of a red-hot platinum filament which emits positive ions. The presence of small traces of halogen vapors increases the emission of positive ions markedly. It is this increase in emission that is measured to indicate the presence of a leak. One problem associated with early models was instability of the "no-signal" setting. How well this problem has been obviated by the use of new electronic circuitry depends on the manufacturer of each particular detector. These detectors operate in the micron range—the region from a few microns to 200 μ, or somewhat more, covering most commercial models. Consequently, they are usually used in the backing line of the diffusion pump (where used). Concentrations of halogen gas as low as 0.2 parts per million of air (0.2 ppm) can be measured or leaks of about 10^{-6} std cc/sec can be detected. This is comparable to the sensitivity of a hydrogen-Pirani gauge system (with selective absorber).

Halogen leak detectors can be very useful on systems (or components) where leaks no smaller than about 10^{-6} std cc/sec are to be detected. The equipment involved is simple and not bulky. Also, this device is considerably less expensive than a helium leak detector and can be used effectively within its limitations. However, certain precautions must be taken in using the detector, some of which are common to other leak-detecting methods. The presence of halogen containing material in the surrounding air, such as vapors from carbon tetrachloride and chloroform, and even cigarette smoke, can lead to spurious results. The same is true for rubber, greases, and other materials in the system that have absorbed halogen containing substances. It must be kept in mind that Freon is heavier than air and, therefore, leak hunting with it should proceed from bottom to top of the system (or component). Overexposure to halogen may necessitate flushing the sensing head with halogen-free air. If this is not done, it may take days for the detector to recover.

15.11   The Helium Leak Detector

This device is the most sensitive form of leak detector using a probe gas that is generally available commercially. It is basically a form of mass spectrometer in which ions of different mass are separated from each other according to their mass. Ions of a particular mass are selected and measured. Emphasis is placed here on helium ions, with helium as the probe gas. However, the ions of other probe gases, such as argon, neon, or hydrogen could be selected. This alternative may become one of the solutions to the problem of leak detection in countries where helium is in short supply. A helium leak detector is sometimes referred to as an MSLD (mass spectometer leak detector). However, this designation could refer to a mass spectrometer using any probe gas. The elements of a typical helium leak detector are shown in Fig. 15.4. Gas molecules enter the ion source, where they are ionized by electron bombardment. The ions formed are speeded up by appropriate grids and voltages, and an ion beam is formed by suitable slits. The beam passes through the analyzer system, where it is split into separate beams of ions, according to mass. Usually a magnet is used to split up the beam. However, electric fields or a combination of

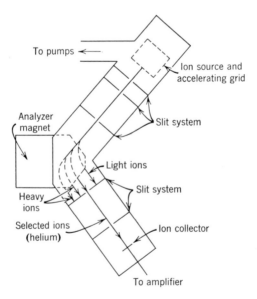

Fig. 15.4   Elements of a helium leak detector.

electric and magnetic fields can be used. The ions of desired mass are then selected (usually by a slit system) and allowed to fall on an electrode (target) connected to a suitable amplifier. The current gives a measure of the amount of gas (probe gas) present in the system since it is the ions of this gas that have been selected.

A number of commercial helium leak detectors are available. However, the principle of operation is essentially the same for all of them. They differ primarily in physical appearance, accessories, and operational procedures. The use of a specific commercial model is discussed later. Many of the procedures noted are applicable to other commercial units and differences can be resolved by reference to the literature of the manufacturer involved. First, the general methods of leak detection with a helium leak detector are discussed. It should be noted that these general methods are applicable to many other types of leak-detecting schemes.

The typical helium leak detector contains: (1) a vacuum pumping system (mechanical and diffusion pumps) for pumping down the analyzer and associated lines, (2) a cold trap for pumping condensable vapors, (3) an appropriate vacuum coupling for connecting a standard leak (or the object to be tested), (4) flanges for connecting objects or systems to be tested, (5) valves for controlling the pumping down of the test object, or a "hood" (such as a bell jar) around the test object, and for special methods of testing ("accumulator" valve), (6) a leak indicator, which is usually a meter, with an audio signal added (optional), (7) vacuum gauges, and (8) appropriate electronic circuitry and controls. The leak detector pumping system could be used to evacuate the test object (or a surrounding hood). However, even with adequate valving, this method is not satisfactory. Without the addition of more valves and lines, the pumping has to be done slowly through the throttle valve so as to avoid damage to parts such as gauge and pumps. The usual procedure is to use an auxiliary mechanical pump. A capacity of 5 cfm is adequate for most purposes. Of course, where a complete system is being tested, the pumps are already available. In this case, the leak detector is usually connected to the backing line ("backing space" technique). Some commercial models incorporate an auxiliary mechanical pump.

15.12   Sensitivity of Helium Leak Detectors

Leak detector sensitivity is specified in two ways: (1) by the smallest detectable helium concentration in air at a specified source pressure [in parts per million (ppm)], (2) by the minimum partial pressure of air which, if changed to the same pressure as the search gas, would produce the minimum detectable indication, e.g., an indication of three times random noise in 3 sec (see below). The second definition is commonly applied to helium leak detectors. Manufacturers often use another form of this definition, viz., the smallest pure helium leak which can be detected at a specified source pressure (usually atmospheric), and under specified test conditions. This is often called the *smallest leak detectable* (SLD) and is given in units of throughput, such as std cc/sec, μcfh, and μl/sec. Another term that is often used is *minimum detectable leak* (MDL), which is the smallest leak that can be unambiguously detected by a given leak detector in the presence of noise or background (see below) *or* the product of the minimum detectable pressure change and the pumping speed at the detector. These definitions are used interchangeably here, with the same types of units. However, it must be kept in mind that helium flows more readily through a leak than air. Both smallest and minimum detectable leaks will be used here for helium. To convert the values to the corresponding values for air, simply divide by 2.7. The sensitivities of most commercial helium leak detectors are expressed in terms of helium. It must be kept in mind that air leaks will reduce the sensitivity (smallest or minimum detectable leak) since they result in a reduced helium partial pressure. Consequently, sensitivity checks should be done with a minimum of additional vacuum components, and with assurance that all parts are "tight." Manufacturers supply *sensitivity calibrators*, which are also called *standard leaks* or *calibrated leaks*. Such leaks can be obtained to cover various leak sizes. Some commercial models include a leak as a part of the system, while with others models, leaks are supplied as separate units. The separate leaks can be attached to the detector by suitable vacuum couplings. The most sensitive standard leaks are usually small sealed containers of helium at atmospheric pressure with a very small leak built into the container (usually glass or quartz). Larger leaks are usually made by flattening metal capillary tubing and soldering it inside a larger metal tube.

The maximum sensitivity is limited by the characteristics of the electronic circuitry (usually the amplifiers) used to measure the current due to helium ions. This limitation is caused by *noise* and *drift*. Noise shows up as a random fluctuation in the output of the electronics and will be indicated by an erratic motion of the needle of the detector or *output meter*. Drift shows up as a gradual wandering of the output meter needle and is expressed in scale divisions per minute. For the purpose of discounting any insignificant fluctuations in the reading, the measurement is usually made over a period of an hour or so under steady conditions, an electrical recorder often being employed. The drift for most commercial leak detectors is usually less than ½ of 1 percent per minute of full scale on the output meter. This type of drift ordinarily introduces no particular problem, although it may be necessary occasionally to "zero" the output meter while leak hunting. The combination of noise and drift gives a *smallest readable deflection* (in terms of scale divisions), which is also called the *minimum detectable signal* (MDS). The smallest readable deflection is specified by the manufacturer for the most sensitive setting of the leak detector (output meter). The term *background* is also used in connection with leak detectors. This refers to the spurious output (scale divisions) due to the response of the detector to other gases than the probe gas (helium). The background may be inherent in the detector or accidental. It must be subtracted from the reading indicated on the output meter in order to get the true reading of the leak.

Methods of calculating the sensitivity (smallest leak detectable) and the size of the unknown leak are now shown here. The smallest or minimum leak detectable (MLD) is given by:

$$MLD = \frac{\text{smallest readable deflection} \times \text{standard leak rate}}{\text{leak rate meter reading}}$$

Sometimes the phrase *sensitivity calibrator leak rate* is used instead of standard leak rate. In either case, the units of throughput are used (μcfh, std cc/sec, etc.).

EXAMPLE 1

Suppose the smallest readable deflection is 2% of full deflection (0.02) on the 1 scale of the output meter (a reading of 0.02) and the standard leak rate is 0.002 μcfh. If the meter reading is 2 on the 3 scale, then the minimum detectable leak is

$$MLD = \frac{0.02 \times 0.002}{2} = 2 \times 10^{-5} \ \mu\text{cfh}$$

$$= 2 \times 10^{-10} \ \text{std cc/sec}$$

since 1 std cc/sec = about 100,000 μcfh. This is a typical sensitivity for a commercial helium leak detector.

It should be noted that the minimum detectable signal (or smallest detectable signal) is often taken as three times the mean peak-to-peak fluctuations of the output meter needle, averaged over ten fluctuations. The signal-to-noise ratio is then 3.

The size of an unknown leak can be determined by using the following relationship:

$$\text{Unknown leak rate (size)} = \frac{\text{standard leak rate} \times \text{unknown leak reading}}{\text{reading with standard leak}}$$

The unknown leak reading and the reading with the standard leak are given in scale divisions (same scale).

EXAMPLE 2

For accurate results the unknown leak should be placed in the same port as the standard leak and under identical conditions. Suppose the unknown leak gives a full scale deflection on the 10 scale, i.e., a reading of 10. If the same standard leak is used as in the previous example, the standard leak reading is 2 and the leak rate is 0.002 μcfh. Then, the unknown leak rate is

$$\frac{0.002 \times 10}{2} = 0.01 \ \mu\text{cfh} = 1 \times 10^{-7} \text{ std cc/sec}$$

In this example the value given is for helium. To obtain the air leak value, divide this value by 2.7.

15.13   Response and Clean-Up Times

High sensitivity is only one important factor in hunting leaks. The time involved in finding leaks is also important. This time is determined by two factors, response time and clean-up time. *Response time* is the time it takes for a leak detector to yield a signal output equal to 63% of the maximum signal attained when tracer gas is applied indefinitely to the system under test. It is usually desirable to keep this time under 3 sec. The *clean-up time* is the time required for a leak detector to reduce its signal output to 37% of the signal indicated at the time tracer gas ceases to be applied to the test system.

If the requirements of short response and clean-up times are not met, the leak-testing process is delayed to a large and sometimes intolerable extent. As an example, consider the case of a section of weld being probed at a constant rate. If the response time is poor, the leakage indication will appear some time after the probe has moved well beyond the leak, and the probe will then have to backtrack slowly until a second signal is obtained. However, this second signal cannot be observed distinctly until the first signal has been removed, or "cleaned up." Therefore, the clean-up time is equally as important as the response time. It should be noted that response and clean-up times are characteristics of the test system as a whole, including the part being tested and the auxiliary pump. They are not merely functions of the leak detector alone. The leak detector responds almost immediately to any changes in helium concentration that occurs at the throttle valve. In a test system, the delay in response is due to the time necessary for the helium concentration to build up at the throttle valve. The delay in clean-up is due to the time necessary for the pump to remove the helium from within the volume of the part being tested. A test object having a large volume will cause long delays though a high speed pump will reduce the delays. The important characteristic of the system is the ratio of volume to pumping speed (V/S).

Suppose a volume of V l is being evacuated at a speed of S l/sec. The relationship between the response time T_1 and clean-up time T_2 is then given by

$$T_1 = T_2 = \frac{V}{S}$$

NOTE: When V is in liters and S is in liters per second, T is in seconds. When V is in cubic feet and S is in cubic feet per minute, T is in minutes.

EXAMPLE

A part having a volume of 2 l must be leak tested, and the desired clean-up time is 2 sec. What pumping speed is required?

The solution is as follows:

$$2 \text{ sec} = \frac{2 \text{ l}}{S}$$

and

$$S = \frac{2 \text{ l}}{2 \text{ sec}} = 1 \text{ l/sec}$$

It should be noted that the speed required is the effective speed at the leak, which may be much smaller than the speed of the vacuum pump itself, due to the limiting effect of the connecting lines. The conductance of a connecting tube or pipe can be obtained from Fig. 3.7 and the required speed can then be obtained by using eq. 2.9 or Fig. 2.10. It is important to use short lengths of large diameter tubing to cut down on response time (also clean-up time). When a volume is pumped by both an auxiliary pump and the leak detector pumps, the speed of the former is used since it is generally much greater than that of the latter. The use of molecular pumping speed is usually satisfactory.

When the test object is pumped directly into the leak detector (first pumping down to a few microns with an auxiliary pump), the time of response will be determined by the pumping speed of the instrument itself. The speeds of the pumping system on most commercial leak detectors are such that reasonable response times can only be obtained with volumes of a few liters. However, even on large systems, where time is a secondary factor, this method yields maximum sensitivity, since there is no auxiliary pump to "rob" the helium. Of course, in this method the pumping speed used is that of the leak detector, corrected for losses in pumping lines.

15.14   Vacuum Testing

Certain general methods for hunting leaks are common to all helium leak detectors. However, it should be kept in mind that these methods can be used with many other types of leak detectors. Common methods are:

1. Vacuum testing.
2. Pressure testing.
3. Pressure-vacuum testing (inside-out).

Vacuum testing (see Fig. 15.5a) involves having the leak detector pumped down (diffusion pump on and liquid nitrogen in the trap) and then pumping down the object to be tested with an auxiliary mechanical pump. Once the test object has been pumped down to a few microns, it can be connected to the leak detector by suitable valving. The outside of the test object is then sprayed with a fine jet of helium (*probe test*). A tube having a fine opening at one end (*the probe*) is used to direct the stream of helium. Such probes are obtainable commercially, although it may be necessary to attach a tip with a smaller opening. A probe with an on-off valve is useful since by means of such a valve helium loss is avoided while not testing and the background of helium does not build up as fast as it would otherwise. The probe is attached to a cylinder of commercial helium (about 99.99% pure). Of course, a pressure regulator is needed on the cylinder. Also, a flow meter is often very useful since the flow can be adjusted to a value that has been found to be satisfactory for the type of leak hunting involved. The actual flow rate will vary considerably, depending on the size of object being tested and the exact nature of the test. For relatively small objects, the flow rate is sometimes adjusted by detecting the flow on the lips, the cheeks, or ears.

Certain general precautions should be observed in vacuum testing. Some of these are:

1. Use well-designed test connections. The connections can be a major source of difficulty on account of leakage, excessive helium contamination, and dirt.

   a. Use minimum amounts of elastomers and plastic. Since rubber tubing is particularly bad because it absorbs helium, which is difficult to remove by pumping and also eventually becomes contaminated with other materials, use as short a length as possible.

   b. Use a good, low vapor pressure vacuum grease sparingly. Apply a light film to gaskets and wipe off the excess. Large quantities of grease act as a source of helium contamination and also as a dirt catcher. The system may become so dirty because of excess grease that a good vacuum cannot be attained.

2. The need for cleanliness of the system as a whole, including the test object, cannot be overemphasized. The larger the system, the more important is cleanliness. A cold trap in the test object or in the connecting line is useful in the interest of cleanliness.

Certain points should be observed in carrying out the probing methods. Some useful techniques are:

1. Probe from the upper parts of the test object to the lower. In this way the helium, which rises, will flow back only over areas already tested.

2. When testing individual joints, time may be saved by using a generous flow of helium from a flexible rubber tubing ($\frac{1}{4}$ in. ID). When a leak is indicated, its exact location can be determined by means of a finer probe. On the other hand, with large leaks in the system enough helium may enter the system to saturate it and the leak detector for a while. With many small leaks, diffusion of helium to these small leaks may make leak detection difficult. The use of a fine probe will limit narrowly the area covered by helium. The detector output will be a maximum when the probe is directly over the leak.

3. A very large leak will give an indication even when the probe is some distance from it. Locate the leak and repair it (permanently or temporarily). Vacuum putty may be used for temporary repairs.

4. When a point appears to leak, but does not give a consistent response, a large leak in some other location is to be suspected.

5. To distinguish between two possible points of leakage close to one another, mark one of them (say with tape). Using a fine probe and a minimum flow of helium will help.

6. Different types of joints sometimes give the same typical leakage indication—a delayed and slow build-up of the leak signal and a very slow clean-up. The signal may even stay constant for some time. Such indications are usually due to porosity, flanges with flat gaskets, and rubber tubing joints. The behavior is due to long leakage paths and accumulation of helium (at atmospheric pressure) in crevices and pockets. Similar effects are produced by leaks in volumes that are behind constrictions or that are otherwise being pumped slowly.

7. Test subunits before incorporating them into an assembly or system. Then only joints between units require investigation.

With large test objects (several hundred liters) the response time may be quite long. It can be reduced by using short, large diameter connecting lines and adequate pumping speed. A large mechanical pump is often used. This results in a considerable reduction in sensitivity since this pump will divert helium from the leak detector. A better method is to pump the test object with a separate diffusion pump–mechanical pump arrangement. This simply amounts to testing a separate vacuum system. Two methods are used to connect the leak detector to the system:

1. Connecting it to the system near the pump connection or to the pump line near the system.

2. Connecting it to the backing line, i.e., the line between diffusion pump and forepump ("backing space" technique).

Method 1 is used when the pressure in the system is greater than the pressure in the detector while method 2 is used when this situation is reversed. A valve in the leak detector (the *throttle valve*) is used to control the pressure in the leak detector. The remarks above regarding precautions and probing techniques are generally applicable to a complete system. Method 1 is not as sensitive as method 2, since the system pumps remove helium, which does not reach the leak detector. In method 2, all helium entering the system passes through the backing line.

In the case of the "backing space" technique the leak detector can be "teed" into the backing line (the line between forepump and diffusion pump). However, a rotary forepump removes gas by a "batch process," which results in pressure fluctuations. Therefore, it is best to use a second diffusion pump in the backing line (backed by a rotary pump) with the leak detector connected to a backing volume between the two diffusion pumps. A valve between the backing space and the second diffusion pump is used to build up the backing pressure, which results in an increased sensitivity for leak detection. This may amount to a factor of 100.

It should be noted that with large volumes to be tested not only is a mechanical–diffusion pump system desirable, but also leak detectors that have an automatic balance control are particularly useful. These leak detectors respond only to a change in helium signal rather than to the presence of a quantity of helium. Thus, the leak detector automatically balances to 0 signal unless the concentration of helium changes. The result is a great reduction in the time spent in locating large leaks. With small systems (a few liters) a mechanical auxiliary pump is adequate. It is desirable to include a valve above the auxiliary pump and a valve between the leak detector and the test object (usually part of the leak detector). This makes it easy to first pump down with the auxiliary pump and then pump only with the leak detector (where possible), or to adjust the pumping speed of both pump systems. Often a gauge on the test object is useful in determining when the leak detector can be used. Also, a cold trap in the test object (or connecting line) is useful in minimizing contamination of the leak detector. Another method prevalent in vacuum testing is the "hood" technique. In this method, the test object is connected through appropriate valving to the leak detector and an auxiliary pump (as in the probe technique). However, in this case the test object is covered by a hood which is filled with helium. This is an overall test and does not pinpoint leaks. The hood can be removed and the probe technique used to locate individual leaks. In many cases the method is used as a "go—no go" technique in production testing. Test objects showing leaks are simply thrown out. Any type of hood, such as a bell jar, can be used as long as it can be filled with helium. Thin plastic sheet is useful for covering the test object or part of it. This is particularly true of parts with irregular surfaces. The helium indication with this method will depend on the size of leak(s), the amount of helium in the hood, the pressure inside the test object, the pumping speed used, the pressure inside the leak detector, and the length of time since application of helium to the hood. Consequently, specific rules of procedure must be set up for each installation where hood testing is used. The method can also be used for testing water lines. With the unit evacuated, a helium hose should be attached to one end of the line, the other end being open. After enough time has been allowed to displace all air (a matter of seconds to minutes, depending on size and length of the line), the open end is capped and the helium pressure is raised to a safe value for the line being tested. The use of high pressure makes it possible to detect very small leaks.

The sensitivity of the hood method can be increased by using the *accumulation method.* The test object is evacuated and isolated by a valve from the auxiliary pump while it is being covered with helium in the hood. The higher the helium concentration and the longer the test object remains in the hood, the more helium will pass through a given leak and accumulate in the test object. Suppose a vacuum gauge is included in the system between the leak detector and the test object. If there is a considerable rise in pressure during the "soaking" period, a gross leak exists. If this is not the case, the leak detector valve can be opened rapidly and at a uniform rate. A detectable leak will show up as a rise in the output signal to a peak, followed by a decline as the accumulated helium is pumped out by the leak detector. This accumulation method can also be used in connection with the probe technique. Most commercial helium leak detectors are provided with a valve at the top of the diffusion pump (the *accumulator valve*). Suppose a leak is suspected in part of a test object but a reliable reading cannot be obtained. In such a case, if the accumulator valve is closed, with helium being played on the suspected part, a leak will show up as a gradual increase in the output signal due to the accumulation of helium. However, the greatest value of the accumulator method is in its use with hoods.

15.15   Pressure and Pressure-Vacuum Testing

Pressure testing involves pressurizing the test object (or system) with helium or a helium-air mixture and then looking for leaks by sucking helium escaping through leaks into the leak detector (see Fig. 15.5b). A *sniffer probe,* also called a *pressure probe* or *sampling probe,* is used to go over the surface of the test object. This probe can be of various forms of design. A suction cup is useful for drawing in helium from a leak without pulling in too much surrounding air. The method is intrinsically less sensitive than vacuum testing because of air being drawn in with the helium. A valve is needed in the leak detector or the sampling line to regulate the flow of air and helium into the leak

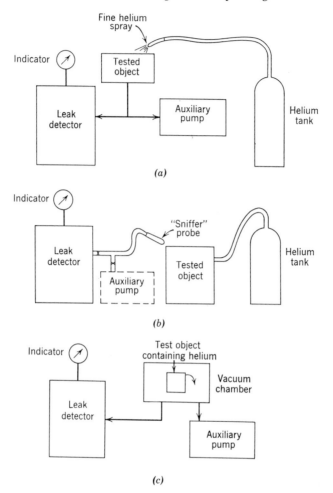

Fig. 15.5 (*a*) Vacuum testing. (*b*) Pressure testing. (*c*) Pressure-vacuum testing.

detector. Clearly, this method is simply a variation of the general method of pressure inside–indicator outside (Section 15.6). Its sensitivity is comparable to that of methods involving a vacuum gauge with suitable probe material and to the sensitivity of the halogen leak detector, i.e., around 10^{-6} to 10^{-7} std cc/sec. The manner in which the method is used can have considerable effect on the sensitivity.

The helium pressure that can be used will depend on the nature of the test object or system. When gaskets and fragile parts are involved, it may not be possible to use more than 1 or 2 psig. Small, rugged components can be pressurized to around 80 psig (or more). The sensitivity increases as the pressure is increased. The method is useful for vacuum components that will not stand internal vacuum. The "sniffer" is usually a small orifice, its size determined by the allowable pressure in the leak detector. For maximum sensitivity, the orifice should be large enough to admit a large sample of gas from the vicinity of the probe. Practically, the orifice must be rather small to maintain a low enough pressure in the detector. As an example, the orifice might be a tube 6 in. long with an inside diameter of 0.001 in. Rather small samples of the atmosphere at the probe tip are admitted to the detector. This results in a sensitivity of around a thousand times less than in vacuum testing. The response time is important in this method. The longest practical length of probe line (hose) is about 10 ft for a sensitivity of 10^{-6} std cc/sec. The test area should be kept free of helium from sources other than the test object. This is usually not too serious a problem since helium diffuses rapidly in air and rises. As in vacuum testing, the testing should proceed from top to bottom.

Certain points regarding pressure testing can be summarized as follows:

1. The flexible hose to the sniffer preferably should be of the metal bellows type. However, rubber tubing (preferably neoprene) is often used. Use as short a hose as possible.

2. Sometimes a specially shaped sniffer probe is required. This can be made by flattening the end of a piece of copper tubing. Needle valves are also used to vary the suction through the sniffer. A long needle with slight taper is desirable. The valve must be at the end of the sniffer so there is no volume between the valve and the point where the sample enters the sniffer.

3. Sometimes the flow through the sniffer is too large to be handled by the pumping system of the leak detector. Accordingly, an auxiliary pump is used (see Fig. 15.5b). Alternatively, a special sniffer with small flow can be used. A long hose makes it necessary to use an auxiliary pump because of the large amount of outgassing of the hose. Also, a long hose results in a large response time, due to the long path.

4. When pressure testing large objects, it is preferable to use a short hose and to move the leak detector rather than to use a long hose and a stationary leak detector.

5. If it is desired to use several atmospheres of helium, and if the test object is large, considerable helium will be used. To avoid loss of

helium, "pressure up" with some gas such as nitrogen, for example, by combining 10 atm of nitrogen and 1 atm of helium. Although the helium is diluted, the increased flow rate more than offsets this dilution.

Some special procedures that might be followed in pressure testing are the following:

1. Connect a suitable length of flexible hose to the sniffer, with the other end connected to the outlet of the leak detector.

NOTE: If an auxiliary pump is used, then a tee is used to make connections to the sniffer, the leak detector, and the pump, with a valve in the pump line. This pump is turned on before going to step 2 below.

2. Carefully open the leak detector throttle valve until the pressure indication is correct (leak detector already pumped down to operating conditions).

3. Put leak detector into operational condition for testing.

4. Test by passing end of sniffer slowly over suspected points of leakage (or proceed slowly from top to bottom of test object).

5. After sniffing in one location, the probe may use up all the helium issuing from a leak there. Remove the probe until a detectable concentration of helium has built up again around the leak.

6. Keep test object out of drafts so helium can build up in a concentration around a leak.

7. Increase the sensitivity by restricting the slight suction of the sniffer to the neighborhood of the area under test. A rubber fitting (even a small piece of rubber tubing) over the end of the sniffer will act as a "suction cup."

The pressure-vacuum testing method (see Fig. 15.5c) involves combining pressure testing with vacuum testing so that parts can be tested with the high sensitivity inherent in vacuum testing. The part is pressurized with helium (sealed) and placed in a vacuum chamber, which is often a bell jar. The chamber is usually pumped down by an auxiliary pump and is also connected to the leak detector. The vacuum container should be sealed to a base plate with an O-ring (or L-gasket) and should have a minimum volume so as to shorten pump-down time. Specially built metal units are usually preferable to glass bell jars. Vacuum grease is not needed to seal the container to the base plate. In this method, the helium flowing into the vacuum chamber shows no dilution so high sensitivity results.

If a leak is indicated during the test, its location may be determined by removing the chamber, attaching a sniffer connection to the pumping port, and pressure testing the port while it is under helium pressure. The pressure-vacuum method is recommended for leak testing parts having long tubing connections. Vacuum testing such parts results in long response and clean-up times. The method is also recommended for checking parts which must be back filled with helium (filled with helium and sealed with the helium inside), such as many electronic instruments and components that are hermetically sealed. A valved manifold is used to admit helium to the part or parts (back fill). A modification of this method can be used to check parts that will not stand atmospheric pressure. The valved manifold is arranged so that the part and the chamber are pumped down simultaneously. Helium is then admitted to the part at a pressure that will not damage it. Leak testing can then proceed.

15.16   Commercial Helium Leak Detectors

Commercial helium leak detectors will differ in physical appearance and in constructional details. However, the same basic principle of operation is used and the detailed operating procedures are much the same for various models. Figure 15.6 shows a typical commercial leak detector. This is a unit manufactured by Veeco Vacuum Corporation, New Hyde Park, Long Island, New York, and is designated as Model MS-9AB, Production Leak Test Station. This model is most suitable for semiautomatic leak testing of small parts. However, it can be adjusted for manual operation.

The main features of the MS-9AB are shown in Fig. 15.7. It will be noted that the test object is set up for a hood test. It could be connected to either of the two side inlet flanges (TP 1 or TP 3) for vacuum testing, or, alternatively, a sniffer probe could be connected to one of these flanges for pressure testing, with the other flange blanked. In the second case, the quick vacuum coupling would be plugged or used for a standard leak (called SENSITIVITY CALIBRATOR in this brand of detector). The test, air, and pump valves, together with the vacuum controller and roughing pump, comprise the features making automatic operation possible. In the Model MS-9A these features are left out and a connection is made directly from the manual throttle valve to the tube joining the side inlet flanges. In this case an auxiliary mechanical pump must be provided for vacuum or pressure-vacuum

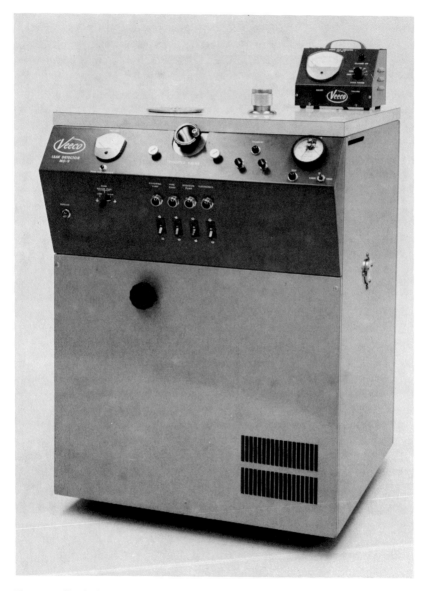

Fig. 15.6 Typical commercial leak detector. (Veeco Vacuum Corporation, New Hyde Park, Long Island, N.Y., Model MS-9AB.)

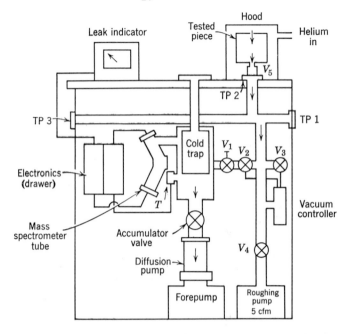

Fig. 15.7   Schematic of MS-9AB.

$$V_1 = \text{manual throttle valve}$$
$$V_2, V_3, V_4 = \text{automatic solenoid valves}$$
$$V_5 = \text{quick vacuum coupling}$$
$$T = \text{cold cathode tube}$$

leak hunting, whereas in the Model MS-9AB the roughing pump serves this function.

Attention will be directed here to the MS-9A type of leak detector since it is very commonly used in small laboratories and vacuum plants. It is generally necessary to test various sizes of vacuum components (or systems). This makes it difficult to set up any type of semiautomatic operation. The initial setup of the leak detector should be made in close cooperation with the manufacturer's representative. He will usually advise setting up an auxiliary pump with appropriate pumping lines and valves for pumping down the test object. In the hood method the user will have to provide a suitable hood and means for connecting the test object to the leak detector. The pressure testing method will require the acquisition of a suitable sniffer probe and a flexible connecting line. Assume now that the leak detector has been set up and tested.

STARTING FROM TOTAL SHUT-DOWN

The switches for FOREPUMP, DIFFUSION PUMP, and ELECTRONICS should be in the OFF positions, the THROTTLE VALVE should be CLOSED, and the ACCUMULATOR VALVE should be OPEN. The power cable should be plugged into a 115 v, 60 cycle source and the cord for the portable LEAK INDICATOR plugged into the chassis. Specific steps to be followed are:

1. Turn on FOREPUMP switch.

2. When the forepump stops gurgling (about 2 min), turn on DIFFUSION PUMP.

3. After about 15 min, turn on ELECTRONICS. This will also turn on the DISCHARGE GAUGE.

Turning on the electronics too early will contaminate the gauge. If the pressure reading is off scale, shut off the electronics switch and wait a few minutes. After this time the pressure meter should be on scale. When the pressure is 1 μ or less, add liquid nitrogen to the trap (about one-quarter full). Use a funnel and allow a large clearance between it and the fill pipe or trap so that the displaced air can escape from the trap, thus avoiding violent bubbling of the liquid nitrogen. After about 3 min add sufficient nitrogen to fill the trap. The pressure should drop to 5×10^{-5} Torr or less in a few minutes.

NOTE: Do not obstruct the filler opening of the cold trap.

4. Push FILAMENT ON button. This should light up the filament pilot lamps on the front panel and on the LEAK INDICATOR. (The filament is automatically turned off if the pressure rises above about 0.3 μ).

SENSITIVITY CHECK

This check should be made before leak testing and at the end of the day.

1. Close throttle valve and remove plugs from PORT and SENSITIVITY CALIBRATOR. Insert SENSITIVITY CALIBRATOR into test port (TP 2, Fig. 15.7). Check that the valve on the sensitivity calibrator is open so it can be pumped down at the same time the manifold is pumped.

2. Push FILAMENT ON button. Close valve on SENSITIVITY CALIBRATOR and zero the LEAK INDICATOR METER by means of its zero ADJUST knob. If necessary, use the COARSE BALANCE control on the auxiliary control panel inside the pull-out drawer. Check emission current by pressing

EMISSION button. The emission should be between 2 and 4 mamp. If this is not the case, you will find that detailed instructions for adjustment are given by the manufacturer.

3. Rough down the port manifold by using an auxiliary pump and slowly open the THROTTLE VALVE while maintaining the pressure at 10^{-4} Torr or lower. If the filament goes off (too high a pressure), close the throttle valve slightly and press FILAMENT ON button.

4. Valve off external system. The throttle valve is now fully open and the SENSITIVITY CALIBRATOR is connected to the leak detector.

5. Read the deflection on the LEAK INDICATOR and close the valve on the SENSITIVITY CALIBRATOR.

6. The LEAK INDICATOR should return to zero. (If it does not return to zero, the remaining signal reading is not due to helium from the SENSITIVITY CALIBRATOR and must be subtracted from the original reading.)

7. Record the LEAK INDICATOR reading for the known input of helium. See Section 15.12 for calibration.

8. Close the THROTTLE VALVE.

9. Vent the PORT MANIFOLD; remove the SENSITIVITY CALIBRATOR, plug the end with the red stopper, and open the SENSITIVITY CALIBRATOR valve.

LEAK TESTING

Once the sensitivity check has been made, leak testing can proceed. In vacuum testing, connect the test object to the port desired, "rough" it down, and "throttle" into the leak detector. Spray the test object with helium and look for a deflection of the leak indicator (or listen for an audio signal, where provided). See Section 15.15 for using the pressure and pressure-vacuum methods.

PARTIAL SHUT-DOWN (OVERNIGHT OR WEEKEND)

1. Close THROTTLE VALVE.
2. Turn off ELECTRONICS.

COMPLETE SHUT-DOWN

1. Close ACCUMULATOR VALVE.
2. Turn off DIFFUSION PUMP—note time.
3. Turn off ELECTRONICS.

4. Open THROTTLE VALVE and TEST MANIFOLD to vent the LEAK DETECTOR system up to ACCUMULATOR VALVE.

5. Remove, clean, and replace cold trap.

6. About 15 min after shutting off the DIFFUSION PUMP, turn off FOREPUMP.

7. Open ACCUMULATOR VALVE to vent rest of system.

NOTE: If cold trap is not at room temperature when the system is shut down completely, remove the bucket immediately after venting the system. Decant any refrigerant remaining and wipe off the bucket with a lint-free rag moistened with acetone. Dry the bucket (hot air, etc.) and re-install in the trap, taking care to avoid touching surfaces with fingers (grasp bucket by top flange only).

The specific steps involved in using a commercial leak detector will differ from model to model and from manufacturer to manufacturer. However, the details of operation for each model are usually spelled out carefully by the manufacturer. Maintenance problems are generally minimum. The openings for air circulation through the leak detector should not be blocked and vibrations should be avoided. A maintenance schedule should be set up for changing mechanical and diffusion pump oils. Gasketed joints that are disassembled periodically, such as the flanges for connecting test objects, the liquid nitrogen trap, and the standard leak, should be inspected on disassembly (gaskets and grooves). Spare parts, such as filaments, gaskets, and certain electronic parts, should be kept on hand. The manufacturer can provide a list of parts that are most likely to be needed.

15.17   What Leak Detection System Should Be Used?

No simple answer can be given to this question. Although considerable attention has been given to helium leak detectors, this does not necessarily mean that such detectors should be used in all circumstances. In many cases a simpler system is quite adequate. One general criterion that can be used to determine when a leak detection system other than a helium leak detector is applicable is when the leak to be measured is larger than 10^{-6} std cc/sec. Of course, even when a total tolerable leak much larger than 10^{-6} std cc/sec is allowable, a number of very small leaks (each considerably less than 10^{-6} std cc/sec) may be responsible. In this case it may be necessary to use a helium leak detector or some leak detection system with a comparable sensitivity. Consider the instances in which experience indicates that only leaks of 10^{-6} std cc/sec or larger need be detected.

Several detection methods with the necessary sensitivity could be used in such instances (this is indicated in Table 15.1). The minimum detectable leaks shown are rather conservative values, being applicable to commercially available equipment without extraordinary precautions. However, even the values shown will not be achieved without reasonable care, particularly with respect to reaching stable pressure conditions. Electronic circuitry and outgassing are important problems that must be considered. The pressure ranges shown are very approximate. Only some common probe materials are shown for leak detectors 1, 2, 3, 4, and 5. The sensitivities obtained with thermocouple and thermistor gauges are comparable to that shown for the Pirani gauge, although usually somewhat less. Similarly, cold cathode and Alphatron gauges have sensitivities comparable to that for the thermionic ionization gauge. More attention will be devoted in this section to the halogen leak detector than to other types because it is readily available commercially. This detector can be used if the answer to any of the following questions is "yes."

1. Is the test object presently being tested with helium under pressure?

2. Will a leakage of 10^{-6} std cc/sec or more at 1 atm difference be acceptable?

3. Can the test object stand a pressure of several atmospheres?

4. Is the test object small, or can the area to be leak tested be enclosed in a volume of a few cubic inches?

If question 1 can be answered "yes," then halogen instead of helium can be used with an appropriate halogen leak detector. It is possible to detect halogen concentrations in air as low as 0.2 ppm or a leak of about 10^{-6} std cc/sec. If the answer to question 2 is "yes," then either vacuum or pressure testing can be used. In pressure testing, the halogen leak detector is pumped down to the operating pressure and a sniffer probe is connected to it through an appropriate valve, as in the case of the helium leak detector. The sensitivity with either vacuum or pressure testing is about 10^{-6} std cc/sec. A considerable gain in sensitivity can be obtained when it is possible to pressurize the test object to several atmospheres. At a pressure of 140 psig, a hundred times more gas will leak through than at 15 psig differential and a sensitivity of about 10^{-8} std cc/sec can be achieved. With a small test object or a small area to be tested, increased sensitivity can be obtained by using the accumulation method of leak testing. A small, fairly tight cap or enclosure can be placed around the area to be tested (test object under pressure), with halogen allowed to accumulate for a period of time. The leak detector is then connected to the enclosure to sample the contents.

The methods described above for using a halogen leak detector are essentially the same as those used with a helium leak detector, or variations thereof. As a matter of fact, any leak detector scheme that has its own vacuum system can be adapted to vacuum, pressure, and pressure-vacuum testing. However, the methods using spark coils, discharge tubes, or vacuum gauges generally are used only in vacuum testing. This is a matter of convenience since such devices (except the spark coil) are readily incorporated into a vacuum system as a permanent part of the system. The above detectors (spark coils, etc.) are usually used with the probing technique, although the hood method can also be used. The minimum detectable leaks shown in Table 15.1 for vacuum gauges can be increased considerably by the use of special techniques. The problem with vacuum gauges is drift and instability. Careful attention to electronic circuitry, gauge design, degassing, and elimination of contaminating gases and vapors is usually necessary to achieve high sensitivities. Through the use of specialized techniques, reports have been made of sensitivities comparable to that of a helium leak detector (or even higher). However, such sensitivities cannot be obtained routinely. Also, as the sensitivity increases, the time required to find a leak may increase considerably. At the present time, the palladium barrier ionization gauge appears to offer considerable promise, with a sensitivity of 10^{-8} std cc/sec fairly feasible without extraordinary precautions. The use of a tungsten filament with oxygen (diode or triode) also bears consideration.

Several methods for increasing the leak detection sensitivity are sometimes used. The differential method has already been mentioned for eliminating meter fluctuation. An increase in sensitivity over the single gauge of a factor of 10 is fairly common with this method. Two other methods commonly used to increase sensitivity are:

1. *"Backing Space" Technique.* The effect of the probe gas is increased by having the diffusion pump compress the gas into a dead space by closing a valve in the foreline. Clearly, the gauge must be such that the pressures involved can be read. Pumping the gauge with a separate system and bleeding in gas could be done.

2. *Rate of Pressure Build-Up.* The apparent change in rate of pressure build-up is measured when the probe gas replaces air through the leak.

Although these methods give an increased sensitivity, they suffer from two serious disadvantages:

1. A substantial increase in time for finding leaks.
2. The requirement for highly trained personnel to interpret results.

The helium leak detector is a form of mass spectrometer, which is "tuned" to measure only helium ions. Many efforts have been made to increase the sensitivity of such devices above the nominal value of 10^{-10} std cc/sec of commercial devices. Peters (*Rev. Sci. Instr.*, **30**, 1093, 1959) has reported on a mass spectrometer design claimed to have a sensitivity of 10^{-13} std cc/sec of helium. However, where helium is in short supply it is necessary to have a leak detector with a sensitivity comparable to that of the helium leak detector but using another probe gas. In principle, it should be possible to convert a commercial helium leak detector over to the detection of some other gas. However, in practice this can turn out to be expensive and unsatisfactory. Consequently, considerable attention has been given to other designs of mass spectrometer, although there is no reason why commercial helium leak detectors could not be built to detect other gases. The problems encountered stem from the higher concentrations of other gases in the atmosphere and from the difficulty in separating the desired ions from unwanted ions. Hydrogen and argon have been used fairly extensively with mass spectrometers. However, whereas the normal concentration of helium in air is 1 part in 670,000, the normal concentrations of hydrogen and argon are about 1 part in 10,000 and 1 part in 100, respectively. Consequently, the background is considerably higher with these latter gases and suitable means must be used to compensate for this. Hydrogen has attractive properties in that it diffuses rapidly through small leaks. However, it is flammable, so suitable safety precautions must be taken, such as good ventilation, no open flames, etc. Cossuta and Steckelmacher (*J. Sci. Instr.*, **37**, 404, 1960) have reported on a lens mass spectrometer with a sensitivity of 4×10^{-10} std cc/sec for hydrogen.

Many of the types of spectrometers used for leak detection were originally developed to analyze gases in vacuum systems. No attempt will be made to discuss the various types of such mass spectrometers reported in the literature or available commercially. Only one type, on which considerable work has been done and which is in fairly general use, will be discussed briefly. This device is called the Omegatron and is available commercially, e.g., from Elliott Brothers, Ltd., London, England. Its operation is based on the principle of a cyclotron, in which ions are speeded up by a radiofrequency (rf) voltage and forced to move in curved paths by a magnetic field. Ions of interest can be selected by adjusting the rf voltage and the magnetic field strength. Sensitivities as high as 6×10^{-11} std cc/sec with standard vacuum testing techniques and as high as 3×10^{-14} std cc/sec with accumulation techniques have been claimed for helium and argon. For such high sensitivities, the vacuum in the detector must be good, using bakable components. The sensitivity with hydrogen is lower because of the difficulty in eliminating all hydrogen contamination. At this writing, the Omegatron is used for various special applications in leak detection but has not become competitive with standard helium leak detectors for general purposes.

For detecting leaks in ultra-high (and very high) vacuum systems it might be thought at first that a helium leak detector would be the obvious choice. However, this is not the case (Alpert, *J. Appl. Phys.*, **24**, 860, 1953). The pressure in the leak detector is usually considerably higher than that in the system being leak tested (10^{-9} to 10^{-12} Torr). Consequently, it is difficult to get helium from a leak into the detector. The standard method of leak testing ultra-high and very high vacuum systems is to use the vacuum gauge (Bayard-Alpert type) which is part of the system. Suppose the equilibrium pressure in a system is 5×10^{-9} Torr. The application of acetone, ether, or carbon tetrachloride to a leak may either send the current reading flying off scale to an equivalent pressure above 10^{-8} Torr, or, the leak having thus been temporarily clogged, result in a reduction of pressure to 10^{-9} Torr or less. Another technique for finding leaks is a "shut-off" test (rate-of-rise method), where the vacuum system is isolated from the diffusion pumps and the rate of pressure rise is measured on an ion gauge. Without leaks, the ion gauge will *reduce* the pressure when operated continuously, due to sputter-ion pumping. A rate-of-rise curve is obtained by operating the ion gauge intermittently, taking readings every few minutes. Only a few seconds are required for a reading. The shape of the curve, as compared to that for a tight system, reveals the presence of leaks. This method will also distinguish between real and virtual leaks. Virtual leaks are usually due to volatile constituents which give a constant high pressure, whereas real leaks show a constant rate of rise over a long period of time.

15.18   Some General Rules for Leak Hunting

There are such large differences between various vacuum components and systems that it is not possible to give hard and fast rules for all situations. However, there are a few procedures which find general application.

1. Test components before assembly.

a. When the part can be pressurized, use one of the appropriate testing methods—soap bubble test, chemical indicators, sniffer test with halogen or helium leak detector. The choice of method can be roughly estimated by the maximum leak allowed. Some idea of this can be obtained, in turn, by looking at the tolerable leak for the system as a whole and the number of components involved.

b. When the part cannot be pressurized, set up the part in a vacuum system and use one of the vacuum testing methods (probe or hood)— vacuum gauge with probe gas, halogen or helium leak detectors, etc. Again, the choice of method can be guided by the tolerable leak.

2. Do not look for leaks in a system unless there is fair assurance that leaks exist. A rate-of-rise measurement will give valuable information.

3. Do not neglect the fact that apparent leaks may be due to malfunctioning of pumps.

1. If the pressure does not come down below the Torr range, check for gross leaks such as poorly seated gaskets, cracked parts (in assembly), and partly open valves. Hissing, wavering of a flame, and visual inspection should determine the cause. Make sure the mechanical pump is operating properly.

2. If the pressure reaches the micron range (say, 300 μ) but is still too high to turn on the diffusion pump, make rate-of-rise measurements using isolation techniques to get the general location of the leak.

a. If there is no indication of a leak in the main vacuum system, check the level of oil in the mechanical pump and its condition. Usually this is not necessary with a new system.

b. If there is an indication of a leak in the main system, use a vacuum testing method appropriate to the pressure involved. Often the use of the thermocouple or Pirani gauge is simplest. The whole system can be pressurized but this is usually done only with large systems and is one of the last resorts.

3. After the diffusion pump has been operating for some time (sev-

eral hours or days, depending on the size and nature of the system) and the pressure has not reached the desired operating value, determine whether the pump is at fault. This can be done by taking a rate-of-rise measurement or by measuring the speed of the pump (see Chapter 7). A rate-of-rise measurement will also determine whether virtual or real leaks are the cause of the trouble. Make sure the virtual leaks are not due to vapors from the cold traps by keeping these traps filled. With other virtual leaks it is usually necessary to simply keep pumping. Highly contaminated parts may necessitate dismantling the system. When real leaks are indicated, any of the vacuum testing methods appropriate to the pressure may be used.

a. With small systems, the vacuum gauge–probe gas method or the use of a halogen leak detector are most common.

b. With large systems (several hundred liters or more) it is often necessary to use a helium leak detector to find small leaks.

TESTING A SYSTEM THAT HAS BEEN USED

1. If major modifications have been made to the system, the steps for testing a new system should be followed.

2. When a system has been shut down simply to change working parts (as in a bell jar) or for minor modifications, the testing procedure is very simple. If the pressure does not come down to the operating value in a reasonable time, make a rate-of-rise measurement. This will show whether or not real leaks exist. Once the fact is established that real leaks exist, check the obvious sources of leaks such as gasket joints that have been disassembled and new parts. If gross leaks are present, many of the steps for testing a new system may have to be followed.

15.19   Repairing Leaks

A general rule that should be followed in repairing leaks found during leak hunting is to make permanent repairs whenever possible. Temporary repairs are usually permissible if the system absolutely must be operated for a limited length of time before permanent repairs can be made. A permanent repair means reworking the part involved or replacing the part. Some common sources of leaks and possible methods of repair are indicated below:

1. Static gasket seals.

a. Tighten seal but not too much.

b. If method *a* does not work, shut down the system and examine the gasket and gasket surfaces.

c. Replace any damaged gasket and smooth rough surfaces (fine emery cloth).

d. When the gasket is not damaged, clean (acetone, etc.), coat with light film of good quality vacuum grease (when permitted), and reassemble seal.

e. Do not use sealing materials such as glyptal to stop the leak. This procedure is temporary and the gasket can't be used again.

2. Movable, gasketed seals (Wilson, chevron, etc.).

a. Add a small quantity of good quality vacuum grease to the moving member and operate this member a few times through the seal.

b. If method *a* doesn't work, try tightening the retaining rings on the seal.

c. If method *b* doesn't work, dismantle the seal and examine the component parts. Replace all damaged parts.

3. Flare fittings (and similar metal-to-metal seals).

a. Tighten the compression nut moderately. Too much tightening is likely to twist the tubing passing through the fitting.

b. If method *a* doesn't work, take the joint apart.

c. Try annealing the copper flare.

d. If method *c* doesn't work, use a thin coating of a suitable sealant (glyptal, etc.) on the surfaces that make contact.

4. Soldered, brazed, and welded joints.

a. For temporary repair, use a cement. This method works for small leaks where the diffusion pumps are in their operating range. A "thin" cement (clear glyptal, Eastman Kodak Resin 910, etc.) can be used for the smallest leak. For somewhat larger leaks use a "thicker" cement such as red glyptal.

b. For permanent repair, rework the joint or replace the part.

5. Leaks through metal parts.

a. For temporary repair, use a cement as in the case of soldered, brazed, and welded joints.

b. For permanent repair, rework the part involved (as appropriate) or replace the part.

c. Peening the leak is not reliable.

6. Glass-to-metal seals.

a. A temporary seal can be made with a cement or wax (heat part appropriately).

b. Rework wax seals. Such seals are not recommended.

c. For permanent repair, replace the seal.

7. Glass-to-glass joints, cracks, and pinholes in a glass system.

a. For small leaks, use a cement (at room temperature) or heat the glass and apply a suitable wax (picein, sealing wax, etc.).

b. For a large leak, rework the glass or replace the part of the system where the leak is located.

Quantities from Kinetic Theory of Gases

Table A.1 Some Constants and Conversion Units

N_0 = number of molecules in a mole (Avogadro's number) = 6.02
$\times 10^{23}$

k = Boltzmann's constant = 1.381×10^{-16} dyne-cm/deg

R = gas constant = 8.314 cal/°C/mole

1 atomic weight unit = 1.66×10^{-24} g

1 poise = cgs unit of viscosity = 1 g/cm/sec

1 microbar = 1 dyne/cm^2 = 0.745 μ Hg

Table A.2　Physical Properties of Some Gases and Vapors

Gas	M	$10^{23}\,m$	$10^{-4} \times \bar{v}$ (0°C)	$10^{-4} \times \bar{v}$ (25°C)	$\dfrac{10^{-4} \times}{\sqrt{\overline{v^2}}}$ (0°C)	$10^8\,\sigma$	λ 15°C, $1\,\mu$ Hg	$10^3\,K$	η	D
H$_2$	2.02	0.335	16.93	17.70	18.38	2.74	8.97	0.416	87	1.0
He	4.00	0.665	12.01	12.56	13.05	2.18	14.1	0.344	194	0.87
CH$_4$	16.04	2.663	6.01	6.27	—	4.14	3.92	0.072	108	0.29
NH$_3$	17.03	2.827	5.83	6.09	6.33	4.43	3.42	0.051	97	0.27
H$_2$O vapor	18.02	2.992	5.67	5.92	6.15	4.60	3.18	0.055 †	93	0.25
Ne	20.18	3.351	5.36	5.59	5.81	2.59	10.0	0.110	310	0.42
N$_2$	28.02	4.652	4.54	4.75	4.93	3.75	4.77	0.057	173	0.28
Air	28.98 *	4.811	4.47	4.67	4.85	3.72	4.86	0.057	180	0.27
O$_2$	32.00	5.313	4.25	4.44	4.61	3.61	5.15	0.057	200	0.28
HCl	36.46	6.06	3.98	4.15	4.32	4.46	3.38	—	140	0.21
A	39.94	6.631	3.81	3.98	4.13	3.64	5.06	· 0.039	220	0.25
CO$_2$	44.01	7.308	3.62	3.79	3.93	4.59	3.18	0.034	145	0.20
Hg	200.6	33.31	1.70	1.77	1.84	4.26	—	—	—	0.18
Electron	5.49×10^{-4}	9.1×10^{-5}								

* Calculated from density of 1.293×10^{-3} g/cc at 0°C and 760 mm Hg.
† At 100°C.

M = molecular weight
m = mass of a molecule, g
$\bar{v}$ = average speed, cm/sec
$\sqrt{\overline{v^2}}$ = root mean square velocity, cm/sec
σ = molecular diameter, cm

λ = mean free path, cm
K = thermal conductivity at 0°C, cal/cm/sec/°C
η = viscosity at 15°C, micropoises
D = relative diffusion into air, arbitrary scale

M, m, $\bar{v}$: From S. Dushman, *Scientific Foundations of Vacuum Technique*, 2nd ed., J. M. Lafferty, editor, John Wiley and Sons, New York, 1962, p. 16.
$\sqrt{\overline{v^2}}$ Calculated from $\bar{v}$ values.
σ, λ, K, η: From Kennard, *Kinetic Theory of Gases*, McGraw-Hill Book Co., New York, 1938, pp. 149, 180.
D: From Guthrie and Wakerling, *Vacuum Equipment and Techniques*, McGraw-Hill Book Co., New York, 1949. (By permission of the copyright holder, U.S. Atomic Energy Commission.)

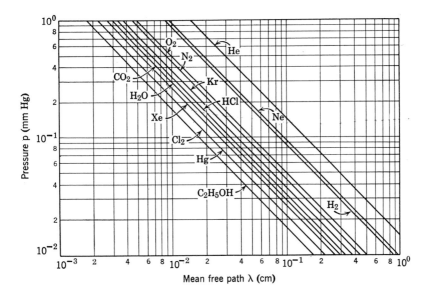

Fig. A.1   Mean free path as a function of pressure.

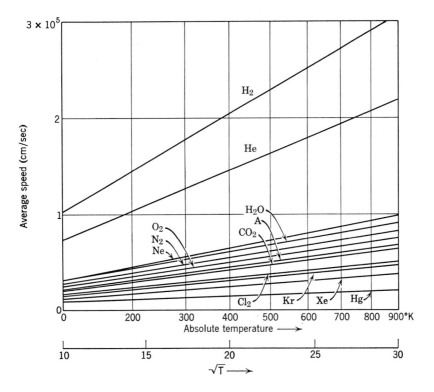

Fig. A.2   Average speed of molecules as a function of absolute temperature.

Mechanical, Ejector, and Diffusion-Ejector Pump Oils

Mechanical pump oils are generally petroleum oils. Paraffin oils are the most common. The various commercially available oils will differ in their vapor pressures and resistance to vapors. The manufacturers of mechanical pumps should be consulted regarding the proper oil to use.

Ejector pump oils need not have as low vapor pressures as diffusion pump oils. Examples of such oils are the Convoil and Convachlor oils (Consolidated Vacuum Corp.) and Narcoil 10 (see Table B.1).

Diffusion-ejector pump oils are much the same as ejector pump oils. However, certain diffusion pump oils are suitable in the same applications where diffusion-ejector oils are used.

Table B.1   Some Vapor Pump Oils

Trade Name	Supplier *	Chemical Composition	°C for Given Pressure (mm Hg)		Pressure at 25°C (mm Hg)	Reference †
			10^{-2}	10^{-5}		
Amoil	C.V.C.	*i*-Diamyl phthalate	93	22	1.3×10^{-5}	1
Amoil S	C.V.C.	*i*-Diamyl sebacate	114	43	1×10^{-6}	4
Apiezon A	J.G.B.	Mixture of hydrocarbons	110	37	2×10^{-6}	3
Apiezon B	J.G.B.	Mixture of hydrocarbons	127	50	4×10^{-7}	3
Apiezon C	J.G.B.	Mixture of hydrocarbons	160	77	1×10^{-8}	5
Arochlor	M.C.C.	Polychlorinated biphenyls	93	27	8×10^{-6}	3
Butyl phthalate	C.V.C.	*n*-Dibutyl phthalate	81	18	3.3×10^{-5}	1
Butyl sebacate	C.V.C.	*n*-Dibutyl sebacate	—	—	2×10^{-5}	8
b-S	—	Dibenzyl sebacate	155	82	4×10^{-9}	2
D.C. 702	D.C.	Semiorganic mixture of silicones	—	—	2×10^{-7}	7
D.C. 703	D.C.	Semiorganic mixture of silicones	153	83	5×10^{-9}	5
D.C. 704	D.C.	Single molecular species	—	76	4×10^{-8}	6
Litton oil	L.I.	Petroleum hydrocarbon	132	57	1.4×10^{-7}	3
m-Cr	—	Tri-*m*-cresyl phosphate	141	67	9×10^{-8}	2
Narcoil 10	N.R.C.	Chlorinated diphenyl	—	—	2×10^{-5}	7
Narcoil 40	N.R.C.	Di-3,5,5-trimethyl hexyl phthalate	—	—	10^{-7}	8
Octoil	C.V.C.	Di-2-ethyl hexyl phthalate	128	54	2.3×10^{-7}	1
Octoil S	C.V.C.	Di-2-ethyl hexyl sebacate	142	68	2×10^{-8}	3
p-Cr	E.K.C.	Tri-*p*-cresyl phosphate	144	71	2×10^{-8}	2

* *Suppliers:*

C.V.C.:   Consolidated Vacuum Corp., Rochester, N.Y.
D.C.:   Dow-Corning Corp., Midland, Mich.
E.K.C.:   Eastman Kodak Co., Rochester, N.Y.
J.G.B.:   James G. Biddle Co., Philadelphia, Pa.
L.I.:   Litton Industries, San Carlos, Calif.
M.C.C.:   Monsanto Chemical Co., St. Louis, Mo.
N.R.C.:   National Research Corp., Cambridge, Mass.

† *References:*

1. K. C. D. Hickman, J. C. Hecker, and N. Embree, *Ind. Eng. Chem.*, **9**, 264 (1937).
2. F. H. Verhoek and A. L. Marshall, *J. Am. Chem. Soc.*, **61**, 2737 (1939).
3. Metropolitan Vickers, 1946.
4. From constants deduced by S. F. Kapff and R. B. Jacobs of Distillation Products, Inc.
5. A. Herlet and G. Reich, *F. Angew. Phys.*, **9**, 1, 14 (1957).
6. A. R. Huntress et al., *Vac. Symp. Trans.*, **104** (1959).
7. H. M. Sullivan, *Rev. Sci. Instr.*, **19**, 1 (1948).
8. Manufacturer's data.

Quantities and Relationships
Useful in Vacuum Practice

Table C.1 Symbols and Units

a	length of side of rectangular or triangular duct, in.
b	length of side of rectangular duct, in.
L	length of duct, in.
D	diameter of circular pipe or aperture, in.
B_1	perimeter of pipe, in.
B	total perimeter of annular passage (inner + outer), in.
A	cross-sectional area of duct or aperture, in.^{2}
V	volume of vessel, l
V'	volumetric flow, l/sec
λ	mean free path, cm or in.
T	absolute (Kelvin) temperature
M	molecular weight
ρ	density of a gas, g/cc
η	coefficient of viscosity, dynes/sec/cm^2
$\bar{v}$	average velocity, cm/sec
v_p	most probable velocity, cm/sec
v_r	root mean square velocity, cm/sec
P	pressure, μ Hg
P'	time rate of pressure = dP/dT, μ Hg/sec
P_1', P_2'	time rates of pressure corresponding to pressures P_1, P_2
P_1	upstream pressure, μ Hg
P_2	downstream pressure, μ Hg
$\bar{P}$	average pressure in a pipe, μ Hg
Q	amount of gas flowing by a certain plane at constant T, μl/sec
S	pumping speed, l/sec
U	conductance, l/sec
W	impedance, sec/l
m'	mass of gas striking unit area per unit time, g/cm^2/sec
Y	correction factor in rectangular pipe (viscous conductance)

508

Table C.2 Some Vacuum Formulas

Mean Free Path (λ)	$\lambda = 8.524\eta/P\sqrt{T/M}.$ $\lambda = 1.91/P$ in. $\lambda = 4.86/P$ cm.
Average ($\bar{v}$), most probable (v_p) and root mean square velocity ($\sqrt{\overline{v^2}} = v_r$)	$v_p = 12{,}900\sqrt{T/M}.$ $\bar{v} = 1.128v_p.$ $v_r = 1.224v_p.$
Mass of gas striking unit area per unit time (m')	$m' = 58.32 \times 10^{-6}P\sqrt{M/T}$ g/cm^2/sec
Pumping speed (S)	$S = Q/P$ l/sec
Impedance (W) and conductance (U)	$\dfrac{1}{W} = U = \dfrac{Q}{P_1 - P_2}$ l/sec
Series impedances and conductances	$W = \sum_i W_i;\quad \dfrac{1}{U} = \sum_i \dfrac{1}{U_i}.$
Parallel impedances and conductances	$\dfrac{1}{W} = \sum_i \dfrac{1}{W_i}.\quad U = \sum_i U_i.$
Pumping speed through impedance (W) of pump with speed (S_p)	$\dfrac{1}{S} = W + \dfrac{1}{S_p} = \dfrac{1}{U} + \dfrac{1}{S_p}$
Reynolds number (Re)*	General: Re $= D\bar{v}/\rho\eta$ Air at 20°C, circular pipe: Re $= Q/226D$
Transition from turbulent to viscous flow, air at 20°C*	Turbulent: $Q > 5 \times 10^5 D$ Viscous: $Q < 2.5 \times 10^5 D$
Transition from viscous to molecular flow, air at 20°C*	Viscous: $\bar{P}D > 200$ μin. $\lambda < D/200$ Molecular: $\bar{P}D < 6$ μin. $\lambda > D/3$
Viscous conductance, thin, small aperture, air at 20°C, $P_2/P_1 \leq 0.1$*	$U_v = 129$ l/sec $= 100D^2$ l/sec (circular aperture)

Table C.2 (Continued)

Molecular conductance, thin, small, aperture, $\lambda \gg$ aperture *

$U_M = 75A$ l/sec
$= 59D^2$ l/sec (circular aperture)

Aperture pumping speed *

$S = U(1 - P_2/P_1) = U$ for $P_2/P_1 < 0.1$

Molecular flow through large aperture *

$U_M = 75AA_0/(A_0 - A) = 59D^2D_0^2/(D_0^2 - D^2)$ (circular aperture)

Reduces to $75A$ or $59D^2$ for $A_0 \gg A$

Aperture pumping speed
$= [U(1 - P_2/P_1)A_0]/(A_0 - A)$
$= UA_0/(A_0 - A)$ for $P_2/P_1 < 0.1$

Viscous conductance, long circular pipe, air at 20°C *

$0.35D^4\bar{P}/L$ l/sec

Viscous volume flow, long circular pipe, air at 20°C *

$0.25D^4[(P_1 - P_2)/L]$ l/sec

Viscous conductance, long rectangular duct, air at 20°C *

$0.36a^2b^2\bar{P}Y/L$ l/sec

a/b:	1.0	0.9	0.8	0.7	0.6	0.5	0.4	0.3	0.2	0.1
Y:	1.00	0.99	0.98	0.95	0.90	0.82	0.71	0.58	0.42	0.23

Viscous volume flow, long rectangular duct, air at 20°C *

$V' = 0.36a^2b^2Y[(P_1 - P_2)/L]$ l/sec

Y (see above)

Impedance of any short pipe *

Add impedance of aperture to impedance of long pipe

Molecular conductance, long circular pipe, air at 20°C *

$U_M = 6.5D^3/L$ l/sec

Molecular conductance, long pipe of noncircular cross section *

$U_M = 61.8(KA^2/B_1L)$
$K = 1$ for pipe of circular cross section

Molecular conductance, annular aperture, air at 20°C *

$U_M = 75AK$ l/sec
$K =$ shape factor

$U_M = 33.3AK/LB$ l/sec
K = shape factor

Molecular conductance, annular passage, air at 20°C *

Values of shape factor—annular aperture or passage

D_1/D_2	K		D_1/D_2	K
0	1		0.707	1.254
0.259	1.072		0.866	1.430
0.500	0.866		0.966	1.675

D_1, D_2 (see above)

Molecular conductance, long rectangular duct, air at 20°C *

$$U_M = \frac{16.6a^2b^2}{(a+b)L} \cdot K \text{ l/sec}$$

Values of shape factor—for long rectangular duct *

b/a	K		b/a	K
1	1.108		0.200	1.297
0.667	1.126		0.125	1.400
0.500	1.151		0.100	1.444
0.333	1.198			

Molecular conductance, long equilaterally triangular duct, air at 20°C *

$2.58a^3/L$ l/sec

Molecular flow at bends and elbows (for each bend allow for an effective length in the range indicated by the inequality) *

$L_{axial} < L_{eff} < L_{axial} + 0.11D$

Pumping speed of pump of speed S_p near base pressure P_0 *

$S = S_p(1 - P_0/P)$ l/sec

Pump speed formula, metered-leak method *

$S_p = 760V'/AP$ l/sec

Pump speed formula, rate-of-rise method *

$S_p = [(P_2' - P_1')/(P_2 - P_1)]V$ l/sec

*Guthrie and Wakerling, *Vacuum Equipment and Techniques*, McGraw-Hill Book Co., New York, 1949, pp. 246–247. (By permission of the copyright holder, U.S. Atomic Energy Commission.)

Table C.3   Combined Viscous and Molecular Flow

(2 in. diameter (ID) pipe)

P (Torr)	C_V (l/sec)	C_M (l/sec)	C_T (l/sec)
1	3910	43.3	3953
10^{-1}	391	43.4	434
10^{-2}	39.1	44.0	83
10^{-3}	3.9	47.2	51
10^{-4}	0.39	51.9	52
10^{-5}	0.039	52.9	53
10^{-6}	0.004	53.0	53

C_V = viscous conductance
C_M = molecular conductance
C_T = total conductance

EXAMPLE OF CONDUCTANCE (IMPEDANCE) CALCULATIONS

Straight-through thimble trap.   Four inch nominal diffusion pump. Molecular flow.

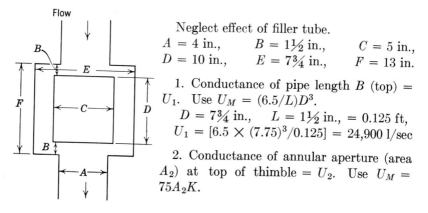

Neglect effect of filler tube.
A = 4 in., B = 1½ in., C = 5 in.,
D = 10 in., E = 7¾ in., F = 13 in.

1. Conductance of pipe length B (top) = U_1. Use $U_M = (6.5/L)D^3$.
D = 7¾ in., L = 1½ in., = 0.125 ft,
$U_1 = [6.5 \times (7.75)^3/0.125] = 24{,}900$ l/sec

2. Conductance of annular aperture (area A_2) at top of thimble = U_2. Use $U_M = 75A_2K$.

Area of housing = $\pi/4 \times E^2 = \pi/4 \times (7¾)^2$
Area of thimble = $\pi/4 \times C^2 = \pi/4 \times (5)^2$
Area $A_2 = (\pi/4)[(7¾)^2 - 5^2] = 27.5$ in.^{2}
$75A_2 = 75 \times 27.5 = 2060$ l/sec
$C/E = D_1/D_2 = 5/7.75 = 0.64$, $K = 1.2$ (see Table C.2),
$V_2 = 1.2 \times 2062 = 2472$ l/sec

3. Conductance of annular passage between thimble and housing = U_3. Use $U_M = 33.3A^2K/LB$.

$A = A_2 = 27.5$ in.^{2}, $K = 1.2$ (as in calculation 2),
$L = D = 10$ in. $= 0.833$ ft,
$B = \pi \times C + \pi \times E = \pi(5 + 7.75) = 12.75\pi$,

$$U_3 = \frac{33.3 \times (27.5)^2 \times 1.2}{0.833 \times 12.75\pi} = 900 \text{ l/sec.}$$

4. Conductance of pipe length B (bottom) = U_4.
        Same as for calculation 1, i.e., = 24,900 l/sec.

5. Conductance of aperture $A = U_5$. Use $U_M = [75A_0A/(A_0 - A)]$.
$A_0 =$ area of housing $= \pi/4 \times E^2 = \pi/4 \times 7.75^2$,
$A =$ area of pipe or pump $= \pi/4 \times A^2 = (\pi \times 4^2)/4 = \pi \times 4$,

$$U_M = \frac{75 \times \pi/4 \times 7.75^2 \times \pi/4 \times 4^2}{\pi/4(7.75^2 - 4^2)} = 1284 \text{ l/sec}$$

The conductances are in series. Use $1/U = \sum_i (1/U_i)$

$$\frac{1}{U} = \frac{1}{U_1} + \frac{1}{U_2} + \frac{1}{U_3} + \frac{1}{U_4} + \frac{1}{U_5}$$

U_1 and U_4 are very large (24,900 l/sec) so they can be neglected:

$$\frac{1}{U} = \frac{1}{U_2} + \frac{1}{U_3} + \frac{1}{U_5} = \frac{1}{2472} + \frac{1}{900} + \frac{1}{1284}$$

Solving. $U = 435$ l/sec (say 430). Assume an average speed of 250 l/sec for a diffusion pump attached to the bottom of the trap. Then the pumping speed at the top of the trap (entrance to pipe) is given by $1/S = 1/U + 1/S_p$ where $S_p = 250$.

$$S = \frac{US_p}{U + S_p} = \frac{430 \times 250}{430 + 250} = 158 \text{ l/sec}$$

Loss of pumping speed due to trap $= \dfrac{250 - 158}{250} \times 100 = 37\%$

Appearance of Discharges in Gases and Vapors at Low Pressures

Gas	Negative Glow	Positive Column
Air	Blue	(Reddish)
Nitrogen	Blue	Yellow (red gold)
Oxygen	Yellowish white	Lemon
Hydrogen	Bluish pink (bright blue)	Pink (rose)
Helium	Pale green	Violet-red
Argon	Bluish	Deep red (violet)
Neon	Red-orange	Red-orange (blood red)
Krypton	Green	—
Xenon	Bluish white	—
Carbon monoxide	Greenish white	(White)
Carbon dioxide	Blue	(White)
Methane	Reddish violet	—
Ammonia	Yellow-green	—
Chlorine	Greenish	Light green
Bromine	Yellowish green	Reddish
Iodine	Orange-yellow	Peach blossom colored
Lithium	Bright red	—
Sodium	Yellowish green (whitish)	Yellow
Potassium	Green	Green
Mercury	Green (goldish white)	Greenish blue (greenish)

— indicates no distinctive color given.

From *Lehrbuch der Physik: IV. Eliktrizitat und Magnetismus*, Third Part, p. 373; and Espe-Knoll, *Werkstoffkunde der Hochvakuum Technik*, p. 240 (color in parentheses).

Bibliography

Books and Monographs on Vacuum

Davy, J. R., *Industrial High Vacuum*, I. Bitman and Sons, London, 1951.

Diels, K., and R. Jaeckel, *Leybold Vakuum-Taschenbuch*, Springer, Verlag, Berlin, 1958.

Dunoyer, L., *Vacuum Practice*, translated by J. H. Smith, D. Van Nostrand Co., Princeton, N.J., 1926.

Dushman, S., *Scientific Foundations of Vacuum Technique*, John Wiley and Sons, 2nd ed., J. M. Lafferty, editor, New York, 1962.

Guthrie, A., and R. K. Wakerling, *Vacuum Equipment and Techniques*, McGraw-Hill Book Co., New York, 1949.

Holland-Merten, *Handbuch der Vakuumtechnik*, Halle, Germany, 1950.

Jaeckel, R., *Kleinste Drucke, ihre Erzeugung und Messung*, Springer-Verlag, Berlin, 1950.

Laporte, *Hochvakuum, seine Erzeugung, Messung und Anwendung*, Halle: Knapp, Germany, 1950.

Leblanc, M., *La Technique du Vide*, Librairie Armand Colin, Paris, 1951.

Martin, L. H., and R. D. Hill, *A Manual of Vacuum Practice*, University Press, Melbourne, 1949.

Mönch, G. Ch., *Hochvakuumtechnik*, Rudolph A. Lang Verlag, Possneck, Germany, 1950.

Oetjen, G. W., *Bd. l. Kapitel Vakuumtechnik*, Ullmanns Enzykl. der techn. Chemie 3. Aufl.

Reiman, A. L., *Vacuum Technique*, Chapman and Hall, London, 1952.

515

Wagner, G., *Erzeugung und Messung von Hochvakuum. Aufbau und Betrieb von Hochvakuum-Apparaten,* Franz Deuticke, Vienna, 1950.

Yarwood, J., *High Vacuum Technique,* 3rd ed., Chapman and Hall, London, 1955.

Books on Kinetic Theory and Properties of Gases

Barrer, R. M., *Diffusion in and through Solids,* The Macmillan Co., New York, 1941.

Barton, E. H., *An Introduction to the Mechanics of Fluids,* Longmans, Green and Co., London, 1915.

Bayley, F. J., *An Introduction to Fluid Dynamics,* Interscience Publishers, New York, 1958.

Brunauer, S. B., *The Adsorption of Gases and Vapors,* Princeton University Press, Princeton, N.J., 1945.

Carman, P. C., *Flow of Gases through Porous Media,* Academic Press, New York, 1956.

Cobine, J. D., *Gaseous Conductors,* Dover Publications, New York, 1958.

Daynes, H. A., *Gas Analysis by Measurements of Thermal Conductivity,* Cambridge University Press, Cambridge, England, 1933.

Farkas, A., and H. W. Melville, *Experimental Methods in Gas Reactions,* Cambridge University Press, Cambridge, England, 1939.

Fraser, R. G. J., *Molecular Rays,* Cambridge University Press, Cambridge, England, 1931.

Jeans, J. H., *An Introduction to the Kinetic Theory of Gases,* The Macmillan Co., New York, 1940.

Jost, W., *Diffusion in Solids, Liquids and Gases,* Academic Press, New York, 1952.

Kennard, E. H., *Kinetic Theory of Gases,* McGraw-Hill Book Co., New York, 1938.

Chapman, S., and T. G. Cowling, *The Mathematical Theory of Non-Uniform Gases,* Cambridge University Press, Cambridge, England, 1939.

Lamb, H., *Hydrodynamics,* Dover Publications, New York, 1945.

Loeb, L. B., *Kinetic Theory of Gases,* 2nd ed., McGraw-Hill Book Co., New York, 1934.

McBain, J. W., *The Sorption of Gases and Vapors by Solids,* George Routledge and Sons, London, 1932.

Present, R. D., *Kinetic Theory of Gases,* McGraw-Hill Book Co., New York, 1957.

Schlicting, H., *Boundary Layer Theory,* translated by J. Keslin, McGraw-Hill Book Co., New York, 1955.

Smithells, C. J., *Gases and Metals,* Chapman and Hall, London, 1937.

Some Handbooks and Books Relating to the Vacuum Field

Adam, N. K., *The Physics and Chemistry of Surfaces,* 3rd ed., Oxford University Press, Oxford, England, 1941.

American Institute of Physics Handbook, McGraw-Hill Book Co., New York, 1957.

Cupp, C. R., "Gases in Metals," Chap. 3, Vol. IV, *Progress in Metal Physics,* London, 1953.

Espe, W., and M. Knoll, *Werkstoffkunde der Hochvakuumtechnik,* Akademie-Verlag, Berlin.

Handbook of Chemistry and Physics, Chemical Rubber Publishing Co., Cleveland, Ohio.

Flügge, S., editor, *Handbuch der Physik,* Vol. XII, Springer-Verlag, Berlin, 1958.

Holland, L., *The Vacuum Deposition of Thin Films,* John Wiley and Sons, New York, 1956.

International Critical Tables, McGraw-Hill Book Co., New York, 1928.

Kohl, W. H., *Materials Technology for Electron Tubes,* Reinhold Publishing Corp., 1951.

Leck, J. H., *Chemisorption,* Butterworths Scientific Publications, London, 1956.

Strong, J., *Procedures in Experimental Physics,* Prentice-Hall, Englewood Cliffs, N.J., 1938.

Trapnell, B. M. W., *Chemisorption,* Butterworths Scientific Publications, London, 1955.

Periodicals, Societies, and Conferences

Le Vide, published by Société française des engénieurs techniciens du vide, Paris, France.

Vakuum-Technik, Organ der Arbeitsgemeinschaft Vakuum des Vereines Deutscher Ingenieure [und andere ähnliche Gesellschaften], April 1954–   , Berlin, W. Germany. Vol. 1–2 (1952 to Feb. 1954): *Glas—und Hochvakuum—Technik.*

Vacuum, Pergamon Press, Oxford, England.

International Union for Vacuum Science, Technique and Applications, Brussels, Belgium.

International Congress on Vacuum Techniques, 1st Congress: Namur, Belgium, 1958 [published as *Advances in Vacuum Science and Technology* (Pergamon Press, Oxford, England)], 2nd Congress: Washington, D.C., 1961.

American Vacuum Society, Boston 9, Mass. Formerly Committee on Vacuum Techniques. *Vacuum Symposium Transactions,* 1953–   (*Vac. Symp. Trans.*).

Cambridge High Vacuum Symposium 1947, 1948, 1949 (NRC) French Lick Meeting, May 11–14, 1952.

High Vacuum Convention, Perthshire, England. Ref.: *Chemistry and Industry,* Oct. 1948 (S. Dushman).

Symposium, Institute of Physics, Birmingham, June 27–28, 1950. Ref.: *J. Sci. Instr., Suppl. 1,* 1951.

Vacuum Articles

KINETIC THEORY AND FLOW OF GASES

Warfield, C. N., *Technical Note 1200,* National Advisory Committee for Aeronautics, 1947.

Taylor, J. B., *J. Ind. Eng. Chem.*, **23**, 1228 (1937).

Licht, W., Jr., and D. G. Steckert, *J. Phys. Chem.*, **48**, 23 (1944).

Dickins, B. G., *Proc. Roy. Soc. (London) A*, **143**, 517 (1934).

Knudsen, M., *Ann. Physik*, **6**, 129 (1930).

Blodgett, K. B., and I. Langmuir, *Phys. Rev.*, **40**, 78 (1932).

Jones, H. A., *Gen. Elec. Rev.*, **28**, 650 (1935).

Liang, S. C., *Can. J. Phys.*, **33**, 279 (1955).

Bennett, M. J., and F. C. Tompkins, *Trans. Faraday Soc.*, **53**, 185 (1957).

Ibbs, T. L., *Physica*, **4**, 1135 (1937).

Bardeen, J., *Phys. Rev.*, **57**, 35 (1940).

Langhaar, H. L., *J. Appl. Mech.*, **9**, A-55 (1942).

DeMarcus, E. C., *Atom. Energy Comm. Report K-1302*, Parts I and II (1956);
Parts III and IV (1957).

Davis, D. H., *J. Appl. Phys.*, **31**, 1169 (1960).

Brown, G. P., A. Dinardo, G. K. Cheng, and T. K. Sherwood, *J. Anal. Phys.*, **17**,
802 (1946).

Pollard, W. G., and R. D. Present, *Phys. Rev.*, **73**, 762 (1948).

VACUUM PUMPS

Mechanical

Gaede, W., *Ann. Physik*, **41**, 337 (1913).

Howard, H. C., *Rev. Sci. Instr.*, **6**, 357 (1935).

Eltenton, G. C., *J. Sci. Instr.*, **15**, 415 (1938).

Alexander, P., *J. Sci. Instr.*, **21**, 216 (1944).

Payne, J. H., *J. Franklin Inst.*, **211**, 689 (1931).

Matricon, M., *J. Phys. Radium*, **3**, 127 (1932).

Neumann, R., *Electronics Eng.*, **20** (Jan., Feb.) 1948.

Encyclopaedia Britannica, 14th ed., Vol. **22**, p. 126.

Holweck, F., *Compt. rend.*, **177**, 43 (1923).

von Friesen, S., *Rev. Sci. Instr.*, **11**, 362 (1940).

Blodgett, K. B., and T. A. Vanderslice, *J. Appl. Phys.*, **31**, 1017 (1960).

Siegbahn, M., *Arch. Math., Astr., Phys., Roy. Swedish Acad.*, **30B**, No. 2 (1943).

Eklund, S., *Arch. Math., Astr., Phys., Roy. Swedish Acad.*, **27A**, No. 21 (1940);
29A, No. 4 (1942).

Becker, W., *Vakuum-Tech.*, Oct. 1958, pp. 149–152.

Vapor

Estermann, I., and H. T. Byck, *Rev. Sci. Instr.*, **3**, 482 (1932).

Hickman, K. C. D., *J. Franklin Inst.*, **221**, 215 (1936).

Ho, T. L., *Physics*, **12**, 386 (1932).

Embree, N., *Rev. Sci. Instr.*, **8**, 263 (1937).

Venema, A., and M. Bandringa, *Philips Tech. Rev.*, **20**, 145 (1958).

Blackhurst, I., and G. W. C. Kaye, *Phil. Mag.*, **47**, 918, 1024 (1924).

Zabel, R. M., *Rev. Sci. Instr.*, **6**, 54 (1935).

Amdur, I., *Rev. Sci. Instr.*, **7**, 395 (1936).

Lockenwitz, A. E., *Rev. Sci. Instr.*, **8**, 322 (1937).

Malter, L., and N. Marcuvitz, *Rev. Sci. Instr.*, **9**, 92 (1938).
Hickman, K. C. D., *J. Appl. Phys.*, **11**, 303 (1940).

Getter-Ion

Holland, L., *J. Sci. Instr.*, **36**, 105 (1959).
Carter, G., and J. H. Leck, *Brit. J. Appl. Phys.*, **10**, 364 (1959).
Bayard, R. T., and D. Alpert, *Rev. Sci. Instr.*, **21**, 571 (1950).
Florescu, N. A., *Vacuum*, **10**, No. 3, 250 (1960).
Blears, J., *Proc. Roy. Soc. (London)*, A, **188**, 62 (1959).
Hall, L. D., *Rev. Sci. Instr.*, **29**, 367 (1958).
Holland, L., L. Laurenson, and J. M. Holden, *Nature*, **182**, 851 (1958).
Holland, L., and L. Laurenson, *Le Vide*, Nr. 81, 141 (1959).
Herb, R. G., R. H. Davis, A. S. Divatia, and D. Saxon, *Phys. Rev.*, **89**, 897 (1953).

Some Other Types of Pumps

Vrátný, F., and B. Graves, *Rev. Sci. Instr.*, **30**, 597 (1957).
Jepson, R. L., S. L. Mercer, and M. J. Callaghan, *Rev. Sci. Instr.*, **30**, 377 (1959).
Caswell, H. L., *Rev. Sci. Instr.*, **30**, 1054 (1959).
Grimsehl, E., *Physik. Z.*, **8**, 762 (1907).
Waran, *Proc. Phys. Soc.*, **34**, Part 3, 120 (1922).

Vapor Pump Fluids

Jaeckel, R., *Z. tech. Physik.*, **23**, 177 (1942).
Blears, J., *Nature*, **154**, 20 (1944); *Proc. Roy. Soc. (London)* A, **188**, 62 (1947).
Hickman, K. C. D., *Nature*, **187**, 405 (1960).
Brown, G. P., *Rev. Sci. Instr.*, **16**, 316 (1945).
Latham, D., B. D. Power, and N. T. M. Dennis, *Vacuum*, **2**, 33 (1952).
Martin, C. S., and J. H. Leck, *Vacuum*, **4**, 486 (1954).
Bishop, F. W., *Rev. Sci. Instr.*, **30**, 830 (1959).
Power, B. D., and D. J. Crawley, *Vacuum*, **4**, 415 (1954).
Hickman, K. C. D., J. C. Hecker, and N. D. Embree, *Ind. Eng. Chem., Anal. Ed.*, **9**, 264 (1937).
Burch, C. R., *Proc. Roy. Soc. (London)* A, **123**, 271 (1929).

Pumping Speed

Alexander, P., *J. Sci. Instr.*, **21**, 216 (1944).
Eltenton, G. C., *J. Sci. Instr.*, **16**, 27 (1939).
Howard, H. C., *Rev. Sci. Instr.*, **6**, 327 (1935).
Matricon, M., *J. Phys. Radium*, **3**, 127 (1932).
Dayton, B. B., *Ind. Eng. Chem.*, **40**, 795 (1948).
Korsunsky, M., and S. Vekshinsky, *J. Phys. U.S.S.R.*, **9**, 399 (1945).
Blears, J., *Proc. Roy. Soc. (London)* A, **188**, 62 (1946).
Bureau, A. J., L. J. Laslett, and J. M. Keller, *Rev. Sci. Instr.*, **23**, 683 (1952).

VACUUM GAUGES

Barr, W. E., and V. J. Anhorn, *Instruments*, **19**, 666, 734 (1946).
Klumb, H., and H. Schwarz, *Z. Physik.*, **122**, 418 (1944).
Venema, A., and M. Bandringa, *Philips Tech. Rev.*, **20**, 145 (1958).
Yarwood, J., *J. Sci. Instr.*, **34**, 297 (1957).

Kenty, C., *Rev. Sci. Instr.,* **11**, 377 (1940).

Alpert, D., *J. Appl. Phys.,* **24**, 860 (1953).

Hobson, J. P., and P. A. Redhead, *Can. J. Phys.,* **36**, 251, 278 (1958).

Lafferty, J. M., *J. Appl. Phys.,* **32**, 424 (1961).

Conn, G. K. T., and H. N. Daglish, *J. Sci. Instr.,* **31**, 412 (1954).

Pirani, M., and R. Neumann, *Electron. Eng.,* Dec. 1944, Jan., Feb., and Mar. 1945.

Dushman, S., *Instruments,* **20**, 234 (1947).

Burrows, G., *J. Sci. Instr.,* **20**, 21 (1943).

Hasse, G., *Z. tech. Physik,* **24**, 27, 53 (1943).

Flosdorf, E. W., *Ind. Eng. Chem., Anal. Ed.,* **17**, 198 (1945).

Dibeler, V. H., and F. Cordero, *J. Res., Natl. Bur. Standards,* **46**, 1 (1951).

Alpert, D., C. G. Matland, and A. O. McCoubrey, *Rev. Sci. Instr.,* **22**, 370 (1951).

Beams, J. W., J. L. Yound, and J. W. Moore, *J. Appl. Phys.,* **17**, 886 (1946).

DuMond, J. W. M., and W. M. Pickels, *Rev. Sci. Instr.,* **6**, 362 (1935).

Hughes, A. L., W. J. H. Moll, and H. C. Burger, *Z. tech. Physik,* **21**, 199 (1940).

Webber, R. T., and C. T. Lane, *Rev. Sci. Instr.,* **17**, 308 (1946).

Scott, E. J., *Rev. Sci. Instr.,* **10**, 349 (1939).

Becker, J. A., C. B. Green, and G. L. Pearson, *Trans. AIEE,* **65**, 711 (1946); *Bell Systems Tech. J.,* **26**, 170 (1947).

Morse, R. S., and R. M. Bowie, *Rev. Sci. Instr.,* **11**, 91 (1940).

Schulz, G. J., and A. V. Phelps, *Rev. Sci. Instr.,* **28**, 1051 (1957).

Bleecker, H., *Rev. Sci. Instr.,* **23**, 56 (1952).

Penning, F. M., and K. Nienhaus, *Philips Tech. Rev.,* **11**, 1161 (1949).

Thomas, H. A., T. W. Williams, and J. A. Hipple, *Rev. Sci. Instr.,* **17**, 368 (1946).

Downing, J. R., and G. Mellen, *Rev. Sci. Instr.,* **17**, 218 (1946).

Spencer, N. W., and R. L. Boggers, *ARS J.,* **29**, 68 (1959).

Blears, J., *Proc. Roy. Soc. (London) A,* **188**, 62 (1947).

Bayard, R. T., and D. Alpert, *Rev. Sci. Instr.,* **21**, 571 (1950).

Apker, L., *Ind. Eng. Chem.,* **24**, 860 (1953).

Houston, J. M., *Bull. Am. Phys. Soc.,* **11**, 1, 301 (1956).

Sommer, H., H. A. Thomas, and J. A. Hipple, *Phys. Rev.,* **82**, 697 (1951).

Alpert, D., and R. S. Buritz, *J. Appl. Phys.,* **25**, 202 (1954).

Edwards, A. G., *Brit. J. Appl. Phys.,* **6**, 44 (1955).

Reynolds, J. H., *Rev. Sci. Instr.,* **27**, 928 (1956).

Nelson, R. B., *Rev. Sci. Instr.,* **16**, 55 (1945).

Leck, J. H., and A. Riddock, *Brit. J. Appl. Phys.,* **7**, 153 (1956).

TRAPS AND BAFFLES

Morse, R. S., *Rev. Sci. Instr.,* **11**, 277 (1940).

Anderson, P. A., *Rev. Sci. Instr.,* **8**, 493 (1937).

Northrup, D. L., C. M. Van Atta, and L. C. Van Atta, *Rev. Sci. Instr.,* **11**, 207 (1940).

Morse, R. S., *J. Ind. Eng. Chem.,* **39**, 1064 (1947).

Hunten, K. W., G. A. Woonton, and E. C. Longhurst, *Rev. Sci. Instr.,* **18**, 842 (1947).

Alpert, D., *Rev. Sci. Instr.,* **24**, 1004 (1953).

Venema, A., and M. Bandringa, *Philips Tech. Rev.,* **20**, 145 (1958).

Biondi, M. A., *Rev. Sci. Instr.,* **30**, 831(1959).

Harris, L. A., *Rev. Sci. Instr.,* **31**, 903 (1960).

SORPTION, PERMEATION, AND DIFFUSION OF GASES

Deitz, V. R., editor, Vol. 1, 1900–1942, Bone Char Research Project, Inc., 1944;
Vol. 2, 1943–1953, Nat. Bur. Standards, Circ. No. 566, 1956.
Kipling, J. J., *Quart. Revs. (London)*, **10**, 1 (1956).
Smith, R. N., *Quart. Revs. (London)*, **13**, 287 (1959).
Maggs, F. A. P., P. H. Schwabe, and J. H. Williams, *Nature*, **186**, 956 (1960).
Savage, R. H., *J. Appl. Phys.*, **19**, 1 (1948).
Biondi, M. A., *Rev. Sci. Instr.*, **30**, 831 (1959).
Alpert, D., and R. S. Buritz, *J. Appl. Phys.*, **25**, 202 (1954).
Young, J. R., and N. R. Whetten, *Rev. Sci. Instr.*, **32**, 453 (1961).
Smithels, C. J., and C. E. Ransley, *Proc. Roy. Soc. (London) A*, **150**, 172 (1935);
152, 706 (1936).
Barrer, R. M., *Trans. Faraday Soc.*, **36**, 1235 (1940).
Norton, F. J., *J. Am. Ceram. Soc.*, **36**, 90 (1953).

OUTGASSING, VAPOR PRESSURES, AND EVAPORATION

Verhoek, F. H., and A. L. Marshall, *J. Am. Chem. Soc.*, **61**, 2737 (1939).
Honig, R. E., *R.C.A. Rev.*, **18**, 195 (1957).
Law, R. R., *Rev. Sci. Instr.*, **19**, 920 (1948).
Bond, W. L., *J. Opt. Soc. Am.*, **44**, 429 (1954).
Stull, D. R., *Ind. Eng. Chem.*, **39**, 517 (1947).
Berl, E., *Chem. Met. Eng.*, **53**, 130 (1946).
Smith, R. V., *U.S. Bur. Mines Inform. Circ. No. 7215* (Aug. 1942).
Hickman, K. C. D., *Chem. Revs.*, **34**, 51 (1944).
Langmuir, I., H. A. Jones, and G. M. J. Mackay, *Phys. Rev.*, **30**, 211 (1927).
Neumann, K., and E. Volker, *Z. physik. Chem. A*, **161**, 33 (1932).
Santeler, D. J., *G. E. Report No. 58 GL*, 303 (Nov. 5, 1958).
Cloud, R. W., and S. F. Philip, *Rev. Sci. Instr.*, **21**, 731 (1950).
Zabel, R. M., *Rev. Sci. Instr.*, **4**, 233 (1933).
Hogg, B. G., and H. E. Duckworth, *Rev. Sci. Instr.*, **19**, 331 (1948).
Britt, J. R., *NRL Report 3827*, Naval Research Lab., Washington, D.C., July 1951.
Monk, G. W., *Atom. Energy Comm. Doc. No. MDDC-1307*, Tech. Inform. Div.,
Oak Ridge, Tenn.
Atom. Energy Res. Establish. A.E.R.E. GP/R 1875.

GETTERS AND ELECTRICAL CLEAN-UP

Stout, V. L., and M. D. Gibbons, *J. Appl. Phys.*, **26**, 1488 (1955).
Wagener, S., *Brit. J. Appl. Phys.*, **1**, 225 (1950): *Proc. Nat. Conf. Tube Tech.*, 1958.
Della Porta, P., *Vacuum*, **10**, 181 (1960).
Espe, W., M. Knoll, and M. P. Wilder, *Electronics*, **23**, 80 (1950). *Brit. J. Appl.
Phys.*, **6**, 161 (1955).
Leck, J. H., K. B. Blodgett, and T. A. Vanderslice, *J. Appl. Phys.*, **31**, 1017 (1960).
Siegler, E. H., and G. H. Dieke, *Clean-up of Rare Gases in Electrodeless Dis-
charges*, ONR Contract NONR 248, Tech. Report No. VIII.
Young, J. R., *J. Appl. Phys.*, **27**, 926 (1956).
Varnerin, L. J., and J. H. Carmichael, *J. Appl. Phys.*, **28**, 913 (1957).
Holland, L., *J. Sci. Instr.*, **36**, 105 (1959).

LEAK DETECTORS AND TECHNIQUES

Kuper, J. B. H., *Rev. Sci. Instr.,* **8**, 131 (1937).

Ridenour, L. N., *Rev. Sci. Instr.,* **12**, 134 (1944).

Manley, J. H., L. J. Hayworth, and E. A. Luebke, *Rev. Sci. Instr.,* **10**, 389 (1939).

Nelson, R. B., *Rev. Sci. Instr.,* **16**, 55 (1945).

Jacobs, R. B., and H. F. Zuhr, *J. Appl. Phys.,* **18**, 34 (1947).

Nier, A. O., C. M. Stevens, A. Hustrulid, and T. A. Abbott, *J. Appl. Phys.,* **18**, 30 (1947).

Thomas, H. A., T. W. Williams, and J. A. Hipple, *Rev. Sci. Instr.,* **17**, 368 (1946).

Cassen, B., and D. Burnham, *Intern. J. Appl. Rad. and Isotopes,* **9**, 54 (1960).

White, W. C., and J. S. Hickey, *Electronics,* **21**, 100 (1948).

Weber, A., *Glas-u. Hochvackuum-Tech.,* **2**, 259 (1953).

Van Leeuwen, J. A., and H. J. Oskam, *Rev. Sci. Instr.,* **27**, 328 (1956).

Peters, J. L., *Rev. Sci. Instr.,* **30**, 1093 (1959).

Dowling, D. J., *J. Sci. Instr.,* **37**, 147 (1960).

Cossuta, D., and W. Steckelmacher, *J. Sci. Instr.,* **37**, 404 (1960).

Bloomer, R. N., and W. C. Brooks, *J. Sci. Instr.,* **37**, 306 (1960).

Turnbull, A. H., *Atom. Energy Res. Establish. Report G/R478,* Mar. 1950.

Blears, J., and J. H. Lech, *Brit. J. Appl. Phys.,* **2**, 227 (1951).

Minter, C. C., *Rev. Sci. Instr.,* **29**, 793 (1958).

ULTRA-HIGH VACUUM SYSTEMS AND COMPONENTS

Nottingham, W. B., *Phys. Rev.,* **55**, 203 (1939).

Bayard, R. T., and D. Alpert, *Rev. Sci. Instr.,* **21**, 571 (1950).

Alpert, D., and R. S. Buritz, *J. Appl. Phys.,* **25**, 202 (1954).

Lafferty, J. M., *J. Appl. Phys.,* **22**, 299 (1951).

Hagstrum, H. D., *Phys. Rev.,* **89**, 244 (1953).

Becker, J. A., and C. D. Hartman, *J. Phys. Chem.,* **57**, 153 (1953).

Venema, A., and M. Bandringa, *Philips Tech. Rev.,* **20**, 6 (1958).

Alpert, D., *J. Appl. Phys.,* **24**, 860 (1953); *Rev. Sci. Instr.,* **22**, 536 (1951); *Vacuum,* **9**, 89 (1959).

Biondi, M. A., *Rev. Sci. Instr.,* **30**, 831 (1959).

Harris, L. A., *Rev. Sci. Instr.,* **31**, 903 (1960).

Hall, L. D., *Science,* **128**, 279 (1958).

Blodgett, K. B., and T. A. Vanderslice, *Gaseous Electronics Conference,* October, 1958 [see *Bull. Am. Phys. Soc.,* **4**, 111 (1959)].

Axelrod, N. N., *Rev. Sci. Instr.,* **30**, 944 (1959).

Hintenberger, H., *Z. Naturforsch.,* **6a**, 459 (1959).

Pattee, H. H., *Rev. Sci. Instr.,* **25**, 1132 (1954).

Yarwood, J., *Brit. J. Appl. Phys.,* **10**, Sept. 1959 (Conf. report).

Holland, L., *J. Sci. Instr.,* **36**, 105 (1959).

Munday, G. L., *Nucl. Instr. Methods and Instruments,* **4**, 367 (1959).

Index